D1577175

600079184

MOORLAND
MANAGEMENT

MOORLAND
MANAGEMENT

——— FOR ———
AGRICULTURE, CONSERVATION
AND FIELD SPORTS
A Practical Guide

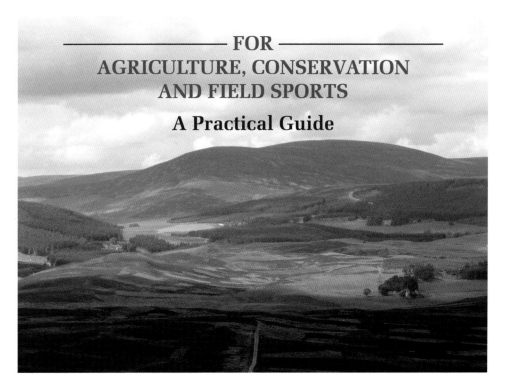

JOHN PHILLIPS

Quiller

First published in the UK in 2012
by Quiller, an imprint of Quiller Publishing Ltd

British Library Cataloguing-in-Publication Data
A catalogue record for this book
is available from the British Library

ISBN 978 1 84689 122 9

Printed in China

Quiller

An imprint of Quiller Publishing Ltd
Wykey House, Wykey, Shrewsbury, SY4 1JA
Tel: 01939 261616 Fax: 01939 261606
E-mail: info@quillerbooks.com
Website: www.countrybooksdirect.com

CONTENTS

ABBREVIATIONS AND ACRONYMS

ATV	All Terrain Vehicle
BASC	British Association for Shooting and Conservation
BP	Before Present
BSE	Bovine spongiform encephalitis
BTO	British Trust for Ornithology
CAP	Common Agricultural Policy
CLA	Country Landowners Association (now Country Land and Business Association)
CMRP	Clyde Muirshiel Regional Park
CROW	Countryside and Rights of Way Act
DAFS	Department of Agriculture and Fisheries for Scotland
DEFRA	Department for Environment, Food and Rural Affairs
EGAS	Eley Game Advisory Service (now Game and Wildlife Conservation Trust)
EU	European Union
FC	Forestry Commission
GCT	Game Conservancy Trust (now Game and Wildlife Conservation Trust)
GWCT	Game and Wildlife Conservation Trust
HFRO	Hill Farming and Research Organisation
HLS	Higher Level Stewardship (England and Wales)
IPC	International Publishing Corporation
ITE	Institute of Terrestrial Ecology
JMT	John Muir Trust
LFA	Less Favoured Area
MA	Moorland Association
MGA	Moorland Gamekeepers' Association
MoD	Ministry of Defence
NE	Natural England
NFU	National Farmers' Union
NNR	National Nature Reserve
NT	National Trust
NTS	National Trust for Scotland
RDC	Red Deer Commission/Deer Commission for Scotland (now SNH)
RSPB	Royal Society for the Protection of Birds
RSPCA	Royal Society for the Prevention of Cruelty to Animals
SAC	Special Area of Conservation; also Scottish Agricultural Colleges
SFP	Single Farm Payment
SGA	Scottish Gamekeepers' Association
SGRPID	Scottish Government Rural Payments and Inspectorate Division
SNH	Scottish Nature Heritage
SNP	Scottish National Party
SPA	Special Protection Area
SRDP	Scottish Rural Development Programme
SRPBA	Scottish Rural Property and Business Association
SSSI	Site of Special Scientific Interest
SWT	Scottish Wildlife Trust
UK	United Kingdom (England, Scotland, Wales and Northern Ireland)
US	United States of America
WAGBI	Wildfowlers Association of Great Britain and Ireland (now BASC)
WANE	Wildlife and Natural Environment (Scotland) Act, 2011
WT	Wildlife Trust(s)
WWF	World Wildlife Fund for Nature

FOREWORDS

BOWHILL, SELKIRK, SCOTLAND

Britain's moorlands have an enduring capacity to arouse the deepest passions and stir the strongest controversies. Man's impact has shaped them over the centuries as generations of farmers and shepherds, foresters, gamekeepers and stalkers have adapted their workplaces to their needs. Yet with their natural beauty and their wonderfully diverse flora and fauna their protection and management has become a source of anguished concern for conservationists, while into the mix come the growing numbers from town and country who identify them as precious resources for leisure and recreation. Trying to ensure that our moorlands can match all these hopes and expectations – above all that they survive for future generations – is one of the great challenges facing all who care about the countryside.

Few people have shown such single minded devotion as John Phillips to teasing out with maximum calm observation and minimum prejudice a practical analysis of the many facets that have to be properly understood if we are to succeed. For my father, a lifelong countryman but one who believed we should never stop learning, John was something of a guru. He had huge respect for his blend of ancient wisdom and current observation, and his underlying philosophy that all of our ambitions for the moorlands could be made to co-exist if we were meticulous in our practices. In particular he believed that hill sheep farming and grouse moors could be two sides of the same coin.

For that to happen, however, we all needed a bible of common and best practice. I have little doubt that in *Moorland Management* my father would have reckoned we had that bible. He would have made it required reading for all those involved on his estates in upland management. I have no hesitation in echoing that enthusiasm, in urging the widest possible dissemination of its ideas and in thanking John for his great service to the cause of our moorlands.

The Duke of Buccleuch and Queensberry

GARROWS

AMULREE

DUNKELD

PERTHSHIRE

For more than fifty years I have been responsible for a small moorland estate in the central Highlands of Scotland. My great interest in the habitat as well as the creatures that inhabit the moor has never diminished but paramount in that interest is the grouse – that most exciting of all wild game birds – coupled with awareness of the need for good management of the moorland for its inhabitants.

It is well established that there is no single factor that makes a successful grouse moor. In addition to the importance of the underlying geology, I believe it is a combination of suitable habitat created by careful heather burning, good keepering, tackling diseases, particularly those transmitted by tick and, last but not least, dedicated and efficient sheep management. These factors, applied with meticulous attention to detail, represent both sides of the (golden) coin mentioned in the Foreword.

It is also a most welcome side-effect that, beyond all doubt, well managed moors harbour a far greater variety of birdlife than those that are unmanaged.

Since Lord Lovat's masterful work *The Grouse in Health and in Disease* was published in 1911, a great deal has been written and much research work undertaken on the subject of grouse and grouse moors. How timely it is that, a century after Lord Lovat's report, John Phillips has now produced a comprehensive and practical guide to moorland management covering all aspects of agriculture, conservation and sport.

John Kemp-Welch

Sir John Kemp-Welch

11

ACKNOWLEDGEMENTS

When a book of this nature is written, it is the sole responsibility of the author. Its contents contain facts, observations and opinions which have been accumulated over a long period which in this case have been whipped into shape by Dave Bruce and Douglas Phillips without whose inputs it would never have appeared.

The author's interest in moorland began in childhood with much time being spent 'helping' on a west of Scotland hill farm. It was nurtured by both formal and practical education in agriculture and wildlife management which led in its turn to an independence of mind in matters of moorland management and a desire to see the wider development of multi-uses on the uplands. The people who helped to realise this objective are too numerous to mention individually and if one cannot mention everybody, perhaps one should mention no one. However, this is unfair as there were critical interactions with individuals which resulted in major changes of course which led ultimately to this book being written. First must be mentioned the paternal schooling in field sports skills which was well-advanced by the age of ten, followed by the influence of staff at the Eley Game Advisory Station, particularly Terence Blank. Contact followed with scientists at many levels in both the UK and elsewhere and Art Lance, Robert Moss, Hugh Reid, Adam Watson and Derek Yalden deserve special mention. In the field of land ownership, the late Duke of Buccleuch was pre-eminent in the support he provided for field study and the testing of ideas. John Blackett-Ord, Hugh Blakeney, Iain Chalmers, Reay Clarke, John Denton, Peter Doran, Andrew Duncan, James Duncan, Michael Fenwick, Robert Howden, John Kemp-Welch, Bill Lithgow, Willie Macgregor, Bob McNeil, Rob Marrs, Jack Meath, Mike Nairn, Hugh Reid, Robert Steuart Fothringham, Peter Straker Smith and Roger Wheater were far-seeing, provided wise counsel, original thought and played key roles. However, nothing happens without on-ground help, which is where numerous farmers, shepherds and keepers contributed so much. Giants were Jamsie Bain, Mervyn Browne, Bobby Fowler, John Grave, Archie Maclellan, Brian Mitchell, Mike Nisbet, Matt Ramage, Michael Skelly and Colin Stewart. One cannot ignore the help from family and friends which included support from Jenny Braasch, John and Sue Hutchinson and Claire Rackham.

The future holds many challenges if moorlands are to survive in the form in which we recognise them so it is to be hoped that future generations show the wisdom and dedication which has been displayed by those who have been professionally active in the field during the past seventy years. Hopefully, this book may help the cause.

John Phillips
Perthshire and Sutherland
December 2011

To S. R. Crockett
On receiving a dedication

Blows the wind today and the sun and the rain are flying,

Blows the wind today on the moors and now,

Where about the graves of the martyrs the whaups are crying,

My heart remembers how.

Grey remembered tombs of the dead in desert places,

Standing stones in the vacant wine-red moor.

Hills of sheep and the homes of the silent vanished races,

And winds, austere and pure.

Be it granted to me to behold you again in dying.

Hills of home! And hear again the call;

Hear about the graves of the martyrs the peewees crying,

And hear no more at all.

Robert Louis Stevenson (1850–1894) Vailima, Samoa

From *Songs of Travel*

Moorlands – An introduction

The popular conception of heather moorland.

Stevenson's poem enjoys pride of place because it contains something for everyone who lives among and loves moorland. The heather in bloom in late August, the snell wind of autumn, the livelihoods of people past and present are all acknowledged. For the owner, moorland is a land asset which needs to be looked after; for the keeper, it is a workplace; for the field sportsman, a place for recreation and holidays. For the farmer or shepherd, it provides their livelihoods and, for the naturalist, it is an extensive and multi-faceted place of great conservation interest. For the general tourist, the sheer wonder of the open spaces it represents provides its own 'Wow!' factor.

Perhaps optimistically, this book too aims to offer something to everyone with a serious interest in moorland. Any book with 'management' in its title is likely to be wide ranging in content if it is going to do justice to its subject but it must be stressed from the outset that no management can take place without clear objectives and that the would-be manager has to chart a path 'from here to there' to deliver them. It matters not whether the intention is to increase ravens in Argyll, choughs in Islay, hen harriers in Orkney, blackgame in Perthshire, quality sheep on moorland or bags of grouse to commercial levels – no management can succeed without a plan. All plans involve a degree of intervention. Sometimes it has to be drastic, sometimes a mere tweak but, for the object of desire to be advantaged, other organisms may have to be disadvantaged – and that sometimes means competing animals and birds being killed.

*Along the crag,
Reeth.*
(ILLUSTRATION:
J. LOWES)

This is such a truism that it is a matter of constant wonder that people connected with conservation are so reluctant to accept it. Many have been brought up on a diet of game and wildlife management on North American lines where no predators are ever killed and the numbers of target species available for man to crop is influenced by the twin prongs of habitat improvement and bag limits. This works well in the United States (US) where land areas are vast and hunters have modest expectations. In France – a predominantly rural country – the policy is essentially the same but the results of *laissez faire* are disastrous. Many hunters associated with a policy of *chasse commune* (go shooting where you like) ensures that there is no game for anyone. No individual *chasseur* (shooter) in France raises a finger to help the brown hare by shooting a fox or the grey partridge by knocking out a crow's nest. As there is no intervention, there is scarcely any wild game for anybody. British managers must heed the lessons of France.

In Britain, we are fortunate to have a tradition of interventionism when it comes to managing species, quarry or otherwise. It is of paramount importance that this philosophy is not eroded. Private landowners by and large know how to manage habitats for game species and are permitted by law to kill common predators. It is, however, demonstrable that certain species of birds of prey – all of them protected – occur in saturation numbers in some locations and jeopardise rural enterprises relying on moorland management and its products. An open and objective debate must take place soon to resolve the problem. Many owners bear the burden of land management responsibly and, on the moors of the north of England for example, create dramatic assemblages of birds for the pleasure of the general public. Some do it better than others but nearly all do it better than quangos. Once the dead hand of officialdom enters the field of land and wildlife management, clear objectives become blurred and enterprises tend to run aground.

The moorlands of Great Britain and Ireland make up the largest area of heather-dominated dwarf shrub heath in the world. There are approximately 1.28

million ha (3.2 m ac) of heather in Scotland, 385,000 ha (950,000 ac) in England and almost 100,000 ha (250,000 ac) in Wales.[1] Although the habitat also occurs on mainland Europe, extending from western France to northern Norway and eastwards into the north German plain, it is often fragmented. The extensive blocks of moorland so typical of the Pennines, the northeast Highlands of Scotland and eastern Ireland are absent. The dominance of a single plant species has created a series of landscape features unlike any other. Ling heather is the main constituent but it grows in association with many herbs, grasses, sedges, rushes, lichens, fungi, mosses and liverworts as well as shrubs and trees and supports a wide assemblage of vertebrate and invertebrate fauna. Heathlands and moorlands vary greatly in character from the coastal heaths at sea level to arctic-alpine moors above 1000 metres (3300 ft) but their common characteristic is that they all have ling heather as a dominant or co-dominant species.

Heather moor as we know it is largely an artificial habitat although it maintains many semi-natural features. In boreal times, more than five thousand years BP, there was much less than there is today although what we see in the twenty-first century covers a much smaller area than it did even as recently as a hundred years ago. About five thousand years ago, moorland was confined largely to a narrow band above the tree line and extended upwards into more varied vegetation in the arctic-alpine zone. Following colonisation by man and the arrival of settled pastoral agriculture, most of the forest was cut down. Grazing animals combined with accidental and intentional fire to prevent the natural regeneration of trees. The process reached its zenith during the last quarter of the nineteenth century; the area of heather moor at that time was greater than at any time before or since.

Ling heather, cross-leaved heath, grasses and sedges; typical moorland constituents. Bar scale -5cm.

The moorland area declined during the twentieth century for a number of reasons. In some cases, domestic and wild animals converted the dwarf shrub heath to acid grassland through over-grazing. In others, heath was destroyed intentionally to support crop-based agriculture. Natural and plantation forest sometimes became established as a result of direct or indirect intervention. What we see today are the remains of what was once a larger area that has since been fragmented by piecemeal changes of land use driven as much by fiscal pressure as any consistent assessment of land capability or landscape design.

Heather moorland is characterised by its open nature. The wide, treeless landscape often extends for many kilometres with the occasional intrusion of small patches of woodland and farmland typically confined to the better soils in the valleys. To many, it is a wild area that should be 'tamed' and made more productive. To others it is an area of remarkable complexity and wildlife interest with an intrinsic worth and uniqueness in the global context of a small group of islands supporting some sixty million people living mostly in towns and cities. It represents an ecosystem which shows some contradictions; in one sense, it is simple on account of the apparent paucity of species which inhabit it while in another it is immensely complex and multi-faceted. It is robust yet sensitive, impoverished yet productive.

It is land that has to be managed by people with the wisdom of Solomon and the patience of Job, a mistress which is slow to forgive a wrong yet quick to yield a bounty if handled with sensitivity and knowledge. In the long term, it rarely responds to unsympathetic treatment with a financial dividend but, if its moods are understood and management cajoles rather than bullies it, the response can be economically dramatic.

While the products of conventional importance are sheep, cattle, ponies, red deer and grouse, management for these species is but part of the story and provides only some of the answers for the long-term maintenance of moorland. If managed sympathetically for them, it supports a wide spectrum of other species as well, many of little or no economic importance on their own but playing a vital part in ensuring the long-term survival of the moorland system.[2]

The past sixty years have seen extensive losses. It is estimated that out of a total of around 4.3 million ha present in Scotland in 1940, twenty-four per cent has disappeared completely, its place taken by forestry, improved grassland, bracken, poor quality 'white' land dominated by purple moor grass and white bent, enclosures for conservation purposes and urban development. Losses are unevenly spread. They are more acute in the wetter, western half of Scotland, notably Dumfries & Galloway, Ayrshire and Argyll and significantly worse in Scotland than in the north of England. In Wales, heather is managed on a scale which reflects the generally smaller size of land holdings and tends to be fragmented as a result.

Quantifying the moorland extent

It has been estimated that there were 3100 grouse moors in Scotland in 1958[3] and 485 by 1991.[4] It was reported in 2010 that there were 140 'core' grouse estates left in Scotland with perhaps a further 160 estates where grouse were shot.[5] Inevitably, definitions change over the years but it is reasonable to conclude that Scotland lost around eighty-four per cent of its grouse moorland in the thirty-five years from the 1950s to 1991 and that moorland units qualifying as grouse moor declined by a further seventy-five per cent (485 to 140) between 1991 and

Table 1.

Scotland's land surface by type.[6]

Type	Area	Percentage
Urban	616,800	8
Woodland	1,310,700	17
Crops & fallow	539,700	7
Grass & rough	5,165,700	67
Miscellaneous	77,100	1
Total	**7,710,000**	

2011, i.e. by three and a half per cent per annum.[7] Scotland lost around ninety-five per cent of her grouse moors in half a century; in the eyes of many, the remainder are being hounded to extinction by the Scottish Government, the Royal Society for the Protection of Birds (RSPB) and Scottish Natural Heritage (SNH) the last of which provides a scientific veneer to the whim of government. The chief strongholds for grouse are now the Grampians and Cairngorms east of the A9, the Monadhliaths and the Lammermuirs in southeast Scotland with small, isolated pockets of good numbers elsewhere.

> Despite the millions of pounds (£11m in 2011) the estates bring in [to rural Scotland for grouse shooting], we have a fanatically anti-grouse-shooting, anti-elitist, anti-landowner Scottish Government which wants to put us out of business.[8]

Estimates of the moorland extent also vary, mainly according to definition. Twenty years ago, it was estimated that twenty-two per cent of Scotland's 'grass and rough' (1.1 million ha) could be classed as heather but that thirty to forty per cent of that (350 to 450,000 ha) was 'in such poor condition that it may disappear over the next twenty years'.[9] Extrapolating from an examination of ninety-two of the 304 estates classified by the Game Conservancy as grouse moor, it appears there are 1,280,000 ha of heather ground on estates totalling 1,836,000 ha.[10] If so, heather makes up about twenty-five per cent of Scotland's 'grass and rough' (thirty-three per cent if heather/grass moor is included).

It is easier to arrive at a figure for England as the Moorland Association holds good data. Heather ground totals 350,000 ha and makes up 149 moor ownerships with a mean moor size of 2349 ha;[11] the area of managed heather has risen by twenty-five per cent (88,000 ha) since the association was formed in 1986.[12] (The figure for Wales is *c.* 80,000. It appears that no hard data are available for the island of Ireland.)

Table 2.

A comparison between England and Scotland gross annual grouse moor income. Grouse shooting in England is worth five times the Scottish total and is sourced from one third of the area. (+ Blackmore 2011, x Gray, 2011)[8]

	Number of moors	Total area of grouse moor (ha)	Gross income (£million)	Income per moor
England	149	350,000	52+	£348,993
Scotland	140	1,100,000	11x	£78,571

Moorland is the last extensive area of semi-natural landscape and habitat left in the British Isles although the substantial areas of arctic-alpine terrain, coastal islands and native woodlands can also be classed as semi-natural. Moors can be productive agricultural subjects that yield a substantial bounty of quarry species to be cropped by sport shooting as well as being valuable sites for nature study and general recreation. How we manage this great resource is a matter of frequent discussion. The decisions we make now and in the years to come will have considerable impact on the future of a unique area and on how we, today's trustees, are judged by those that follow. In our hands lies the responsibility for the conservation and management of a remarkable assemblage of wild life.

Historical perspective

The worth of heather and associated plants that make up dwarf shrub heath has been recognised for centuries even if its productivity is low compared with conventional low-ground agriculture. From the Middle Ages until the last quarter of the eighteenth century, moorlands provided fuel, winter forage and bedding for animals, thatch for houses, hand tools and artefacts. It was a resource people knew how to manage for a sustained yield. In a rural economy based on cattle, moorlands were grazed at all times of year. This led to great vegetation diversity; plant nutrients, scarce in such an infertile environment, were recycled rapidly and held within the system. Nutrient losses were small when the only substantial exports were mature animals. The transhumance or shieling system ensured that grazing was evenly spread across the contours and throughout the year. On the less fertile western coasts, seaweed was an important source of plant nutrients for peasant agriculture (as it is to this day in western Ireland) though not available inland. Rain brought small annual dividends of nutrients in aerosol form, notably nitrogen, phosphorus and potassium which became incorporated into the nutrient cycle. The system may have been marginal but nothing was wasted. Even the seasonal runs of fish into the rivers would have contributed, representing as they did an import of nutrients into the hinterland from the limitless store in the oceans.

A territorial cock in a deer forest.
(PHOTO: N. MCINTYRE)

Since those times, it has been the shooting of grouse and the preservation of game in general that has driven change and influenced attitudes towards heather and its management. It is useful to examine old writings by sportsmen who had time to travel widely and at a leisurely pace.[13] Thomas Pennant visited Scotland in 1769 and 1772 and Colonel Thornton's sporting tour of 1786 covered a cross section of Scotland and the north of England.[14] The accounts of their many adventures helped to popularise 'Caledonia Stern and Wild'. The pioneers were followed by many excellent writer-sportsmen of the Victorian era who described the situation in Britain

19

and Ireland.[15] For the serious student, the many books published before 1939 make an excellent baseline for assessing game productivity and understanding how game and wildlife management has evolved since. Though they typically describe a British countryside teeming with small game of all kinds, some authors were already aware that all was not well. As early as 1840, concerns were expressed for the long-term survival of heather on Loch Lomondside[16] and a commentator wrote in 1895 that:

> Shepherds have done much to change the appearance of our Highland glens and to bring about a complete metamorphosis in the conditions under which grouse live in many districts. An enormous increase in the number of sheep kept in Great Britain has destroyed and is still destroying a vast quantity of heather.

After a brief comment on heather management, the author tackles the topic of predation with a balance and sensitivity not commonly attributed to the Victorians.

> The next point is to supplement a good supply of food for the grouse by waging war against its four-footed and winged persecutors. The modern game preserver has often a genuine sympathy for the wild creatures that are roughly classed together as 'vermin'. It would be a grievous sin, undoubtedly, to extirpate even 'vermin' altogether. People who have no game of their own to preserve are sometimes tempted to say hard things of those who have because they kill out hawks and other high-spirited creatures. Nevertheless, there is a mean to be attained in game preserving as in all other things. Landowners exercise good sense and decency if they tolerate a little 'vermin' both to gratify their own taste and to satisfy the requirements of an interested public. But if they allow their lands to swarm with 'vermin' entirely unchecked, they injure their neighbours and show a want of propriety.

Under certain circumstances, hill sheep can be responsible for heather loss.
(PHOTO: C. MACGREGOR)

A female hen harrier hunting. (PHOTO: RSPB)

It would be useful if, one hundred and thirty years after these sentiments were penned, the anti- and pro-shooting fraternities could soften the increasingly strident rhetoric, bury some of their differences and start to work together for the long-term good of the uplands. Perhaps this is a vain hope; deeply entrenched positions are held on both sides of the argument and there is a depressing failure to attempt to recognise the different points of view.

Whatever the rights and wrongs of the discussions, it behoves all to accept that were it not for the grouse management that has taken place over the past two hundred years, we would be looking at a great deal less heather today. Historically, the shooting of grouse has put a value upon heather which has helped to prevent further extensive swathes being reclaimed to agriculture, converted to plantation forest or subjected to aggressive pastoral farming, activities that would have added many hectares of moor to the category of high forest or acid grassland, all but useless for agriculture, conservation and game alike.

End notes

1   Ratcliffe & Thompson, 1998; Moorland Association, 2010.

2   Phillips & Watson, 1995.

3   Waddington, 1958.

4   McGilvray, 1996.

5   Fraser of Allander Institute, 2010.

6   Wikipedia: Geography of Scotland.

7   Of the 140, 93 are in the Highlands, Grampian and Tayside regions and most of the rest in the Lammermuirs and Southern Uplands.

8   Blackmore, 2011; Gray, 2011.

9   Thompson, 1992; Robertson *et al*, 2001.

10   Fraser of Allander, 2010.

11   Moorland Association, pers comm.

12   Moorland Association, 2010.

13   It is, of course, wise to take some statements from times past with a pinch of salt but many in the scientific community dismiss historical reports altogether on the grounds that they express opinion rather than fact. Such a rash attitude to sound observation, though widespread, has a touch of arrogance about it.

14   Thornton, 1804.

15   Hamilton-Maxwell, 1832; Hely-Hutchinson, 1871; Maxwell, 1922.

16   MacPherson, 1895.

Chapter 1
Heather and its management

A callunetum.

Introduction

Living heather is a dwarf shrub familiar to anyone who visits Britain's moorlands that occurs from sea level up to 1000 m (3000 ft) across the length and breadth of the country. Its morphology varies greatly according to site; in favourable and sheltered places at low altitude it can grow to a height of well over a metre while mature plants on the upper altitudinal limit may be only a centimetre or two high. Some of the variability is genetic in origin.[1] It may grow in association with other shrubs and trees but usually dominates large areas of ground to the exclusion of most other plants. It may live for less than twenty years in places favouring rapid growth or over a hundred where conditions are harsh. It can regenerate from lateral stem buds, by layering and from the seed it produces in copious quantities. Because it grows poorly in shade, it tends to be a precursor to high forest; most of the heather-dominated moorland of Europe is ecologically forest without trees. When undisturbed by grazing or burning, the sward comprises plants of many different 'ages' because of its layering habit but its nutrient productivity there is often low compared with managed swards. It can leave little scope for other, perhaps more useful, plants to colonise and survive long term. There may sometimes be good conservation reasons for managing heather as pure, undisturbed, even-aged stands but they are uneconomic as range for managed wildlife and domestic animals alike.

About seventy-five per cent of the world's heather moorland occurs in Britain and Ireland; Britain currently accounts for about 1.3 million ha (Scotland, 1 million; England, 0.3 and Wales, 0.08).[2] Since 1945, about twenty-five per cent of the total area of heather has been lost to land reclamation for agriculture and afforestation or to bracken encroachment and reversion to 'white' moor caused by over-browsing. Land use changes have often been matters of hot debate.[3] Much of the moorland in northern England and Scotland is managed for grouse shooting: were it not for people paying good money to shoot grouse, the losses would have been greater. As a dominant or co-dominant, heather now covers only thirty per cent of British uplands with some authorities estimating that over half of it is in a poor or suppressed state due to over-grazing and perhaps a further fifteen per cent is at risk due to under-management caused by the lack of a planned burning programme. Between a half and two-thirds of the nation's moorland requires a complete change in management emphasis. A lifetime's experience of a wide variety of sites indicates that the majority of the heather moor present in 1940 but since lost or seriously damaged could be recovered using well established methods of grazing control and scarifying/re-seeding.

From a conservation point of view, heather management, regeneration and reclamation are key topics as the habitat supports a unique assemblage of birds, many of which are rare or endangered in Britain and Europe and are 'Red' or 'Amber' listed.[4] They include golden eagle, peregrine, hen harrier, merlin, raven, red grouse, black grouse, curlew, dunlin and golden plover. Recent published work shows that these species are found in greater numbers on managed moorland and, in some cases, reach their highest recorded densities there.[5] Though it has for many years been a declared intention of successive governments to conserve and enhance Britain's heather moors, the standard of moorland management declined steeply over wide areas during the twentieth century and, particularly in Scotland and Wales, continues to do so. If the decline is to be arrested and reversed, it is necessary first to understand the basics of the ecology and management of heather moorland.

A typical mixture of shrubs, herbs, mosses and lichens growing at high altitude.

23

The generic term for the many combinations of plants in a sward containing a high proportion of ling heather is 'callunetum'. Heather moorland in Britain is highly variable. Rodwell's definitive handbook describing its variability was published in 1991; its classification of the main plant associations is used here.[6] As techniques for the proper management of calluneta are equally varied, the starting point has to be the categorisation of different moor types; each one responds differently to fire and the attentions of the grazing animal. Though the moor manager needs to be able to distinguish the broad types, detailed botanical knowledge is not essential. Of Rodwell's twenty-eight categories of mire and twenty-two of heath, the manager needs to be familiar with nine, only four of which require controlled rotational burning.

Politics, public bodies and moorland management

In Great Britain, countryside issues have always tended to be better understood by right-of-centre administrations. Social-democratic ones seem generally less attuned to the detailed workings of the countryside perhaps because the majority of their members have not grown up there. While countryside matters tend to get a reasonably fair hearing in England, the situation is different in Scotland where over eighty per cent of the members of the Holyrood parliament are members of two left-leaning political parties. This may have a bearing on the long-term future of wildlife management and game shooting in Scotland.

Over the past one hundred and thirty years, there have been considerable changes in land ownership with many huge estates being divided up, often into much more manageable parcels. In 1872, the only significant public owner of land was the Church of England with 891,000 ha, almost all of which was let to agricultural tenants. The nine next largest landowners together owned 1,582,186 ha.

A hundred and forty years later, the emphasis is different; the Forestry Commission (FC) owns 1.05 million ha followed by the National Trusts for England (NT) and Scotland (NTS) (332,000 ha), the Ministry of Defence (MoD) (240,000 ha), the Crown Estate (146,000 ha) and the RSPB (130,000 ha). Pension funds own a further 223,000 ha and water and power companies 200,000 ha, making a total of 2.3 million ha of land in public or quasi public ownership. This figure does not include land covered by National and Regional Park designations where management objectives are often not driven primarily by the profit motive.[7] The two largest private landowners are Buccleuch Estates (97,000 ha) and Atholl Estates (61,000 ha) with a high proportion of both being moorland of relatively low value per hactare.

A large estate is not necessarily managed any better or worse than small parcels of land but land that becomes too fragmented creates problems for game and wildlife and the manager of a large block of land who has clear objectives is more likely to see his plans implemented.

The average landowner can only stay in business if he trades successfully, year in, year out. Landowners must take steps to stay solvent and need in many cases to 'up their game' and improve their corporate image as well.[8] No such constraints appear to influence the managers of land in public ownership whose objectives are often primarily to satisfy the demands of the general public, members of which seek access for recreation in a wide variety of forms. 'Membership' of these bodies may comprise subscribers to RSPB and conservation bodies such as the John Muir Trust (JMT), Scottish Wildlife Trust (SWT) and the National Trusts, shareholders in the public utilities or the general public (FC and National

Parks). The management has to take note of the various interests and do what it can to accommodate them. In many cases, decisions are made without reference to the best interests of the farming community earning a living from the land or of wildlife managers seeking to enhance biological diversity. For example, it is not uncommon to find that perfectly legal predator control is either discouraged or forbidden outright because of the risk of offending public sensibilities.

All too often, boards of management are made up of people from peripheral interests which, in the name of inclusivity, include representatives of community councils, local schools and animal welfare groups. Large and unwieldy committees make it difficult for an executive to exercise dynamic management because of the need to satisfy diverse viewpoints. Because the primary land uses of farming, forestry and shooting are often poorly understood by the general public, it can be difficult for those in charge to manage land in a proactive way even if they are well informed and strongly in favour of existing land uses. Many decisions end up being made on the basis of expediency or political whim.

Regional differences and classifications

The moorland types that Rodwell classified which are particularly relevant to those involved in farming and game shooting are summarised below.

M19   *Calluna vulgaris/Eriophorum vaginatum*

Common parlance: Heather/cotton sedge blanket bog, 'Flow' in the north of Scotland; occurs at all altitudes and varies in appearance; where burning has been frequent and grazing heavy, it is tussocky with sub-dominant heather partly concealed in the sedgy tufts; where grazing has been light, particularly in winter, heather dominates the sward; blaeberry and crowberry are common.

H9   *Calluna vulgaris/Deschampsia flexuosa*

Common parlance: Heather/wavy hair grass moor. Widespread in Pennines and North York Moors, derived from H12 by decades of industrial pollution; when burned, fire sites are initially colonised by grass; in summer, flowering grass stems give moorland a characteristic pinkish appearance if viewed from a distance.

H10   *Calluna vulgaris/Erica cinerea* [9]

Common parlance: Ling/bell heather mixture. Often mixed with fine-leaved grasses and forbs on better quality brown earth soils; bell heather usually dominant after fire but ling takes over as the stand matures; under severe grazing pressure, white bent (on dry ground) and purple moor grass (on ground with seeping water) can take over.

H12   *Calluna vulgaris/Vaccinium myrtillus* [10]

Common parlance: Heather/blaeberry mixture often with crowberry. Typical of large swathes of moorland; blaeberry is often prominent after fire and can dominate on good soils particularly on back-lying slopes; at higher altitudes away from the western seaboard, cowberry[11] can be plentiful particularly on better soils; soils tend to be free draining; plants typical of boggy or moist ground such as purple moor grass, deer and cotton sedge usually absent; heather regeneration from seed can be inhibited due to colonisation of bare peat surfaces with blue algae creating anaerobic peat conditions; heavy grazing over long periods tends to increase grasses, particularly on brown earth soils often allowing bracken to encroach and colonise former heath land.[12]

H13   *Calluna vulgaris/Cladonia arbuscula*

Common parlance: Heather/reindeer moss mixture at high altitude. Typical of exposed ridges and high ground in northeast Scotland above H12; provides summer grazing for deer, habitat for ptarmigan and good breeding habitat for dotterel and golden plover; vegetation kept short by severe climatic conditions; should never be burned. See H16.

H16   *Calluna vulgaris/Arctostaphylos uva ursi*

Common parlance: Heather/bearberry. A high altitude community typically found in the central and eastern Highlands; can have a substantial grass and forb component on rich soils providing good grazing for sheep and deer; due to altitude and exposure, burning should be performed infrequently and with extreme care.

H18   *Vaccinium myrtillus/Deschampsia flexuosa*

Common parlance: Blaeberry/grass moor at altitudes of 400 to 800 m in the central and eastern Highlands. Has much in common with H13; can be important range for grazing animals in summer and carry large breeding populations of grouse; similar to H12 but without the heather component; needs no management.

H19   *Vaccinium myrtillus/Cladonia arbuscula*

Common parlance: Blaeberry/moss at high altitude. Very like H13, occurs in similar places and merits the same lack of management.

H21   *Calluna vulgaris/Vaccinium myrtillus/Sphagnum capillifolium*

This association, although unusual, is included here because it is common in the northwest of Scotland, particularly on steep, back-lying and sunless hillsides; may receive varying levels of grazing, particularly by deer but is sensitive to damage by fire; hardly ever managed by shepherd or keeper; location means it rarely dries out sufficiently to carry a fire; often of significant conservation interest as it consists of long heather which falls over, layers in the sphagnum understory and allows blaeberry to flourish.

The survival of heather

Beyond the boundaries of nature reserves, heather tends to survive poorly unless attention is paid to its economic value. If it is neither grazed nor burned, it becomes colonised by trees. We have kept as much as we have because grouse shooting has justified its management and maintenance particularly on hillsides below 500 m (1600 ft) which is where most grouse shooting takes place.[13] Above this, climatic and soil conditions are marginal for the production of sown crops and grasses and the pressure to change heather moor by cultivation falls. However, the high ground still produces annual growth available for use by both wild and domestic animals though if there are too many of either decline of heather dominance and subsequent site deterioration can take place.

Ten-week-old grouse brood on dry heath. (PHOTO: N. MCINTYRE)

Effect of soils on heather survival and performance

The fertility of any soil affects the nutrient content of the plants growing on it and, in turn, any animals using them for food and shelter. Inevitably, this influences the plants' performance. As a rule of thumb, the more base-rich a soil, the greater the biomass and variety of insect and plant species and the number and performance of vertebrates which live there.

Moorland soils vary widely. A thin, free draining immature soil made up of large particles of a hard, acidic and infertile rock such as granite or quartzite supports a different set of organisms than a mature and stable soil sitting on a base-rich parent material such as basalt or limestone. Even allowing for variables such as altitude, exposure, rainfall and grazing regime (historic and current), it is easy to appreciate the importance of soil quality; much of the variation in the performance of tracts of heather moor in different parts of Britain can be attributed to its influence. The granites, quartzites, gneiss and undifferentiated schists of the north and west of Scotland generally support poorer moorland for grouse and sheep than the limestones of the north of England. Of course, management exerts an important influence but, however well burned and grazed a heather moor overlying granite may be, it is always out-performed by a superficially similar moor overlying limestone or basic rocks such as basalt, diorite, epidiorite and hornblende schist managed to the same level.[14] Under comparable circumstances over a period of years, grouse breeding densities and brood sizes are higher on moors overlying the 'better' rocks and sheep are bigger framed and produce more lambs. Where they occur, red and roe deer grow larger bodies and antlers.

Moors overlying poor rocks are more vulnerable to degradation and fertility loss under moderate or poor management than richer ones with a greater fertility 'credit' and where the natural processes of soil weathering help to replenish the fertility bank. Base-poor soils have a lower fertility base and are readily depleted by clumsy management. They are, for example, vulnerable to nutrient losses caused by over-grazing or leaching after a hot fire that are not readily made up because the parent material is impoverished and weathers slowly. On such 'poor' moors, dramatic changes for the worse in the vegetation mix may occur over

Summer fire burned into scree.

three or four burning rotations (sixty to a hundred years). Poor management degrades the soil, tends to increase podzolisation and may result in erosion of the peat layer which holds most of the nutrients, further reducing productivity.

With the exception of blanket bog, most moorland soils which today support grouse moors and sheep walks have been brown forest in character at some time in the not-so-distant past. When natural, mixed broadleaved forest covered the land below 600 m (2000 ft), the annual leaf fall would have been a mixture of deciduous leaves and pine needles, creating mull humus on top of the A to C soil horizons. Tree roots would have penetrated to significant depths, drawn up nutrients, deposited them annually on the top layer as leaf litter, shattered rock and improved drainage. Leaching would have been almost non-existent and losses from the system small.

Once the trees had by and large been replaced by settled agriculture within the forest matrix (a process well advanced by about 2000 BC), heaths developed extensively on the cleared land. The character of the underlying soil began to change, becoming podzolised and inherently less fertile. Through time, many of the plant nutrients accumulated during the forest epoch disappeared through a combination of leaching and livestock export. As long as cattle were the main grazing animals, the process was slow and the decline almost imperceptible: the significant losses are associated with sheep farming and have taken place over the past two hundred years.

After the development of a callunetum, heather leaf litter made a poor substitute for a large annual deciduous leaf fall; a 'mor' or acidic raw humus would have developed on top of the forest soil. Humic acids released by its decay leached through the underlying soil, dissolving and washing out salts of iron and aluminium even if regular heavy grazing and trampling by cattle slowed up the process and helped to retain fertility in the same way as forest leaf fall. Cattle grazing slowed the rate of degradation and podzolisation. By the second quarter of the nineteenth century, cattle numbers were declining as their place was being taken by sheep in ever larger numbers. The grazing habits of sheep and the vegetation management associated with their husbandry encouraged vegetation change and tended to accelerate the formation of a stratified soil which makes up a typical podzol that was less productive than the brown earth or forest soil it replaced and so moved moorlands inexorably down the pH scale. Only the most base-rich land maintained a semblance of its former character. The process was slow at first but by the end of the Second World War it was accelerating, assisted as it was by subsidies aimed at improving agricultural productivity. Vulnerable districts in the wetter, western side of the country were the first to lose most of their game and wildlife interest as soils rapidly degraded under increasingly harsh burning and grazing regimes implemented by farmers, shepherds and crofters.[15] The degraded areas have become progressively more extensive. Almost all good grouse moors are today located on 'better' rocks in dry districts. Few, if any, that overlie infertile rocks, irrespective of how well they are managed, consistently yield bags that compare with the 'rich' moors. It is not only grouse which reflect this difference: insect-eating birds such as meadow pipits are many times more numerous on moorland sites where the soils have a high pH compared with wet, acidic desert. Anyone who doubts the scale of change need only compare *The Sportsman's Guide* with the situation that prevails today.[16] Fifty estates on which grouse were shot were randomly selected from the 'wet west' of Argyll and fifty others from the 'dry east' of Aberdeenshire. Of the Argyll moors, only six (twelve per cent) shoot grouse today whereas twenty-nine (fifty-eight per cent) do so in Aberdeenshire. Argyll has probably suffered more from fragmentation and afforestation than any other Scottish county except perhaps Dumfries & Galloway; the changes in land use have certainly not helped. The moorland in the western half of the country is generally impoverished with widespread heather loss: purple moor grass and white bent are rife and dominate extensive areas. Deer sedge – a certain indicator of degraded, compacted soil – is a major constituent of the sward over vast tracts of moor. That losses have been less marked in Aberdeenshire has much to do with the tenacious way that heather can better withstand abuse in the drier climate of the east than on equivalent stands in the west.

Rough cattle on rough ground. (PHOTO: D. PHILLIPS)

In the drier half of Britain, where transpiration in summer tends to exceed precipitation, the hard and impervious brown coloured pan which develops in the 'B' horizon is made up of leached and precipitated iron salts. It can impede vertical percolation, producing a false water table which leads to waterlogged soil, stifled roots and lower bacterial activity that can result in a sward dominated by plants of little value like deer sedge and cross-leaved heath. This anomalous situation of extensive areas of waterlogged mineral soil in an otherwise dry district surrounded by well drained heath is common in northeast Scotland. Many thousands of hectares could be improved dramatically and cheaply if owners invested modest sums to break the pan mechanically.

Ross tine plough.

LEFT:
Ross tine plough breaking pan.
(DRAWN BY: D. PULLAN)

BELOW:
Grouse brood on wet heath.
(PHOTO: N. McINTYRE)

Peat makes up a third group of moorland 'soil'. It is generally accepted that it has to be at least thirty centimetres deep to qualify for the name. Peats can, of course be many times that depth; in Ireland, 8 m (26 ft) is not uncommon. The material has been harvested, air dried and burned in purpose-built power stations to make electricity by the state body *Bord na Mona* since 1950 with three million tonnes a year still being harvested from 25,000 ha of blanket bog. The characteristics of deep peat are its permanently waterlogged and virtually undrainable state. It is a hostile rooting medium which is acidic, infertile and short of oxygen. Where it is deep, plants growing on it cannot reach down far enough to draw nutrients from the bed rock so it is of no account what lies underneath. The nutrient content of plants growing on peat is only likely to be enhanced if the peat bog is flushed locally with base-rich water coming from elsewhere. The main nutrients for plants growing on bogs arrive in aerosol form as rain and snow or from particles originating from within the solar system. It has been calculated that an average of 6 kg per ha per year of nutrients arrives from this source compared to the 250 kg of nitrogen (N) (more than forty times as much) which is the normal annual fertiliser dressing on a grass field cut for silage in British lowlands.

Mention should also be made of the effect of glacial drift and flushing. Drift is defined as the importation of mineral material by ice or water from a different area and its deposition on top of other rocks and soils often far from its place of origin. If the imported rocks are base-rich and laid on top of acidic material, they improve that area. On the other hand, a layer of granite debris carried by a glacier and laid down on top of richer rocks will have the opposite effect and make the area less productive, particularly if the layer is thick (> 60 cm). Flushing is the term used for the process whereby water lying on an impervious stratum such as bed rock or a pan emerges in a series of upwellings as individual springs or a spring line. If that water has passed through base-rich mineral, it is carrying dissolved salts and has a fertilising effect on the down slope below the point where it emerges.

In the twenty-first century, red deer are heavily implicated in the process of soil degradation.[17] The lay public is invited to express opinion on the present status of deer by the likes of World Wildlife Fund for Nature (WWF) and RSPB, making the animal increasingly a political football.[18] In Scotland, deer numbers were at their lowest at the end of the eighteenth century; writers recording stalking experiences at that time describe days spent searching but not finding deer in places where counts today indicate a settled population of 1500 or more.[19] Writing in 1845, Charles St John described in arresting detail the sighting, stalk and subsequent killing of a notable trophy animal in Assynt, Sutherland. A shepherd on his way home from church on a Sunday reported the sighting and the chase began on the Monday morning. The 'hart' was finally grassed on the Saturday; the pursuit had lasted nearly a week during which time hardly any other deer were seen.[20] By 1965, there were reckoned to be 150,000 red deer in Scotland. Current counts of groups living on open hill suggest a total of around 350,000 individuals though this is likely to underestimate the national herd because many deer live in plantation forest where they are almost impossible to count accurately.[21] Indications are that numbers have at least doubled over the last forty-five years despite a substantial contraction in available range. As deer are mobile and travel far for palatable forage, management of their range tends to consist of large and frequent fires that further degrade large areas of ground currently used as deer forest. Conservation bodies such as JMT, SNH and WWF

Erosion resulting from high concentrations of wintering stags, Monadhliaths, May 1998.

are currently seeking drastically to change management practices on existing deer forest land by pursuing a policy of creating swathes of scrub woodland to enable the moorland to 'repair' itself and have a rest. However, as many estate owners have come to rely on deer in significant numbers as a source of revenue both on established deer forests and in places where they were all but unknown fifty years ago, it is a policy that can destroy the value of an estate if it is imposed on the owner or, if imposed on adjacent land, can have serious implications for the way he is able to manage his own.

Heather burning – general

The beneficial effect of heather burning on populations of domestic animals and grouse was recognised by the middle of the nineteenth century. Sporting authors of the period all acknowledged in varying degrees its value for grouse and sheep. Before 1850, sheep managers typically burned a tenth of the moor each year, usually in a few big fires, but rotations had become much longer by the third quarter of the century. Those interested in shooting grouse over pointing dogs needed a predominance of long heather to encourage birds to sit and thereby provide a longer shooting season – birds quickly become wild and unapproachable if the heather is very short. By the time of the 1873 grouse disease catastrophe, burning programmes on dogging moors were often covering as little as a hundredth of the moor per year.[22]

Burning is necessary to maintain the open character of moorland; if no burning is done and grazing ceases, most moorland reverts to scrub dominated by birches, Scots pine, rowan and willows. Scrub formation is the first step towards the re-creation of high forest. In parts of northeast Scotland, Scots pine colonises heather rapidly if there is a source of seed from 'granny' trees nearby. There are many examples of land which was open moorland with hardly a tree in sight in the 1950s which are covered with closed canopy pine and birch today. In former times all our most productive grouse moors were covered in forest with red grouse largely confined to ground above the tree line (460 to 615 m, 1500 to 2000 ft), except for pockets of open bog within the forest matrix, areas of open ground following forest fire and on exposed coasts. Grouse today have a much larger range available to them than they had in boreal times at the end of the last ice age some 10,000 years ago. They do not live permanently in woodland; as soon as the canopy begins to close, they desert the area. Regular burning destroys colonising tree seedlings: after a fire the plant succession starts again with heather as the main and pioneer specie.

Grouse need heather shoots of high feeding quality mixed intimately with cover for nesting and protection from predators; achieving this is the main objective of management of moorland for them.[23] They feed on young plants between 10 and 20 cm in height because these tend to be more nutritious and to come into growth earlier than older plants. The ease of feeding is important as grouse like to peck at head height, unlike other gallinaceous birds which feed off the ground.

ABOVE:
Underburned/ ungrazed heath soon reverts to forest.
(NO310022)

LEFT:
Shepherd's fire on Mull 1974.
(NM501401)

33

It is very easy to burn heather: a good dry spring day with a moderate wind is all it needs to achieve hundreds if not thousands of blackened acres but grouse are scarce on moorland treated this way. What is desirable is an intimate mixture of long and short vegetation while maintaining a sward of vigorous heather, defined as plants which are still finely branched and growing densely thus providing a solid stand which excludes light from the soil and prevents the establishment of plants of lesser value. This can be achieved only if fires are carefully controlled and the rotation fairly short. Rotation lengths vary considerably from site to site. If they are too long, plants reach the degenerate stage and the heather sward opens out, allowing mosses, lichens and sometimes even trees to establish and impede the establishment of new heather seedlings. Under most conditions in Britain, the ideal rotation for maintaining heather as a dominant plant is between eight and twenty-five years, the former being typical on dry land at low altitude in the north of England, the latter on wetter ground at higher levels in Highland Scotland.

Summer fire in a deer forest. (NC395565)

Heather morphology

The moorland manager recognises five growth stages of heather with the first, pioneer, occurring in two forms: stem based growth which can start within a week or so of the fire's passing and seedlings which may take several years to appear.

The second is *building* heather, 10 to 20 centimetres (25 to 50 in) long and flowering with the category normally split into *early* and *late building*. Ground coverage is nearly complete with plants at the preferred stage for grouse food.

LEFT: *Blaeberry and heather four months after burning.* RIGHT: *Two years after burning. Stem-based heather regeneration is flowering, the other plants are seedlings.*

Building heather (foreground) and mature (behind).

When the canopy closes, the heather is at the *mature* stage, its length 30 to 40 cm (11 to 15 in). Stems are dense and finely branched; the heather closes up the sward and dominates other plants. If this age class is burned, regeneration is rapid and complete.

Four months after burning mature heather. Rapid stem-based regeneration.

35

If left undisturbed, the stand self-thins and enters the fifth, *degenerate*, stage at which the plants, 50 cm (20 in) or more long, become top heavy. As they fall over, the canopy opens up, light reaches the ground and mosses often grow profusely in the damp semi-shade. When this class of heather is burned, regeneration is usually slow and has to come from seed. The tiny plants are highly vulnerable to sheep and deer grazing and may be pulled out and destroyed. In the extreme, browsing pressure can convert a sward where heather dominates into one which is primarily grasses, sedges and moss in the space of a few years. To maintain heather's dominance, it is important that it is burned no later than the mature stage so that regeneration after fire is rapid and complete.

Degenerate heather on left, mature on right.   *Typical sheep damage on pioneer heather.*

Autumn and spring burning

Although the law allows the burning of heather in Scotland and England for six and a half months of the year, there are long periods when the weather is unsuitable and the vegetation too wet. In practice, the season falls into two parts. October and November for autumn burning and December to mid-April for the so-called spring burning season. In former times, there was robust debate among keepers and managers as to the comparative merits of burning in the two seasons; some were implacably opposed to autumn burning because 'it lets the frost into the roots' while others held that problem areas such as back-lying slopes which often held snow into the late spring could only be burned in autumn. In parts of northeast Scotland, heather burned in autumn usually regenerated better in the following growing season than plants at a similar stage of growth burned in spring. The temperature of the fire and hence the speed and plant density of regeneration depends on the relative dryness of the top growth and the litter underneath when the fire is set.

In a dry autumn, the litter layer can be bone dry and highly combustible; a fire consumes it all, leaving a firm seed bed of compacted peat. In the absence of stem-based regeneration, the heat and smoke affect heather seeds lying on the surface and encourage germination the following season. This is common in areas of degenerate heather where there can be a high fuel load. The old stick heather is green and sappy but burns slowly, the fire moves slowly and consumes everything in its path, leaving the soil surface clean and firm. The scorching effect of the passing fire stimulates seed germination the following year. If degenerate heather is tackled in spring, the litter is soggy even if the top growth is tinder dry. The result is a fierce fire which is hard to control and has little or no effect on the understory of damp moss and litter referred to as fog and its presence inhibits seedling regeneration. Where possible, old heather should be tackled as a priority in a dry autumn as the fires are easy to manage and regeneration is nearly always good. Spring burning can produce hot fires with minds of their own. If the programme concentrates on late building and mature heather which is finely branched with a limited understory, regeneration can be excellent as it comes mostly from lateral buds at the stem bases rather than relying on seed.

In summary, if the litter layer is really dry in autumn, the manager should concentrate on degenerate heather as this ensures that this problem type can be got rid of; it regenerates well from seed. At all other times of year, if the top growth of younger plants is dry enough to burn, the effort should be applied there because regeneration is mainly vegetative, starting within weeks of the passage of the fire, and because cover subsequently is both rapid and complete.

The burning seasons

Historically, heather burning has been controlled by statute.[24] In Scotland, heather burning (muirburn) was permitted between 1 October and 15 April inclusive although, on the authority of the proprietor or the Department of Agriculture and Fisheries for Scotland (DAFS), the period could be extended until 30 April on land below 450 m. Above 450 m, the season ran from 1 October to 30 April, extendable as above to 15 May. There was no provision for extending the season by local licence as there is in England. The Wildlife and Natural Environment (Scotland) Act (WANE), which came into force in 2011, incorporates all previous legislation into one statute and includes provision for the granting of licences to burn heather in September though only at the discretion of SNH.[25]

The legal season for burning heather in the English uplands (Less Favoured Areas (LFA)) is 1 October to 15 April. There is no statutory mechanism permitting extension at either end though, under certain circumstances, a discretionary licence can be (and has been) obtained from the local office of the Department for Environment, Food and Rural Affairs (DEFRA).

In 2008, the Welsh Office decreed that the burning season would be shortened and extend henceforth from 1 November to 31 March. As standards of heather management in Wales have tended to lag behind other districts in Britain in recent decades, taking away October, often one of the best burning months of the whole winter, will do nothing to improve management within the Principality. Burning standards will decline still further and the move will knock yet another nail in the lid on the coffin containing Welsh grouse and moorland birds.

Burning for red grouse in Ireland[26] and willow ptarmigan in Norway has been carried out with mixed results on both the birds and the vegetation.[27]

Heather burning guidance

A welter of information emanates from Natural England (NE) and Scottish Natural Heritage (SNH) on heather burning. SNH has published a succession of 'Muirburn Codes' complete with illustrations and exhortations. The latest Codes of Practice contain contradictory instructions for the guidance of keepers and shepherds; were their recommendations to be implemented widely, they would reduce the area of productive heather moorland throughout the UK.[28] The guidelines seem to have been written by people with the sketchiest experience of burning in the field; practices are advocated which are demonstrably wrong and photographs ostensibly illustrating one point highlight an altogether different one. Recommendations as to suitable conditions for burning are nebulous and, if followed to the letter, would result in many fewer acres being rejuvenated. A lack of professionalism on such a scale helps neither owners, farmers, shooters and keepers on the one hand nor those interested in conservation on the other. There is a crying need for a good 'How to...' booklet on moorland management drawn up by the likes of the Scottish Rural Property and Business Association (SRPBA), the Country Landowners Association (CLA), the Moorland Association (MA) or the gamekeeper organisations for circulation to all practitioners. A clear definition of objectives by interested parties would clear the air and provide a basis for dialogue.

High altitude (600 m, 1800 ft). Very steep, many deer and at risk of gully erosion. Such sites should not be burned. (NN625765).

The burning programme

It cannot be over-stressed that heather burning for farm stock and game is intended to ensure the regeneration of vigorous heather. Heather must be burned no later than its mature phase if it is to regenerate rapidly from dormant stem buds nourished by an established root system. If heather is degenerate when it is burned, the plants die and a sward can only be re-established from seed. This is almost always slow, resulting in patchy cover (particularly if grazing animals are present) and, often results in 'weed' species gaining a foothold. It is *far* more important to get the right acreage burned each year (thus ensuring the continuance of heather dominance and reducing the risk of encroachment by poor quality grasses), than it is to achieve a fine-grained pattern. Such patterns, though good for grouse, must not lead to an unacceptable lengthening of the rotation. Particularly if sheep and deer are present, the longer heather must not become too old to regenerate satisfactorily and pose the risk of the moor reverting to an association of plants of lesser value.

A myriad of tiny seedlings under scrog.

It is useful to examine the normal burning practice on a typical moor. Because keepers know that many small fires support more grouse than a few big ones, they try to get many fires of 0.2 to 1 ha (0.5 to 2.5 ac) burned on their hill each year. If they experience a series of wet springs when suitable days for this work are scarce, they fall behind and the burning rotation becomes longer than the optimum. A simple calculation can highlight the problem.

Consider a typical 'two day' driving moor of 2000 ha (5000 ac) being looked after by one keeper. Assume that examination of the ground shows that some 1600 ha (4000 ac) has to be burned in rotation, the remaining 400 ha (1000 ac) being grassy flushes, bogs, overgrazed places where stock congregate in winter and so on or as high ridges where burning is neither necessary nor desirable.[29] If the moor lies at between 300 and 400 m (1000 to 2000 ft), the heather is likely to reach the mature stage in eight to twelve years at low altitude and perhaps twenty years on the tops, an average of, say, eighteen years. Ninety ha (220 ac) have to be burned every year. If the average fire size is 0.4 ha (1 ac) – and this is a fairly substantial fire, being over sixty metres square – then two hundred and twenty fires have to be burned each year.

According to district, there are typically between fifteen and twenty-five burning days each winter so, on average, the fires have to be done in twenty days, therefore eleven fires of an average 0.4 ha (1 ac) have to be done each day. Since two competent men with swiping and water equipment achieve an average of six fires a day under good conditions, it needs two squads of three or four men to be sure of achieving the programme.

A good pattern of fires show up well in winter.

However, the twenty burning days include Sundays, Easter Day, days when the keeper's wife has to be taken shopping, his children have birthdays, funerals have to be attended and so on. Unless the keeper is single-minded (and probably unmarried), it is impossible to exploit every suitable day – he is likely in practice to need between nine and twelve people in two or three squads. Few estates can turn out even six men at a moment's notice. The result is predictable: the average moor is burned with some small fires as described but also with others, probably accidental, of 2 to 42 ha (5 to 10 ac). The area burned each year rarely reaches the desired one eighteenth of the whole, the moor becomes progressively under-burned and, as the years pass, the proportion of degenerate heather increases.

The tools for burning

A robust aluminium scrubber with weld mesh.

In former years, keepers used heather 'besoms' or brushes made of birch saplings and twigs with a covering of wire netting, the whole bound tightly with soft wire. They were heavy, unwieldy, too short, would blister the horniest of hands and usually broke at critical moments. Beaters made of rubberised quarry belting such as are commonly on display at the edges of tree plantations, were sometimes used. However, neither design is satisfactory under the wide variety of conditions encountered in a season. The heather burner of today must look for more effective burning aids than those archaic designs if he is to take advantage of every available weather window. Despite the regular occurrence of disastrous fires which attract widespread condemnation, many people still try to burn heather using antiquated equipment rather than purpose designed metal 'brushes' which do not break, melt or catch fire at critical moments. No serious heather burner can make a long-term success of the task without tools that are light, strong and well designed.

A head made from steel plate – note the under-hanging chain.

No standard is best for all field situations but there are basic types of kit which show the difference between an excellent tool and an indifferent one. The handle must be long enough for the operator to stand well back from the fire – 3.5 m (12 ft) is not excessive – with a diameter between 30 and 50 mm (1.25–2.0 in). Aluminium pipe sold for mounting television aerials makes an excellent handle. The head should have a scrubber made of a substantial mild steel plate punctured with plenty of 25 mm (1 in) holes – the perforations reduce the weight and make the brush more effective. A loop of chain should be suspended underneath to agitate the burning heather as the brush is pushed backwards and forwards: it flicks the twig stumps and puts out the fire. The head can also be a square frame covered with stout weld mesh which kills the fire by acting like a car radiator, cooling the burning gases to below combustion temperature.

A design that works well as a beater rather than a scrubber and is particularly useful when burning among rocks or on rough ground can be made by fastening a bunch of newly cut birch twigs to the top end of a long aluminium shaft using a couple of Jubilee clips and a series of spikes let into the shaft at right angles. See left of picture on page 40. The twigs are easily replaced when they get broken or dried up following use at a hot fire.

Face masks are a statutory Health and Safety 'must'.[30] No prudent person would ever go to the hill to burn heather without one attached to his belt. It may hang there unused for a whole season but there will be an occasion one day when it is badly needed to avoid a disastrous fire.

The burning squad may vary in number and make-up from day to day. At weekends it may be possible to employ teenagers who, while not being strong enough to wield a full-sized besom, are perfectly at ease with a half-sized one or a light aluminium grain shovel on a long handle. They are useful for dealing with the small back-fires which spark up from time to time and need timely attention. People may scoff at the suggestion of using youngsters in this way but it provides them with valuable experience and allows them to become familiar with the decisions that need to be made in connection with burning. Once competence is acquired, it is never forgotten. In fact, many tasks can be discharged by the young or the elderly. There are always fuel or lunches to be fetched, messages to be carried between squads or, perish the thought, help to be run for if disaster threatens. In recent years, many owners have invested in machines to assist with fire control, typically cutters to create fire breaks and water dispensers to assist directly in controlling fires. Because little physical effort is required to operate them, they can be manned by an older person whose days of wielding a heather besom are over.

A rotavated strip designed as a fire break.

Cutters

A Wessex Scrubmaster cutting a fire break immediately prior to burning.

In stone-free areas, fire breaks can be cut using a swipe, a flail-type forage harvester or a brush cutter. A small cutter can be mounted on the back of a four-wheel-drive quad bike.The technique is widely used on suitable areas of dry heath in east Scotland and the north of England. Those who find what it can do for them become great advocates of cutting *as a critical aid* to better burning. It is useful when labour is scarce and the burning squad faced with substantial beds of even-aged mature or degenerate heather. Cutting has to be done *immediately* prior to burning. The cutter macerates the heather into a coarse, soggy mulch and a fast-moving hot fire dies on contact with the barrier so produced. However, if the mulch is allowed to dry out for a few days, the control strip becomes unreliable as the fire gets into the half-dry blanket and may smoulder for hours. The advantage of the swipe on suitable ground is that a man can start cutting breaks in the morning while the heather is still wet with dew and the burning squad can light the areas so isolated as soon as the dew evaporates. A small squad can cover much ground in a short time as the fires normally need little attention; under really favourable conditions, as many as a dozen or fifteen can be burned by one squad in a day. Unfortunately, the technique is of no value where there are many stones and unsatisfactory if the ground is wet and dissected by sheep drains.

It is impossible to generalise on the topic of which machine is best as circumstances vary so much on different moors. In the eastern part of the country, where the land surface is dry and often level, heather can extend unbroken over hundreds of acres and an eighty or a hundred hp four-wheel-drive tractor can get to most places and handle a large swipe fitted with chains or hardened-steel blades which can cut a swathe up to 1.75 m (6 ft) wide. On wetter ground, smaller cutters mounted on or trailed behind a quad bike work well. These smaller machines usually incorporate a power unit to drive the cutting reel. Redundant forage harvesters with a reel equipped with flails can also be useful and have the great advantage of dissipating the litter by throwing it into the air where it is caught and spread by the wind. Swiping has the disadvantage of leaving a windrow of trash that can take many years to rot and so impede regeneration. By smothering the underlying vegetation, it may even trigger water erosion. Some people feel that the risk of breakages is unacceptably high if there are many stones on the hill but there are places on most moors which are relatively free of stones and where the machine does a good job. Even if only a portion of the moor can be treated in this way, it releases men to tackle more difficult places.

Water – an aid to burning and spray kit

The element entirely antagonistic to fire is water. No keeper can exploit every day that heather can be burned without the reassurance of a copious supply near at hand. Taking a tank of water to the hill is a bit like flying with a parachute – you may never need it but, if you do, you need it desperately. On ground where tractors can go, a modified agricultural crop sprayer with a hand-lance fitted to one of the booms (the other being blocked off) has a useful part to play in dampening down smouldering sticks, litter and peat. It can be used also to lay a trace of water in front of a strong fire. If this is done in a herring-bone pattern, the fire edge can be squeezed in manually and the fire brought under control. Some people have experimented with fire-retardant chemicals which are deliquescent and create a sticky gel of some persistence but they are expensive for everyday use. Others have used foam but the layer often lies on top of the vegetation and allows a fire to creep underneath. In practice, plain water is as good as anything and can be transported to the hill in quantities from as large as 150 gal (675 lt) in a spray tank down to 3 gal (13.5 lt) in a knapsack sprayer. Each operator has his preference, probably based on the available equipment. Most hill vehicles in common use can readily be modified to carry a water tank and a little ingenuity enables most people to have a good supply on hand. An Inverness-shire keeper designed and built a high-pressure water-spraying system for his Argocat in the early 1990s, a design incorporating modifications which has since been widely copied with many keepers. Up-to-date versions can be bought off the shelf. The critical part of any spray machinery is the output pump. It should atomise limited amounts of water – 1 to 2 gallons (4.5 to 9 lt) a minute being typical – into a plume of fine mist. If the water particle size is small, not only does the full tank go further but the fogging effect kills a fire quickly.

ABOVE:
A high-pressure fogger.
(PHOTO: M. BROWN,
DUN-ALISCAIG ESTATE
MANAGEMENT LTD)

RIGHT:
*The AWT400 HP
Muirburn Control
System, which fits
Argo All Terrain
Vehicles, includes a
pump for filling the
tank and a second,
high-pressure, low-
volume pump for fire
control and spraying.*
(PHOTO: M. BROWN)

The problem with water is that it is heavy. A purpose-built 350 lt (80 gal) tank with a mist pump can sit in the well of an Argocat and go virtually anywhere but the payload approaches half a tonne; an agricultural sprayer holding 1500 lt weighs in at well over a tonne. It is obviously important that suitable sources of clean water are identified so that little time is wasted during refilling. Lochs, bog pools and burns should be surveyed in advance to make sure that the machine can get close to the water supply, fill up quickly and depart without getting stuck.

Mapping fires

The development of cheap and easy-to-use global positioning systems has made the accurate recording of fire sizes and patterns easy. Over the years, the kit has become a useful management tool as the history of each piece of heather can easily be recorded. Mapping does not have to be done at the time of burning; the record can be made at any time in the course of the following year.

Mobile phones

Many keepers take mobile phones with them at all times. They are a useful safety aid and can play a part in an emergency. At the lowest level, they are handy at the day's end to warn those at home that it is time to put the tatties on!

The practice of burning

One man burning, 1968. Fine until it is time to put the fire out. (NH781309)

45

Putting a match to the hill can be done by a child, a picnicker, a shepherd or a keeper. Whoever does it, the results can be catastrophic for grouse and moorland wildlife if lit in the wrong place at the wrong time such as late spring when birds are nesting or in high summer when the peat is tinder dry. Properly controlled heather burning is a highly skilled operation that, like all skills, is not learned overnight. The primary objective is to maintain that heather dominance without which there will be no grouse or moorland as we know it. The operation must be planned carefully from the outset so that full advantage can be taken of inevitably limited opportunities.

The first essential is an estate map to a scale of 1:25,000 (2.5 in/mile) on which can be overlaid a rough vegetation map. Allied to a series of walks across the hill, it is easy to identify those areas in need of attention so their treatment can be prioritised. Marking areas with numbered discs is perhaps carrying things a bit far but the principle is sound.[31] Each enthusiastic moor owner and keeper should use his own method to quantify what needs to be done. This is an out-of-season activity though one successful Perthshire proprietor walks up every part of his hill in the first week of the shooting season and combines shooting with a five-minute stop every hour when the ground just traversed is described and recorded in a notebook carried by the keeper. By the end of the shooting season and the likely start of burning, nobody on the estate is in any doubt about the scale of the task to be undertaken. Certain places are earmarked for burning in autumn should suitable weather present itself while other areas are attended to in the spring. Priorities are drawn up; north-facing slopes where snow lies late and are slow to dry out can be identified together with the combination of wind and weather necessary to do the work there. Other, 'easier' ground, perhaps consisting of substantial beds of young heather resulting from a previous large fire, can be considered for treatment when the wind is strong and conditions are too dry to tackle older heather safely. A detailed map can be prepared and hung in the estate office or on the gun room wall indicating immediately where the keeper should go each day, in each season (autumn or spring) and under every combination of wind, time of day and relative humidity.

Identifying suitable weather is an observational skill which is not always well developed with the result that suitable days, particularly in the depths of winter, are missed. An early clue that conditions are becoming suitable, particularly in the wetter west, is when television weather forecasters predict a settled spell of anti-cyclonic weather. In the east, suitable weather can occur even in depressions. Another clue is when fallen leaves suddenly swirl and rustle in a light breeze, indicating low levels of atmospheric humidity. This often happens when the weather is overcast and keepers' thoughts are elsewhere. The keeper should put his staff on 'red alert' for a possible start perhaps two days hence. As a prelude, he probably tries a fire or two in mid-afternoon on his own and, if conditions prove satisfactory, notifies his helpers to be ready for the following day. That evening, he checks the weather forecast again to note the likely wind speed and direction and may well check the detailed forecasts available from the local weather centre situated at the nearest airport or large city.[32] When the next day dawns, he has a clear idea where he is going and there is no dithering or wasted time. It goes without saying that he has already set out depots of diesel fuel in strategic places, made up suitable brushes for fire control and put fire-lighting kit together including a small funnel and a supply of plastic bottles for applying diesel to the embryonic fire. Much time can be saved if the 'face' of the fire gets going quickly – a knapsack sprayer or a pressure pump filled with diesel

Controlled burning 1971. See p.227. (NO396420)

is a worthwhile investment. A portable gas bottle is another option but is uncomfortable to carry. While face masks are a legal essential, gloves should never *ever* be worn by anyone working close to a fire because the risk of the spontaneous combustion of clothes in severe heat is very real. The sensitive backs of the hands provide an early warning of overheating. Not long ago, a keeper suddenly burst into flames when working at a hot fire. His clothes had some spilt diesel on them but, because he was wearing gloves and a face mask, he did not realise how hot his clothes had become. He was seriously burned and lucky to get away with his life.

A skilled man gets the feel of conditions each day by lighting a small fire on a site with an obvious break in front. If the fire goes slowly and burns well, he feels confident enough to tackle a more difficult place nearby. If the initial fire

47

takes off with a roar and a rattle, he knows that caution is the order of the day and concentrates on areas where he is burning downhill into sheltered places where the fire is easier to control. Depending on circumstances, he lights additional fires, leaving one, two or three men with each. Occasionally a real bonanza situation occurs in spring. If a hill face is flecked with small snow drifts, the sun is bright and the wind strong, several years of under-burning can be corrected in a short time. The whole hill can be fired and the flames allowed to ramble about among the snow flakes. When the thaw is complete and the snow has gone, a beautiful pattern of patches of heather of various ages mixed up with the burned ground emerges. Under good conditions, it is possible to have several fires going at once but the man in charge must always keep a weather eye open; conditions can change in the space of a few minutes. As a rule, they become progressively more difficult and dangerous because wind speeds tend to increase as the day goes on before they die down in the early evening. Later in the burning season, when the sun is strong, conditions can change from really sleepy to extremely dangerous in the space of half an hour as the air warms up, overnight dew evaporates, relative humidity falls and the wind picks up. All big fires start as small ones and the really big ones are invariably lit after lunch.

And talking of lunch… one often sees all burning activity stop for an hour or even more at midday. There is no need or excuse for this; when conditions are suitable, fires must be burning all the time. People must learn to eat their sandwiches on the move at this time of year; there are plenty of other occasions when leisurely picnics, happy gossip and jolly banter do not intrude on critically important work.

As the day does on, the man in charge anticipates changing conditions. In the morning, he has burned fires uphill on gentle slopes when the heather was damp from overnight dew and consequently slow to burn. In the afternoon, fires are started in situations where there is no chance of their widening out and becoming a major conflagration. He remembers that people get tired as the day goes on and that there are no medals to be earned if helpers half kill themselves at a fire that should not have been lit in the first place. Neglecting this point makes casual workers understandably reluctant to come out the following day.

Showing how scattered snow patches can facilitate burning. Drumochter, March 2011. (NN618798)

Burning uphill, even on a gentle slope, is always by far the most risky task as, in addition to the reasons noted above, wind speeds increase with altitude. A fire which was docile at noon can be a raging monster by 3.00 pm. Under these conditions the worth of the prior planning is shown as the squad can move to a more sheltered place where fires can be run along and down a slope where control is easy. Burning uphill should only be done when the vegetation is damp, the wind light and burning conditions borderline. The need for caution is particularly true of steep ground as it is often subject to wayward winds created by updraughts which blow the fire from side to side. Such variables can create problems; light winds are always dangerous as they make control tricky, particularly if the vegetation is dry.

A classic example of a poor choice of site with wayward winds created by updraughts blowing the fire from side to side. The 'pair of trousers' pattern is the inevitable result. (NN617800)

Most people do their best work when the wind is strong and true – it does not normally matter how strong it is if the right site is chosen. There are few things more satisfying to an enthusiastic and knowledgeable heather burner than a fire lit on a dry day with a true wind behind it which takes off with a roar and the planned break – sometimes far in front – kills the head so that the fire goes out almost before the back fires have been put out. Often, in a really strong, true wind, little needs to be done to control the sides of the fire as the flames get blown out by the strength of the draught. It is depressing to read in official handbooks on burning that it should never be done in winds in excess of Beaufort 4. The edict is as ill-informed as it is misleading.[33]

Breaks – natural and artificial

It is impossible to extinguish the head of a fire that is going well. All that operators can hope to do is to control the sides closely and gradually pinch them in while letting the head look after itself. Though the competent man lights a

49

fire only if he is certain that the planned break in front will stop it, there are many cases every year where this most elementary of precautions is not taken. The planned stopping point does not do its job, the barrier is jumped and a larger-than-planned fire results. Breaks come in many guises, the two most reliable being fires burned that same season and snow patches. Young pioneer heather, provided it has a limited amount of dead sticks or 'scrogg' left over from a previous fire, will stop a hot fire which burns into it. Normally, a deep-gullied burn side slows up a fire considerably or stops it altogether but stone dykes, hill roads and narrow watercourses can lull people into a false sense of security. Because the moss and old sticks are usually tinder dry, a previous fire with much litter left on it is one of the least reliable and most dangerous breaks there is. Once a fire gets a hold of that kind of material, almost nothing short of added water puts it out. A fire that burns obliquely into a road or dyke is less likely to jump than one which meets the obstacle at right angles though it is easy to reinforce the break with a little care. Before lighting the downwind fire, the edge of the road or the dyke back should be lit on its upwind side for a stretch at least twice the expected width of the downwind fire. The upwind fire is back-burning and needs an attendant for as long as it takes to burn back three to six metres according to the dryness of the day and the length of the vegetation. When the downwind fire is lit and rushes off to hit the upwind one, both suddenly die from lack of fuel, leaving the sides and fringes at either end of the back burn to be put out. The technique can be used to reinforce any break of doubtful reliability such as a wide sheep path, a narrow green damp flush or a small shallow stream.

Back-burning

Back-burning (burning against the wind) and double-burning (re-burning a former fire) are specialised methods for use on difficult material, on problem sites, or under specific circumstances. When conditions are very dry and apparently dangerous and particularly if the wind is strong, back-burning can be used to deal with difficult beds of ancient heather. Heather of this age class can only regenerate properly from seed but seedlings do not establish in a thick litter layer, making it essential that as much as possible of the mat is consumed. Because it is hard to burn it properly with a downwind fire, back-burning was often used by keepers in the last quarter of the nineteenth century. Many were faced with huge areas of degenerate material as grouse driving, particularly in the central and eastern districts of Scotland, increasingly took over from the shooting of grouse over dogs with its attendant (and desired) long heather. Back-burning is a slow business but it does consume old heather cleanly and leads to good seedling regeneration. Because it is so slow, it has fallen out of favour but results are usually excellent provided the material is indeed degenerate and the estate can call on plenty of labour. Under suitable circumstances of wind and weather, a long stretch of heather can be lit on its downwind edge and the fire allowed to burn back for thirty to forty metres. With competent and confident men, a back fire burned like this can do a good job – but operators must always be aware that a change of wind can spell serious trouble. It is a mistake to burn late-building or mature heather with a back fire as the material often has little litter to insulate the soil. The heat can kill the roots and may even set fire to the peat.

Many conservation interests advocate a no-burning policy for sometimes extensive areas of degenerate heather on the grounds that it maintains areas which are valuable for rare plants and birds, conserves mosses and lichens and

Back-burning to reinforce an existing but inadequate break. (NO396420)

avoids the perceived risk of a hot fire destroying the litter and the top layers of peat. It is sometimes argued that degenerate heather plants fall over, layer and produce new plants from adventitious buds which are morphologically young, thus regenerating itself. However, it is virtually unknown for degenerate heather to become so dry within the legal burning season that the litter burns cleanly and the peat becomes damaged. It may well happen in an (illegal) summer fire but it is rare indeed for damage to be sustained in a fire burned by competent people at the correct time of year. There may well be a case for not burning in certain restricted areas but the widespread enforcement of a no-burning policy would reduce rather than conserve the area of productive moorland in the medium to long term and make it much less attractive to sheep and deer. Unless livestock numbers are adjusted downwards at the same time, it would lead inevitably to the over-use of better managed, more accessible and more palatable vegetation elsewhere. It would certainly result in increased grouse territory sizes overall, lower breeding densities and reduced bags from the neglected moorland with, in turn, a decline in profitability for the moor owner.

Many hot fires have occurred in morphologically young heather with disastrous results. Many cool fires have had equally negative effects on degenerate heather. Generalisations are dangerous!

Double-burning

As often by accident as by design, operators double-burn sites which have formerly supported degenerate heather. They start a fire nearby and run it into an area of old, rank heather stubble that was burned with a cool fire a year or two before where much litter, moss and scroggy material remains. In the spring, as the stubble is often tinder dry, its value as a fire break is delusory. However, fire burning into such a site usually consumes all the surface material and dried up fog, leaving a fine, grey ash. The result is a perfect seed bed with no loose litter overlying the soil that almost always results in a dense sward of seedlings in the next growing season. The more widespread use of double-burning on moors with a history of neglect is to be encouraged.[34] Re-burning old fire sites consisting of a forest of grey sticks normally results in excellent regeneration. The fires may be hard to put out but the modern heather burner has no excuse for not having a water dispenser to ensure control.

In short, there are no good reasons for not tackling neglected moors which carry much degenerate heather and revitalising them by good burning practices. Places can often be seen where good-shaped fires burned under cool conditions in degenerate heather have led to a series of grey strips or patches on the hill that remain unproductive for years because the unburned litter inhibits seed regeneration and degenerate plants do not regenerate from stem bases. Grass often colonises the bare ground as any tiny heather seedlings are pulled out, destroyed or browsed flat by sheep and deer. Lovat pointed out that cool fires reduced rather than increased food for grouse – he was a strong advocate of hard burning on short rotations with numerous fires. If his policy is followed, the problem of slow regeneration (or none at all) following the burning of degenerate heather eventually becomes insignificant as the degenerate heather itself becomes scarce.

Side-burning

Side-burning is a modified form of back-burning that can be useful on degenerate heather when conditions are dry and the wind true. Instead of burning a strip into the eye of the wind, the long edge of the strip is all kindled at once and the fire allowed to creep sideways. The heat produced and the speed with which the fire covers the ground are intermediate between a back fire and a downwind one.

Fire pattern and design

Another good general burning pattern, pictured in summer.

Burning patterns on heather moor are as individual as hand writing. The attitude of management and the approach of the keeper to the critical task of burning can be read like a book; the evidence is there for all to see in shades of black, grey, brown, green and yellow. What is done is done, be it good, bad or indifferent, and cannot be concealed. Everyone has his own idea of what a 'well burned' moor should look like but, all too often, the task is done by a farmer whose orders to his shepherd are summed up by 'proceed and blacken' or by the nervous, inexperienced, lazy or understaffed manager whose hill shows little sign of a systematic programme. Obviously, some patterns must be better for grouse and grazing animals than others. So, which is best for supporting a large population of birds year in, year out?

Burning patterns in former days

When cattle were the main grazing animal and before the development of grouse shooting, burning had a low priority because heavy cattle stocks broke up the heather sward and kept the hill in a productive state. In the late 1700s, sheep became more important economically and detested by the crofters because they represented a changing agricultural order that threatened their traditional way of life. One of the main sources of friction between the cattle-based crofter and the sheep-based immigrant farmer was that the latter 'burned the hill'.

During the second half of the eighteenth and the nineteenth century, burning went through three distinct phases.[35] Although it improved the ground for sheep in the short term, the practice was recognised as bad for the carrying capacity of the hill for cattle. Prior to about 1850 and certainly before the expansion of grouse shooting in the second half of the century, burning was performed by farmers who had only the welfare of sheep in mind. It became almost universal as sheep numbers increased in the middle of the century as owners and sportsmen came to recognise that it benefited grouse as well though shooters found that, while a heavily burned moor supported more grouse than an unburned one, it also yielded smaller bags over a season because of the difficulties of approaching birds due to the lack of cover. It entered a second phase after about 1850 as shooting increased in economic importance and substantial rents began to be paid for 'splendid grouse ground – fifteen hundred acres of heather without a single break!' For sheep, unburned ground was poor to the point of worthless so many farmers rented the shooting as well as the grazing, getting thereby a free hand to burn as they liked. They could create better conditions for their sheep stock while relieving the sportsman of a hefty shooting rent for ground that was in fact indifferent for grouse.

Shooters recognised that unburned or under-burned ground was poorer at supporting a large population of grouse and that disease outbreaks occurred more frequently on it. The bumper grouse years of 1872–73 were followed by a spectacular collapse in numbers caused by disease. A great deal of money was tied up in grouse at the time with heavy investment in lodges and estate facilities; the owners who leased out moors and the tenants who rented them both got a rude shock when they found that 'the golden bird' was at very low numbers and that cash flows were seriously affected. In the aftermath of the collapse, burning came to be recognised as a cornerstone of sound grouse moor management because well-burned moors suffered less and recovered more quickly from the effects of disease.

The third phase ran from about 1875 to the outbreak of the 1914 war – the

Golden Age of grouse and grouse management. Just about everyone made it their business to improve their moors, a policy made easier by an abundance of cheap labour. The period saw the achievement of optimum, fine-grained burning patterns; it was accepted that many small fires yielded more grouse than a few large ones but that the right *percentage* of the moor had to be burned. During the first half of the nineteenth century, approved wisdom had been to burn a tenth of the moor each year, the amount normally done by the farmers. Between about 1850 and the 1873 collapse, the area burned annually by the new managers often fell to less than a hundredth of the total but the successful moors of the last quarter of the century were those where ten to fifteen per cent was burned annually *in small strips and patches*. The policy was adopted by the most successful owners such as Rimington Wilson at Broomhead, The Mackintosh at Moy and Lord George Scott at Langholm. It is no coincidence that these moors all held records for one-day bags at one time or another. Notwithstanding the unreliability of bag records as an absolute measure of numbers, they must have been exceedingly well stocked with birds to achieve the totals they did. Although they are the stuff of legend, many others were doing nearly as well as a brief glance at contemporary game books and letting particulars shows.[36]

It would be a mistake to think that neighbouring moors were all managed to the standards of the best burned ones. The quality of management varied as much then as it does now; some would have shown a coarser burning pattern than others nearby. What mattered was the predominant standard; if three out of four moors in a glen, strath or dale were really well managed, the fourth would have been carried by its neighbours just as one sees today in parts of northeast England and southeast Scotland. If the habitat is good over the majority of a district, then the standards of vermin control and the intensity of population cropping (which affects subsequent bird quality) become of equivalent importance in delivering big bags consistently. Not every moor which delivers a brace of birds to the acre today is burned to the standard of Broomhead, Moy and Langholm a hundred years ago but, where neighbouring moors are also well burned, keepering is thorough, stock management first class and shooting pressure rises and falls in line with breeding performance, a sound foundation is laid for the following year's production.

Plenty of young heather resulting from a vigorous fire programme associated with tight keepering yields bountiful bags of grouse. North Yorkshire.

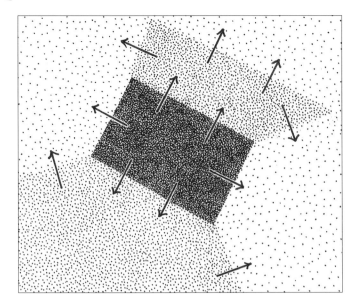

Emigration, immigration and the Sink Effect. Grouse move according to the arrows and tend to level up their breeding density. If the arrows are reversed, the grouse level their general population by in-filling. The same principle holds good for predator dispersal.
(DRAWN BY D. PULLAN)

Lovat drew up a strategy for burning designed to quantify the task and calculate how it could be done in the time available. The main points were:

- To burn old heather in strips fifty yards wide and let the strip run as far as the fire takes it;

- To burn average foot and a half heather in strips and patches between quarter and half an acre (0.1 to 0.2 ha);

- To burn patches and strips on the steep faces of the wintering ground in small blocks of not more than a quarter to a tenth of an acre each (20 to 30 m squares);

- To burn the burnsides, knolls and nesting grounds of grouse in even smaller plots;

- To burn the wet flow ground in big patches of one to ten acres (0.4 to 4 ha) so as to cover the flow ground every six to nine years;

- To burn the high ground with a northern exposure in large three acre (1.2 ha) blocks;

- To burn good broad strips round each of the boundaries [presumably to prevent damage from a neighbouring fire];

- To treat specially those portions of the moor that have a tendency to revert to grass [though he did not elaborate on how this was to be achieved].

He described by way of example a 4000 acre (1600 ha) moor being burned by four parties per day, fifteen to twenty-one burners *in toto*:

> The burners will have to fire twenty patches of one to five acres, say a total of fifty acres, 300 patches of from one fourth to one acre, making, say, 150 acres, 200 patches of from one fourth to one tenth of an acre, say fifty acres, i.e. about 500 burns with a total of between 250 and 200 acres (and)…it will require about ten days or twenty half days to get through the work, calculating that each party burns an average of fourteen patches a day.

Today's manager, even on a moor of this relatively modest size, would find such a programme daunting. The author does not know of a single moor nationwide on which a policy of such detail and complexity is enforced. But neither does he know many moors nationwide that produce *consistent* one day, let alone seasonal, bags on the scale of Abbeystead, Broomhead, Langholm, Moy and Ruabon early in the twentieth century. Lovat was not the only authority who insisted on burning in this manner. Rimington Wilson, Broomhead and Scott, Langholm both described their burning policy and practices and many others recognised that fragmentation of habitat was the key, though many of them would not have been happy with Lovat's toleration of 18 in (30 cm) heather on the hill.[37] One in particular noted that:

> 'Heather should never be so long that a dead bird is at all likely to be lost in it' and that 'you should burn the moor so often that there will be no long and dry heather anywhere with the exception of the tops of the hillocks… For all the rest – all will be young heather, though varying…from the very tender shoots…up to heather of such a height that it just gives a grouse food and comfortable cover.'[38]

As a grouse stands 30 cm (12 in) at full stretch and 13 cm (5 in) or less when crouched, 30 to 35 cm (12 to 14 in) heather is amply long for their needs. It also suits sheep – shepherds managing heather for them prefer the longest plants to be no higher than the tops of their boots or approximately 25 cm (10 in). Heather of this class regenerates well following fire as it is in the 'mature' growth category. Pioneer plants appear on the stem bases in the summer following burning and can provide a small amount of nutritious and palatable browse for lifestock the following winter.

The old literature argues that it is of the first importance to keep the heather young. If it gets to the degenerate stage, trouble lies ahead because degenerate heather does not regenerate nearly as well as vigorous mature plants. A sound burning policy attacks old heather even if it means that fires have to be wider than Lovat's fifty yards or the Banchory team's recommended forty.[39] Only a moor with a lot of young heather can support large numbers of birds consistently *and* maintain heather in the long term (see Chapter 2, *The management of hill grazings*). The problem can be defined as the imperative to burn a tenth to a fifteenth of the moor each year in fires small enough and narrow enough to avoid imposing limits on available grouse territories. As territorial cocks do not defend ground more than twenty metres from escape cover, the maximum fire width should not exceed forty metres. Lovat is cited here as the authority most familiar with the highly productive moors of his times but studies suggest that, although his fine subdivision is highly desirable, it may not be essential if keepering and stock management are both of high quality.[40]

The turn of the century saw intense debate about how to grow more grouse. There were two schools of thought about the shape of fires and what constituted the best for grouse. It was noted in 1903 that:

> On very many moors that we see, and even in some instructions that we read, the method employed or inculcated is to burn the heather in long strips. This is dead against the experience that has been painfully learned on some of the best managed moors… It is very much better to burn in frequent small patches all over the moor… It is well that the patches be allowed to burn out into little bags and recesses… They are far better thus than when ruled out with mathematical precision, for each of these recesses is a sheltered playground for young birds. Looking from a distance at a stretch of the moor at Moy, it has all the appearance of a patchwork counterpane, with the

constant burned patches. This is the first great maxim for the treatment of a moor – burn often, in small patches. *On no account* (says a great authority…) *in strips.*[41]

Here we have a dilemma. Patches large or small with irregular edges increase habitat diversity and the edge effect but few managers can burn in tiny patches these days because they cannot cover sufficient area in the few available days – their programme quickly falls behind what is needed to maintain the correct percentage of young heather. In an imperfect world, strips are almost inevitably central to a burning programme designed to improve heather productivity on a whole-moor basis. Whether narrow or wide depends on circumstance but fires should be allowed to ramble along their edges to avoid those straight lines that are, rightly, condemned. In this context, swiping firebreaks around patches of heather to be burned may be less valuable than many think. Swiping and burning can help to increase the acreage of young heather mixed with old, but the fire edges lack those small-scale indentations which characterise strips and patches burned by a squad of people. Strips with swiped edges are open to criticism on landscape grounds as well.

An aspect of a strip fire which is often overlooked is the contrasting regeneration which occurs frequently along its length. A strip forty yards wide running along a hillside for half a mile (0.8 km) invariably passes over soils with differing moisture characteristics and through heather of different age classes. Five years on, the site usually shows an extraordinary variety of regeneration; some places are covered with a dense growth of finely branched pioneer or early building material, others with bleached grey sticks and an understory of moss. After another five years, the early building is approaching the mature stage whereas the grey stick ground shows the occasional pioneer heather plant but not much else. A substantial long strip fire can be colonised by a variety of age classes in a surprisingly short space of time.

In the early stages, a network of strips can also be decisive in preventing runaway fires in a way that no amount of pocket handkerchiefs can attain. The fate of Tilquhillie (NO721941), a small but productive moor near Banchory, provides a timely lesson. Seemingly well burned with many small, isolated fires, it provided a satisfactory yield of grouse in the 1960s. One spring, the keeper lit a fire which got away from him and destroyed a substantial proportion of the moor before it was brought under control. The small patches all failed to break its progress as the fire burned round them, leaving the keeper powerless. Had the moor been burned with a skeletal framework of long strips, they would have prevented the fire from burning such a big area. The moor was destroyed for grouse and sold shortly afterwards for forestry. Fireproofing a moor by strip burning is particularly critical in places used intensively by the public. The Peak District National Park and the North York Moors both suffer from summer fires on a regular basis – if the management adopted a vigorous policy of burning a latticework of fires, particularly close to the places where many people go, much heather and attendant wildlife would be better conserved.

Over the period of a heather rotation, burning plenty of long strips allied with as many patches as possible is as near optimum as we can get. When an enthusiastic keeper takes on the burning of a neglected moor, in practice he *has* to start with strips though he will over a period of about five years end up burning patches because the strips that have been his norm become shorter by the year until the moor assumes the appearance of the grouse-related diagrams (see p.62).

Lovat's point that it takes several years for heather to regenerate is true of

RIGHT:
The effect of a May fire, 1970. (NH935275)

BELOW:
Three people, three fires, no problems.

degenerate heather but not of late-building and mature heather which may well grow out to 30 to 35 cm (12 to 14 in) in less than the calculated time of up to eighteen years. To keep it 'young', the rotation may have to be reduced. In practice, a moor subjected to intensive strip treatment is almost impossible to over-burn. Fires run only a short distance before encountering a break; even if a fire becomes larger than intended, the fast rate of regeneration of young heather means that the damage is repaired in a year or two. Let no one on a well-burned moor ever be tempted to sit back and say, 'we have burned enough'. It is the author's view that, provided fire sizes are well controlled, the scourge of moor management in our times is not over-burning but under-burning.[42] Lovat summed it up by saying:

> It is very doubtful if too much burning can ever be done in any season provided the areas of the fires are reduced in size as the patching and stripping of the moor progresses. By large fires a moor can easily be burned out, but no old patch is so small that a smaller patch cannot be taken out of it and the moor be thereby improved.

Hopefully, this chapter has demonstrated the need for a multitude of fires of restricted size allied to a typical rotation of no more than fifteen to eighteen years. Lower slopes may be burned perhaps once every ten years and higher ground once every twenty or more but there should be no degenerate heather anywhere on the moor if grouse, sheep and the conservation of the moor itself are the objectives. If the moor is designated as a Site of Special Scientific Interest (SSSI) or owned by someone keen on merlins or mosses, burning will, of course, have to be restricted in places but, that aside, degenerate heather remains the curse of moorland nationwide.

Many people would class this as over-burning. Misty Law 1999.

As long as red grouse are an important objective, an intensive programme with the emphasis on many small fires of irregular shape and size is the way to better bags. There are very few moors nationwide that even approach, let alone reach, the standards of burning that were the norm a century ago. It is for this reason alone that grouse have declined to their present levels on many moors and throughout whole districts. The remedy for improvement is in the owners' hand; the result is as certain as dawn follows night.

It is fine for the student to read of what seems to have been general burning practice in Scotland at the turn of the nineteenth century but what did it actually look like? Was an ideal pattern as assessed by Lovat (Moy, for example) anything like the sort of reference points we use today such as Candacraig, Crosbie, Garrows and Langholm in the 1970s and additionally, Byrecleuch, Clune, Kettelshiel, Leadhills, Misty Law, Monzie, Priestlaw and Whitfield in the 1980s and 1990s? What happens if you try to reconstruct the appearance of a piece of moor treated in the manner described?

Unfortunately, there are few good photographic records of what moorland looked like a century ago. Obviously, Lovat's multitude of tiny fires could only have been done on a moor that had already been well burned. It is questionable whether fires as small as a tenth of an acre are even practicable, particularly in long heather, though it is possible to burn in this way if there are plentiful breaks for stopping fire and a wide choice of late building/mature heather to treat. The diagrams on p.62 illustrate the effect of a five-year crash programme on the worst of situations, an under-burned moor dominated by mature and degenerate heather and carrying a sheep stock.

When taking on an under-managed moor, it is important to try to 'catch up' as quickly as possible and crucial to implement a vigorous burning programme in the first year if pressure from grazing animals is not to disadvantage the grouse by removing all the young nutritious heather shoots before spring.

If the target burning rotation is ten to fifteen years (typical of medium altitude moorland in the eastern half of the country), twenty per cent of the ground needs to be burned in year one and another twenty per cent in year two before easing off to ten per cent per year for the next three years. Five years into the programme, seventy per cent of the moor will have been treated. The aim is to

Burning patterns in 1968 and 1988 showing a collapse in standards over a twenty year period. (NJ350145)

give grazing animals initially as much choice of range as possible. Whatever Lovat *et al* may have thought of them, one can achieve this with strips because they can be dispersed across the moor and run through heather beds of varying ages to influence the rate and mix of regeneration.

Table 3.

A burning programme based on Lovat's figure of c. 500 *(actually 530) fires on a 4000 acre moor.*

The first five years		
Year	Percentage burned	Running total
1	20	20
2	20	40
3	10	50
4	10	60
5	10	70

Though seventy per cent of the moor has been burned by around year six, intensive muirburn (diagram 4) shows one hundred and ten patches of the original tall heather on the 104 ha (257 ac) of our imaginary slice of hill, hardly any larger than 0.4 ha (1 ac). At the end of year six, the manager needs to take decisions. If the heather burned in year one is coming back strongly and looks like reaching 30 to 35 five cm (12 to 14 in) by year ten, a ten-year rotation should be put in hand and seven to ten per cent of the moor be burned every year in years six to ten. If it seems prudent to go for a twelve-year rotation, six to seven per cent a year should be burned in years six to twelve. For a fifteen year rotation, the rate falls to just over four per cent per year.

Table 4.

Initiating a burning programme.

	Grouse & conservation		Sheep and deer	
	Area burned (%)	Fire numbers	Area burned (%)	Fire numbers
Fig 2	20	31	24	1
Fig 3	20	59	26	1
Fig 4	30	72	25	2
Totals	70	162	75	4

The procedure is like an aircraft taking off – at lift off, it starts by climbing on maximum power until it achieves a safe height at which point the pilot levels off and reduces engine thrust. Notionally, this point is the end of year five. From then until it reaches cruising height (the end of the first rotation), considerable effort has still to be applied but nothing like as much as during take off. Just as keeping an aircraft at cruising height is straightforward and fuel effective, so the manager who reaches his 'cruising height' can consider the luxury of Lovat's beloved tiny fires but *not until then can he indulge in the luxury of a 'cruising speed'*.

By the end of the rotation, no old heather which was degenerate at the start of the programme should remain (except for fires burned just before the new programme – heather there will be older than the rotation length). We have seen that small fires in degenerate heather rarely get hot enough to remove the 'fog' and permit regeneration from seed. As it is difficult to limit the size of a hot fire, the ground should initially be opened up with plenty of long strips over a substantial area run through a variety of age classes and vegetation types. When regeneration commences, there is a better mosaic of plant heights. The material burned in years one and two is likely to regenerate over a period of years depending on the morphology of the plants burned. The keeper can perhaps ease off his effort *for a time only* while heather burned in the initial years covers the ground with dense, finely branched plants.

61

Intensive and extensive muirburn programmes

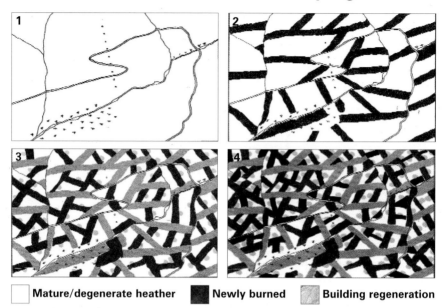

☐ Mature/degenerate heather ■ Newly burned ▨ Building regeneration

Burning patterns optimised for grouse (above) and sheep (below) aimed at regenerating heather. Sketch 1 shows a typical stretch of mixed moorland at the start of the programme. Sketch 2 shows the area at the end of the first year with extensive areas of mature/degenerate heather burned in different ways. Sketch 3 shows how it might look after three or four years, sketch 4 after six to eight years.

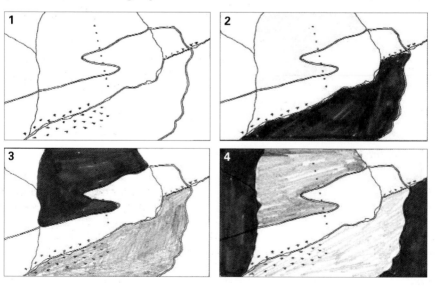

Garrows in winter 1995. (NN813410)

A hot clean fire burned in degenerate heather.

If the theoretical pattern at six years is reversed so that the white patches (old heather) are lumped together and considered as young heather and recent fires (the rest of the matrix being old), many would congratulate the keeper on a fine job. He has, after all, burned about a third of the hill in six years. However, because the heather on the white patches was too old to regenerate from stem bases, there is a strong likelihood that on at least half of the newly burned patches (110 patches on 104 ha or twenty per cent of the moor) there would be no vegetation of worth growing on them. The remainder is attracting grazing deer, sheep and hares to a point where the regenerating heather has little or no

63

Blaeberry and heather seedling regeneration five months later (see previous page).

chance of establishing a sward and increasing in length. If the hill is set stocked at a typical rate of one ewe to the ha (2.5 ac), the area growing quality food available on each hectare is no more than 0.2 ha (0.5 ac), a stocking rate on the young heather that is far too high to ensure the rapid establishment of a tight sward by regeneration from seed. Under these conditions (which are very common nationwide), the grouse response is disappointing. There may even be no response whatever in consequence of the hard grazing because the grazing animals have pre-empted the grouse by browsing off all the valuable new growth.

If seventy per cent is burned in five years, sheep can move freely about the hill, using the strips as walkways as well as grazing sites – the choice of available forage is considerable. There are over 50 ha (130 ac) of burned ground on the area on which, because the fires were hotter and of adequate size, regeneration will be better. Perhaps as much as half the hill is supplying good forage giving each ewe access to 0.6 ha (1.5 ac) of burned ground. The heather clearly has a better chance of recovery with plenty of choice of food for grouse and, crucially, with no patch of young heather being further than fifty metres from a patch of old for cover, so almost all the area is suitable for grouse to occupy and the risk of losses to predators reduced.

The extensive muirburn diagrams show a system which creates seventy per cent of burned ground in the initial five years.[43] Though there is plenty of feeding heather, the fires are too wide for grouse because much of it is too far from escape cover. The pattern is, however, excellent for grazing. Burning substantial fires on a short (ten-year) rotation maintains a vigorous sward of heather and restores the sort of pattern that might have been burned when the hills were managed primarily for sheep in the first half of the nineteenth century. For all it is not good for grouse, it is good for maintaining heather moorland. Perhaps more important, it is a burning pattern typical of much of the heather moors of northern Scotland. Owners who use it have only themselves to blame if they get fewer grouse than expected because breeding densities reflect the fineness of the mosaic created by burning.

Experiments conducted by the Banchory team at Kerloch moor in the 1960s showed that many small fires created conditions for a large grouse breeding stock.[44] The burning was not on Lovat's minute scale but the trials did show that well-burned moorland could support a pair of breeding grouse per 5 to 6 acres (2 to 2.5 ha) as Lovat had stated sixty years earlier. Can Lovat's system be made practicable for today's manager? In the light of what we now know about territorial requirements and bird behaviour (about which Lovat knew little), the answer is probably 'yes'. To do consistently well, grouse need all types of heather in their breeding territory ranging from pioneer to mature. A way of assessing this can be devised by superimposing a quadrat ring of about 2 ha (5 ac, the size of a typical grouse territory) on the map (see picture).[45] As the ring usually impinges on four or five separate pieces of tall heather, the edge effect is substantial and cover and food are intimately mixed. It may well be that such fine subdivision is unnecessary – we do not know how coarse patterns can be before densities start to decline. Though substantially larger than Lovat's tenth of an acre minima, the smallest fire in the diagrams is about three-quarters of an acre (0.3 ha); it is hard to envisage anything smaller being achievable regularly nowadays. It may well be that a policy of fragmented burning patterns would push average breeding densities up to Lovat's 'safe stock' levels and make the bags of yesteryear more credible but any trials would have to be done on a moor surrounded by other well-burned ones, not on a small, isolated moor surrounded by inferior habitat. The evidence suggests that, provided keepering is good *and the heather is young and regenerates rapidly*, burning patterns can be quite coarse. See p.54. What definitely does not work is the combination of coarse burning patterns, indifferent heather regeneration and moderately effective keepering that describes the situation prevailing over extensive areas today, particularly in Scotland.

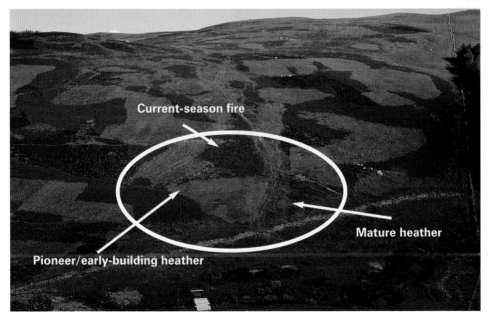

A robust burning pattern illustrating how a notional grouse territory of around 2 ha (5 ac) mixes different growth stages. (NN956095)

Before leaving the subject of what can be learned by extrapolation from old records, it can be said with confidence that grouse managers generally burn on too long a rotation for the good of heather regeneration and hence for grouse in the long term and that fires tend to be either too big or too small. A multitude of fires burned year in, year out, spring and autumn creates optimal conditions for wildlife generally including grouse, hares and other herbivores.[46] Sheep thrive on a finely grained burning pattern and browse the hill more evenly if age classes are intimately mixed.[47]

Cutting and nutrient content

Swiped heather at Vivod, north Wales.

Many who proffer advice on moor management advocate cutting in place of burning. The thinking is that cutting is not weather dependent, enables old material to be removed to make way for new growth provided the site is stone free, avoids the risk of fires getting out of control and removes the issue of smoke nuisance while running no risk of damaging specific areas or the sensitivities of those visiting the hill.[48]

The inexperienced and the fire-shy often advocate the use of cutters but, as any knowledgeable person can explain, the results are invariably a disappointment (see Table 5 opposite).[49] Heather moors are not like hay meadows or fields of ripe barley. They are invariably more bumpy; the cutter is as often slicing through the plant many centimetres above the ground as it is burrowing into the earth and carving up its roots. Inevitably, it does an uneven job that affects subsequent regeneration; machinery breakages are common and can be expensive to repair. The operation creates a swathe of cut material containing plant nutrients and minerals locked up by the shrubs over many years that lies on the top of the cut stubble, inhibits vegetative regeneration and may even be thick enough to prevent all plant regrowth. As the material, which is usually rich in cellulose and lignin, does not lie in close contact with the underlying soil, it is slow to decompose. Its woody nature means that it takes many years to rot and release the nutrients and minerals which are therefore unavailable to regenerating plants. Degenerate heather can regenerate only from seedlings but the weight of

Table 5.

Results of a pilot study comparing cutting with burning in Scotland; the nutrient content of re-growth three years after treatment. The table has no pretensions to science because no chemical analyses were carried out before treatment and the areas were not fenced to exclude browsing – the findings are merely indicative. However, there is clearly a difference in chemical content between treatments. Casual observation indicated that the difference was recognised by hares and sheep as they always preferred to browse the regenerating plants which followed burning.

	Place	Geology	Soil	Treatment	Nitrogen (%)	Difference (%)	Phosphorous (%)	Difference (%)
1	NN924305	Slate/ Phyllite	Mineral	Burned	2.1	16.7	0.140	7.7
				Cut	1.8		0.130	
2	NT121561	Andesite lava	Mineral	Burned	1.8	12.5	0.140	7.7
				Cut	1.6		0.130	
3	NT605630	Landeilo/ Caradoc Shale	Mineral	Burned	1.6	6.7	0.130	0.0
				Cut	1.5		0.130	
			Peat	Burned	1.5	25.0	0.120	27.7
				Cut	1.2		0.094	
4	NS320600	Basalt	Mineral	Burned	2.0	42.9	0.120	48.1
				Cut	1.4		0.081	
			Peat	Burned	1.7	30.8	0.110	44.7
				Cut	1.3		0.076	

trash on the litter surface usually means that seedlings are unable to establish because the soil is out of reach of their limited root run. In the course of time, the mat dries out and becomes a serious fire hazard. The defoliation of established plants can encourage gully erosion. In short, cutting heather is a bad practice with limited value for the moor manager.

The nutrient content of heather that regenerates after cutting is lower than that of heather that grows after burning. Small-scale trials have shown that the nitrogen content of heather regenerating after fire was higher in five of six samples than that from cut sites. The phosphorus content on burned sites was the same (1 site) or better (4 sites) than on cut ones. Even cursory field observation shows that sheep and hares have little interest in heather growing on places where the heather has been cut – both species much prefer to browse heather which has been burned. This is unsurprising as it is higher in both protein and minerals due to the deposition of ash following the fire. That said, cutting is useful in immediate association with burning. A flail cutter or a spindle carrying fast moving rotating chains mounted on a powerful four-wheel-drive tractor can clear a track and throw a windrow of sodden, macerated green material to one side through which a fire cannot run.

The repair and restoration of moorland

The definition of 'moorland' depends on who is defining it. To the purist, it is basically land dominated by ling heather in association with a wide variety of moorland plants. To many, 'the moors' are just open space – hills and mountains – which roll away as far as the eye can see. As a generalisation, any unenclosed upland not dominated by forest (either natural or planted) falls into the category: it has many grades. Heather dominance is not essential; extensive areas of the moors of West Yorkshire and Cumbria are dominated not by heather but by blaeberry, grasses, sedges and bracken while the moors of low-lying Moray are dominated by heather but often support substantial areas of self-sown pine and birch scrub particularly where they are under-grazed. A management strategy that aims to make a moor a productive place for a wide variety of birds and animals must take these variations into account.

A programme for re-creating productive heather on the open moor from ground covered in self-sown scrub has to start by cutting as many trees as possible followed by hard burning on a short rotation to get rid of as much rank heather and tree seedlings as quickly as possible. Much of the heather is likely to be degenerate with a strong-growing understory of mosses and lichens and is unlikely to burn well the first time it is fired. The mosses die and dry out after a year or two; a second fire should ensure a clean, firm seed bed suitable for heather seedlings. The second fire must not be tackled without a water bowser as dead and dry heather sticks burn vigorously and are almost impossible to extinguish without water. Burning kills tree seedlings up to thumb thickness but anything much larger is either not killed (Scots pine) or, if the top growth is killed, emerges as a rosette of regrowth at the basal stem buds (birch). A hot fire which creates good conditions for heather seed also creates good conditions for tree seedlings; it is common to see a dense sward of them following a hot fire in the absence of grazing domestic stock. Unless moorland is properly burned *and* grazed, there is unlikely to be any long-term improvement in its productivity. Burning or grazing badly may even make matters worse and result in an increase in encroaching shrubs particularly in the absence of domestic stock in winter. Managers who seek to return a kind of scrub-dominated vegetation to grouse moor should be braced for criticism from conservation bodies who seek to turn all low-lying heather moor into diverse scrub woodland irrespective of the management objectives of the owner.

If an area supports many ticks, hard burning helps to reduce them as it reduces the mat in which they over-winter. However, critics of burning (who may well be right in some circumstances) say it affects the wildlife interest of

A swiped strip makes an efficient fire break.

Fire has to be hot to kill scrub encroaching on moorland.

the moor by reducing 'biological diversity'. Burning may cause the site to become poorer for black grouse and roe deer but better for golden plover, worse for hen harriers but better for blue hares. The fire prescription described here is for practitioners who want to maintain or increase the acreage of heather. Heather moors have evolved over decades and generally have a history of under-burning with many, particularly on the eastern side of the country, under-grazed as well. Addressing problems with burning and grazing improves the moor for grouse and moorland birds. People who want everything on an area must manage it with sensitivity and a sharp eye for detail. If they want to encourage blackgame, they must leave patches of scrub at the moor edge and along stream sides, patches of mature but not degenerate heather dispersed widely over the hill and rushy bogs to provide habitat for insects for chick rearing. If the objective is to increase native woodland, no fires should be burned and no domestic livestock allowed on the moor, summer or winter. Within a decade, the area will have ceased to be moorland with its typical assemblage of birds and mammals but will instead be home for blackgame, woodcock, roe deer – and sheep ticks.

Golden plover territories require short vegetation for nesting. (PHOTO: N. MCINTYRE)

Ground that reverts readily to scrub woodland is often infertile with a soil derived from base-poor glacial debris. Livestock farming based on stock rearing may not be profitable; crops other than grass or a forage crop of rape or turnips are all it is likely to carry in today's economic climate. It can be difficult for the farmer or landowner to muster enough mouths to utilise vegetation effectively and so affect the moor in a positive way. The decline in productivity has been associated with the decline in cropping on small-scale subsistence farms on the lower slopes of many of the hills in east and northeast Scotland. Long burning rotations and a paucity of livestock, particularly in winter, have allowed the extensive encroachment of scrub and trees. Wide areas of traditional heather moor are now at serious risk from this process but it is, nevertheless, generally welcomed by conservation bodies such as the wildlife trusts and SNH because it offers a short cut to achieving the Scottish Government's target of covering twenty-five per cent of the national land area with woodland cover by 2020. Were that to be achieved, many would see it as a major act of environtalism vandalism that later generations may come to rank with the "The Coming of the Sheep" or "The Clearances."[50] Incidentally, the target is unlikely ever to be achieved because the FC is clear-felling land to sell to the utilities for wind farms almost as fast as other land is being planted.

Burning a grass heath merely maintains grass dominance. (NS911219)

Typical of the Southern Uplands and northwest England are 'white' moors which have been grazed and burned in a manner that allows poor quality grasses and sedges to take over and suppress dwarf shrub heath, usually to the detriment of the wildlife, domestic animals and people who depend on it for a livelihood. The large-scale restoration of such moors to a grass heath made up of a mixture of heather, blaeberry and grasses is a desirable objective and not one necessarily as difficult to achieve as may seem.[51]

The first step is to examine the sward, preferably in late summer when the heather is fairly obvious but before blaeberry has been subjected to autumn grazing by sheep. What does it look like? Of what does it consist? There are three distinct categories though sites are likely to show a combination of two or more. The treatments needed to restore heather are different for each one.

Mid-nineteenth lazy-bed cultivation which has since reverted to hill land. (NS836192)

Type A – Peat soils

On waterlogged blanket bog, any 'grass' is probably not grass at all but cotton sedge – provider of useful 'draw moss' in spring. In the more fertile hollows away from thick peat and where the soil is not waterlogged, purple moor grass may be present in quantity. White bent often dominates drier slopes.

Heather is present but is sub-dominant and in danger of extinction if grazing is hard. The commonest ericaceous plant is cross-leaved heath, of no value to grouse or domestic animals. Cotton sedge often grows in dense tufts with heather and blaeberry apparently growing out from within. If the area has been heavily over-browsed for decades, ling heather may have been extirpated though blaeberry may still be common; there are few or no fine-leaved grasses. Managing this ground to favour ling heather involves no more than limiting winter grazing and banning burning. Heather, being a woody shrub which grows incrementally, will over-top and partially suppress sedge and the less vigorous cross-leaved heath in a few years. Total suppression of the cotton sedge is undesirable because the 'catkins' and stem bases, which grow early in spring, provide valuable food for deer, sheep and grouse. Light winter grazing keeps the heather co-dominant. Cotton sedge responds well to fire so, if the heather does over-top and suppress it, a fire soon restores the balance.[52]

LEFT:
Hare's tail cotton sedge catkins ("draw moss") in flower.
(PHOTO: D. DUGAN)

BELOW:
Purple moor grass a year after spraying with dalapon. The green plants are blaeberry.
(NJ126324)

Type B – Acid grassland. Purple moor grass and white bent

The moor is covered in coarse grass 20 to 30 cm (8 to 15 in) long with few other plants in significant quantity. If the site is wet or low lying, the grass is probably purple moor grass or white bent if it is better drained. Both can present intractable problems as they thrive when subjected to fire and over-grazing.[53] They are of value to livestock only for short periods in summer and

71

have little or no nutritional worth the rest of the year. Purple moor grass in particular competes strongly with heather which is also liable to be suppressed by winter grazing. A common prescription that is still supported officially which is designed to favour heather is either reducing or completely clearing farm live-stock. It is a gravely mistaken policy.[54] With no animals to eat the grasses in summer, the purple moor grass becomes so vigorous that it over-tops the heather. When it dies back in September, it falls over, smothers and kills the heather. Over a century ago, in an effort to further heather dominance, the management at Langholm erected:

> Several enclosures on grassy land where, after the stock was excluded, heather grew well and game swarmed; it became necessary, however, after the heather was established, to introduce a light winter stock. These enclosures were abolished after the last Great War (1914–18).[55]

The original programme presumably involved summer grazing only and no winter stock to allow the heather to re-establish. Summer grazing, particularly with cattle, is ideal. If heather is present as a sub-dominant, easing winter browsing allows it to regain some dominance over a year or two as with Type A. Where heather is absent, more drastic steps may be called for involving the use of gramini-cides such as Fluazifop in July/August followed by burning off the dead grass, scarifying the surface and, if necessary, the harvesting and spreading of heather seed capsules from nearby flowering stands.[56] In the winters following, a carefully controlled grazing regime has to be enforced with no grazing at all in the first year and light stocking only for several years until the heather becomes established to the point where it competes with, and ultimately dominates, the grasses.

Type C – Fine-leaved grassland

The area is lawn-like; dominated by productive, heavily grazed, fine-leaved grasses such as wavy hair grass, fescues and bents together with herbs like eye-bright, wild white clover, yellow tormentil, heath bedstraw, violas and daisies; places cultivated or limed within the past hundred years are easily identified. It often supports moles because the richer soils provide good conditions for earth-worms and may show signs of cultivation (old lazy beds are a give-away and nettles a sure sign of an earlier dwelling as they typically grow on farm and house middens).

A heather seedling from this site after three seasons' growth.

Successful heather re-seeding. (NS628312)

Fine-leaved grassland is perhaps the most difficult site type upon which to restore heather. The soil is often fertile with a pH at or above that at which heather thrives. Sites attract grazing animals and birds because of the vascular plants that grow there and supply a protein-rich bite in spring and early summer for sheep and grouse alike; even where overall domestic stocking rates are low in winter, they remain high enough locally to suppress heather and blaeberry growing nearby. Extensive areas can be treated as Type B above; under normal conditions, they comprise only a small proportion of the moor.

The heather/bracken interface

Bracken is nobody's friend and a scourge on hill land.[57] It colonises the better drained ground formerly occupied by crofts, woodland, heather and grassland. Once established, it can form a dense canopy which kills out all the vegetation underneath. There is some evidence it is carcinogenic which makes it even more disliked. It is estimated that over eight per cent of Britain's land surface (nearly two million ha) is dominated by bracken and that the area is expanding.[58]

Cut bracken being baled for cattle bedding. (PHOTO: W. HUTCHINSON)

As early as 1808, when sheep farming in the Highlands was still in its infancy, bracken was identified as a widespread problem.[59] Useless for fodder and of limited value for thatch, it was never seen as a plant to be encouraged but the industrious subsistence farmers of former times did find uses for it. It was cut, dried and stored for bedding farm animals; its fertility added value to the composted animal dung and household refuse which enhanced the growth of staple crops; its ash was a component of soap. Harvesting at peak growth in July kept it in check while, in a grazing economy based predominantly on cattle, trampling curtailed its spread.

Bracken does not grow on waterlogged ground so does not encroach onto heather swards on blanket bog (Type A). Heather on a fertile site (Type C) is likely to be under pressure from browsing and its competitive edge weakened as a result. Such places are commonly where bracken occurs. If heather at the degenerate growth stage is growing against a bracken bed, it will regenerate only from seed and thus slowly. The bracken on the other hand is unaffected by the fire and has a head start on account of its rhizomatous growth habit so it is likely to take the heather's place. Bracken is opportunistic. It does not compete well with vigorous heather but is quick to take over where the heather has been weakened by over-browsing or allowed to become degenerate. If the heather canopy opens up, it allows the prothalli to establish and start new colonies. The main causes of the increase in bracken acreage nationwide over the last century are poor heather management and livestock husbandry. Perceived wisdom is that where heather is growing alongside, against or in mixture with bracken, it should not be burned: this is incorrect. Sooner or later, heather becomes degenerate and either allows bracken to encroach, colonise and over-top it when the sward becomes senescent and opens up or the keeper or shepherd burns it because it is old which means it loses the little competitive edge it had. There is little doubt that bracken has benefited widely from this 'Catch 22'; the moorland manager is damned if he does... and damned if he does not.

Bracken encroaching on heather growing on well-drained glacial debris. (NC719535)

The solution is simple. Think always of managing the moor as a whole and as the diverse entity it is. Some bits need water; some need scrub control; some need to be protected against rabbits; others need hard burning. And there are bits where heather and bracken coexist and the heather has reached the stage where it ought to be regenerated by burning. The keeper should go ahead and burn it but must treat the bracken fronds the following year because their main competitor has gone. *Re-creation and maintenance of the heather sward is the primary task and must govern site management.* Problems posed by competitors are a secondary consideration.

Fire running into bracken facilitates its spread.

Hill cow grazing and trampling among bracken. (PHOTO: D. PHILLIPS)

RIGHT: *Manage heather regardless of bracken: a hot fire rapidly leads to dense regeneration from seedlings.*

Bracken control

Bracken control is widely practised. Cutting and crushing to weaken and destroy top growth used to be commonplace but chemical sprays are now the main weapon. They can be applied aerially or by wheeled vehicle: application by air, though the most effective method for rough ground, is expensive but vehicles are restricted where the ground is steep and/or rocky. Spot treatment of regrowth can be done by knapsack sprayer or weed wiper. The chemical asulam, marketed as Asulox, is effective when sprayed at full frond emergence in July and August before the onset of large scale translocation of nutrients to underground rhizomes. It kills docks but is non toxic to other vegetation. There is a current proposal from Brussels to ban the sale and use of asulam in 2012 – a retrograde step in terms of moorland management.[60]

Glyphosate (trade name Roundup) is an inexpensive and effective general herbicide but it cannot be sprayed unless the canopy is unbroken because it kills green plants on contact. It can be applied by weed wiper when the emerging fronds are at the 'crozier' stage of growth in late May/June; a repeat application on regrowth in August further weakens the plant. If carried out systematically over three years, bracken can be all but exterminated. An All Terrain Vehicle (ATV) or quad bike can pull a weed wiper easily.

Though kills as high as ninety per cent are often achieved if bracken is sprayed at the right time of year under favourable conditions, the results of many bracken control programmes often fall short of expectation or fail completely because the operation is not planned properly. If the initial treatment is not

75

followed through, the investment is wasted; as a rule of thumb, twenty per cent of the budget should be set aside for post-spray treatments including killing regrowth, seeding heather or establishing grass. Spray-year plus one reveals pieces missed by the contractor and some fresh fronds emerging on treated ground. It is tempting to tackle the regrowth there and then but this is the wrong tactic. Spray-year plus two shows more regrowth and growth from rhizomes which were dormant at the time of the primary treatment and so did not receive the chemical but have since broken dormancy. This is the moment to spray again using a knapsack or weed wiper. In spray-year plus three, there should be few or no emerging fronds; management can move on to the next stage, the re-creation of the heather.

Former bracken sites

In treated places which carried a heavy growth of bracken, there is often a thick layer of undecayed bracken litter on top of the soil even three seasons after spraying. It overlies a structureless black humus which is often mineralised and fertile and can provide a good rooting medium for heather. Without intervention, the area will be colonised by foxgloves, sheep's sorrel and Yorkshire fog, none of which are of much value for grouse and sheep. Heather seedlings need a firm base before they can get that root hold which is the first step to forming a sward; they cannot establish in thick, unstable bracken litter.[61] If the rooting medium is loose and puffy, it dries out; any seedlings in it die. The bracken litter has to be removed or broken up to create micro-niches suitable for heather germination and establishment. The litter can be burned off if weather conditions permit but, failing that, a part-set of agricultural harrows pulled behind a quad bike can tear open the continuous litter cover and display bare earth.

If an area has been under bracken for many years, the heather seed bank may be depleted or completely absent. The easiest way to import seed is in the form of cut heather in October before the capsules have ripened fully and shed their seed. Well-flowered building and mature heather can be cut by forage harvester and the trash scattered on the ground at the rate of 100 gm per sq m. For small areas, donor heather can be cut with a strimmer or Allen scythe. Seed treated with an aqueous solution of phenol can be sprayed on with a hydroseeder.[62] If the area is small and the litter loose and mobile, it is worth laying spruce branches on the top of everything to act as a stabiliser. Sheep must be excluded and rabbits, where numerous, reduced to very low levels.[63] Exclude them both and the battle is all but won.

Heather restoration using prepared seed

Prepared seed has certain advantages although it is expensive. The details of the process of preparing the seed are a closely guarded secret but in broad terms are as follows. The capsules are harvested, sieved, dried and milled to release the seeds which are as fine as dust. They are then heated and treated with a lye made by bubbling heather smoke through water. By simulating a fire, the liquor stimulates germination. The seed can be bulked up with sawdust or sand and spread like fertiliser, or suspended in water and either hydroseeded from a helicopter or sprayed using ground machinery.

Using treated seed can be risky as most of it germinates soon after it is applied. If a dry spell follows, seedling mortality can be high and the end result a disap-

Heather seed harvester made by Norman Close, Kettlewell.

Heather capsules and light trash ready for sowing.

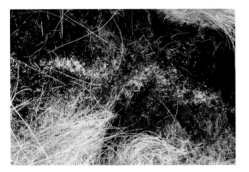

Heather seed bulked up with sawdust and sown by scarifier.

A home-made seeder mounted on a TTS10 scarifier.

pointment. Heather has developed a reproductive strategy which involves the production of vast numbers of tiny seeds (up to 800,000 per sq m per annum) that retain their viability for decades, ensuring that some seeds germinate whenever conditions are suitable. Sooner or later, a cohort of germinating seedlings encounters suitable conditions, a proportion gets a root hold and enough survive conditions such as drought, frost heave or surface disturbance by animals to establish a sward. Even if only five to ten plants per sq m survive, they flower, set seed and act as nurses to a second generation of plants growing alongside the initial establishment. The process can be hastened by sympathetic grazing management.

In 1993, the RSPB treated substantial areas of 'white' moor on their Lake Vyrnwy reserve in mid-Wales. Sheep were removed and seed sown directly onto bare soil created by scarifier. Critics may argue that the result is unconvincing but, in defence, no herbicides were used at the beginning to remove grass competition so the heather has always had to struggle. The result of about twenty-five per cent heather cover after seventeen growing seasons is a triumph under the circumstances. The heather/grass mixture is a great deal more valuable for game and wildlife than the unrelieved white moor it replaced. See p.78

While developing the process of using prepared seed, researchers discovered a sciarid moth whose larvae destroyed heather seedlings at the cotyledon and first-leaf stage and seed germinations could be a total loss.[64] Though no work has been done to find out if it is a significant pest, it is reasonable to suppose that, if

Heather seed sown in 1993 as it looked in June 2011 (Lake Vyrnwy estate, mid-Wales).

a heather braid is sparse, perhaps after the burning of degenerate heather, losses to the midge larvae could adversely affect plant density subsequently.

Restoration of over-browsed heather

The easiest ground to restore is 'white' moor with plenty of suppressed heather as an understory. Because affected areas are often subject to forms of tenant right, it can be almost impossible to control browsing levels, particularly in winter. It used to be thought that heather would recover automatically if the sheep were taken off but those who did so found that, as often as not, de-stocking was only successful if the problem grass was white bent.[65] Clearing sheep completely off areas supporting purple moor grass actually made matters worse because the ungrazed grass smothered the suppressed plants in the autumn. Managers soon concluded that restoring suppressed heather was more efficient than scarifying and seeding, techniques which are appropriate only when heather is completely absent. The problem of restoring over-browsed heather came into sharp relief because farmers could rarely do without the affected land in winter – which was when the damage was being done. They needed it for wintering ewe flocks, the very activity that caused the damage in the first place.

Some felt that if as little as ten per cent of the moor distributed in plots could be protected against winter damage, some long heather could perhaps be salvaged and the damage mitigated. The Joseph Nickerson Heather Improvement Foundation, set up by Sir Joseph, a passionate heather enthusiast and grouse shooter, made money available to test the idea of a series of 1-hectare protective plots covering 10% of the area on four sites in Scotland.[66] It proved easy to persuade graziers to collaborate because the trials involved the loss of only ten per cent of an area generally regarded as sacrificial in any case. Rylock fences excluded sheep for the six months October to March, the plots being open to sheep in summer. The technique worked well as long as there were no rabbits or hares. If either were present, they browsed the current season's growth over winter and negated the initiative. In the best sites, as many as two pairs of grouse would take territory based on each plot and enhance the breeding stock by one or two pairs per ten ha. Though it was never widely adopted, the Nickerson system proved useful in the wider quest for techniques for micro-managing critical areas of moorland.

In summary, restoration and repair of moorland attracts the attention of all managers at some time. The methods of achieving a satisfactory result are well understood but can lead to significant expense. Some people eagerly undertake dramatic operations involving cultivations and herbicides but the money can almost always benefit at least twice the area if it is directed at the root causes of the heather regression – in most cases, too many animals on too small an area during the hard days of winter, or an inadequate burning programme.

Drains and drainage

Heather grows best on dry, free-draining soils and, under ideal conditions, excludes all competitors. Obvious examples of its dominance can be seen on well-drained glacial moraines and fluvio-glacial sands in dry districts, particularly where soil pH is 4.5 or less. Less well-drained soils of similar acidity do not carry dominant heather; other plants maintain a competitive edge and, particularly under heavy grazing regimes, suppress it wholly or in part.

Two categories of heather moor can be distinguished in this context; first, the well-drained northern upland heath typical of low-rainfall districts or steep ground where heather typically makes up over seventy-five per cent of the vegetation and, second, blanket bog where heather may contribute as little as twenty per cent to the sward. Both types have been subjected to varying degrees of drainage during the last one hundred and fifty years.[67]

Waterlogged soils provide poor conditions for root growth. Because they are slow to warm up in spring, the onset of aerial growth is delayed and the growing

Open drains in deep peat often have limited effect on heather performance.

season is shorter than it is on well-drained sites. Because of the wet and airless conditions, bacterial and fungal activity in the soil is low and plants are less vigorous. If an analogy is made with the cultivation of agricultural crops, it is probable that the nutrient quality of the plants is lower as well. In short, heather with a waterlogged root system is unhappy and often a bad colour; it can survive in a semi-aquatic medium but it does not thrive. If the water bathing its roots is base rich (pH ≥ 5.5), the heather may be killed altogether. Attempts were made over many years to remedy situations by cutting ditches 60 cm wide and 45 cm deep (2 ft x 1.5 ft) to run water away rapidly from the heather roots. For forty years from the late forties until 1988, substantial exchequer support was available to farmers who wanted to reclaim moorland although no study of its efficacy was ever conducted – it was just assumed that hill land would benefit from drainage in the way arable land did. This is not the case. People failed to comprehend the hydrological characteristics of peat soils compared with mineral ones. Drawing drains through peat obviously results in significant runoff but the only studies to examine where the water came from were performed by An Foras Taluntais at Glenamoy in Co. Mayo. The 'drawing' effect of an open drain in peat is limited because the combination of sphagnum moss and peat is strongly retentive of moisture. To lower the water table in blanket bog, drains have to be so numerous and so deep that the operation becomes hopelessly uneconomic.

Very large areas were tackled in many districts – thousands of hectares of wet heath in the Flow country northwest of Lairg in Sutherland collected farm improvement grant subsidy and were burned, limed, fertilised and sown with grass seed mixtures scattered on the surface and trampled in by sheep. For a few years, the resulting bright green sward increased sheep productivity substantially but rushes soon encroached and the ground became sour and unproductive. Wet and boggy ground with a high pH favours the *Limnaea* snail which is a host for the liver fluke, a parasite of considerable economic importance affecting sheep and cattle. The false dawn ended when farmers were unable to raise sheep profitably and the land ended up being afforested – again at taxpayer's expense. The traditional treatment of 'fluky' ground consisted of draining, often coupled with the application of 'bluestone' (copper sulphate) which is highly toxic to snails. Although the evidence suggests that moor draining is not worth doing, farmers and keepers used to feel, and many still do, that it confers benefits even if they are small. In the years following open drainage of blanket bog, there is often improved growth locally; heather growing within a metre or so of the sides of a drain on flat ground or on the downhill side of a drain in sloping peat often shows more vigour and flowers better. The improvement obviously affects only part of the ground (perhaps ten per cent if drains are drawn at twenty-metre intervals and twenty per cent at ten-metre intervals) but it is likely to be reflected in improved nutrient content of the vegetation and may help sheep and, more probably, grouse do better because we know that grouse take only two to three per cent of the heather available to them even when they are at high density and living on heather-dominated moorland.[68] On blanket bog, where heather tends to make up a smaller percentage of ground cover, minor improvements in quality might well provide grouse with the choice of better food in winter and earlier growing food in spring, both likely to lead to better breeding.

It is generally a waste of time to drain heather which is growing satisfactorily on the drier upland heaths except where, as is not uncommon, there is lateral water seepage caused by a cap of blanket bog on the top of a hill or a spring line which is oozing moisture. The water trickles down onto lower slopes, often over

Contour drains drawn in this way on a 1:200 decline are stable. Alston Moor, Cumbria.

Ross tine ploughing. (NJ021448)

a wide front; if there is too much for the substrate to absorb, the lower ground becomes waterlogged. A contour drain can be drawn to lead the water into the nearest watercourse. If the drain is dug accurately on a 1:200 decline along the contour, it poses little danger of erosion caused by rapid runoff.

In low-rainfall districts, an impervious iron pan can develop often at a depth of less than a metre which, by impeding vertical drainage, causes surface water-logging. The remedy is to break the pan with a strong tine pulled by a crawler tractor.[69] The improved soil aeration results in higher biological activity and a better supply of nutrients to the heather roots. An advantage of the technique is that the ground is not obviously scarred by open drains although an occasional boulder may be brought to the surface. Access for wheeled vehicles is scarcely affected and the risk of erosion is small. As yet, it is not known how long the effect lasts but areas treated in 1973 still show enhanced heather growth and flowering nearly forty years later. The improvement may well be semi-permanent.

Linear improvement in heather growth following tining in 1973. Tulchan, Strathspey – 1999.

One major effect of drawing drains on heather ground which is often over-looked is that the inverted turf produces a partial firebreak. It is often possible to run fires through drained land with excellent effects on the fire pattern. Drains are unlikely to stop a fire running at right angles into them but often stop a fire spreading laterally. Fires burned by shepherds and allowed to ramble about leave patches of long heather interspersed with burned pieces on ground which has been open drained.

Draining has disadvantages; blanket bog often supports uncommon communities of plants and invertebrates which may be severely damaged or even destroyed by drains; bogs are important for grouse chicks as they provide readily available surface water and good insect habitat.[70] Hen grouse often move their broods to boggy places when the chicks are tiny. Bogs consisting of a complex of pools and hillocks as in the Flow country in north Scotland are important sites for nesting waders; if the pools disappear, much of conservation value goes with them. The manager still has a wet bog; perhaps a little drier than it was but a bog all the same. Recognising this, some managers of drier moors have dammed up watercourses and drains and enriched them with ground limestone to create flush areas where insects thrive.[71] In parts of north England, the activity has become something of a cult, perhaps because public money is available for doing it. Preventing or repairing gully erosion on a wet heath is one thing but changing a good dry heath into a wet one in the name of increasing insects for wader and grouse chicks is to be deprecated.

Bog pools are important for grouse chicks and moorland waders.

There is good evidence that grip blocking may change hydrology by slowing up water loss, increase the production of tipulids and, perhaps, stabilise carbon emissions stemming from the natural oxidation of aerated peat. On the other hand, if heath is allowed to become waterlogged heather dominance is reduced. The moor manager should intercept seepage water if it is flowing onto and saturating extensive areas of otherwise dry heath down slope but should *not* draw networks of open drains in wet heath. To avoid erosion, drains should be drawn accurately on a 1:200 decline. Done properly, the watercourse will scour and keep itself clean and neither gully nor silt up.

Draining as a management technique is now out of favour. There are still situations where it is necessary but the wholesale draining of large areas has been questioned on economic as well as conservation grounds. We are unlikely to see the widespread return of the drainage plough to heather moorland.

A 1:200-slope contour drain with a sound shingle bottom cut in 1995 shows no erosion sixteen years later. (NN913355)

83

RIGHT:
Cotton sedge catkins are an important food source for mammals and birds in spring.
(NH667795)

BELOW:
Rushes: the common end-product of moorland reclamation.
(NO652802)

Cotton sedge is a main constituent of blanket bog communities and an important plant in spring for both sheep and grouse as it is the first plant to come into growth. The catkins are eaten eagerly by grouse from emergence in February onwards and the flowers and stem bases are drawn and eaten by sheep, providing ewes with a good protein supplement and high levels of calcium and phosphorus, important in the weeks leading up to parturition and lactation. Hill farmers have long recognised the worth of 'draw moss'; the late Ben Wilson of Troloss, Lanarkshire, in his day the foremost blackfaced sheep breeder in Scotland, used to say that draw moss was 'better than linseed cake' as a nutrient supplement in the latter part of winter. Over-draining of blanket bog can reduce its value for sheep by reducing sedge dominance. As cotton sedge is fire resistant, it increases its dominance with frequent burning at the expense of heather which struggles because of the wet ground. (See this chapter, *Peat soils*.)

Erosion

The natural erosive forces which have shaped land forms in Britain since the beginning of time are water, wind, ice and snow. Land managers' interests are not well served if land erodes as it can lead to unsightly scarring, floristic impoverishment and soil and fertility losses. The two critical anthropogenic influences are the burning of fires and the grazing of domestic and wild animals, both of them the subject of constant scrutiny by conservation bodies which share a widespread hostility to deer and sheep. Some managers have certainly created some bad situations but the large majority aims to achieve thick ground cover in the form of healthy dwarf shrub heath as it tends to stabilise soils, increase transpiration and reduce runoff rates. If that is properly carried out, everyone is happy.

Every sort of bad practice. (NS902153)

Opposition to the grazing animal can go too far. SNH extols the virtues of braided rivers as a natural feature at, for example, Glen Feshie, seemingly unaware that they arise because the grazing of many animals within the catchment accelerates runoff! With the objective of increasing native woodland from the existing 1900 ha (4700 ac) to 4000 ha (10,000 ac), the Red Deer Commission (RDC) and SNH reduced deer numbers from 34 per sq km to 8.6 per sq km over the five years 2000 to 2004 while claiming in the press that the increase in native woodland would improve conditions for salmonids in the River Feshie.[72] Braided rivers and high salmonid numbers are mutually exclusive.

Braided river, Glen Feshie. (NN852025)
(PHOTO: SNH)

A gully in Glen Tilt photographed in 1968 (TOP), *1988* (BOTTOM LEFT) *and 2011* (BOTTOM RIGHT) *showing natural healing.* (NN921730) *Erosion can lead to something of a witch-hunt with managers at risk of criticism from conservation bodies. A gully can be scoured by a natural event but, even in the presence of many deer and sheep, it heals up in time.*

Heather pests

Heather beetle

Heather beetle is present on most moors in most years but populations irrupt every now and then without warning.[73] The tell-tale evidence of damage is the sight of red heather in late summer; in extensive outbreaks, thousands of hectares can be affected. Heather leaflets and bark are chewed by the larvae, the plants are defoliated, lose sap and die, particularly if they are old. Winter browse for sheep and deer and the choice of food for grouse are reduced, often significantly. Though recognised as 'frosting' or blight well over one hundred and fifty years ago,[74] the beetle was not discussed regularly in the nineteenth century literature, suggesting it was not of much concern to moor managers at that time.[75] Outbreaks appear to have become either more frequent or more severe than they used to be.

Beetle damage. Note the ragged appearance of the plants.

Why it has become a pest of economic importance over the past fifty years is a matter for debate. The reason probably relates to changes in heather management and the make up of the callunetum with changes in atmospheric and soil conditions playing a part. Some observers are quick to blame climatic change but there is no hard evidence to support the claim. Atmospheric pollution may be a contributory cause in some locations, as may local changes in weather patterns. Pollution can enrich vegetation, including heather, raise the nutrient level of plants and so enhance the performance of the beetles feeding on it.[76] The changes in food quality may affect the rate of growth of the larvae and perhaps also its survival rates.[77]

In Europe, the beetle has destroyed thousands of hectares of lowland heath which have then been taken over by purple moor grass.[78] The loss is associated with changes in grazing and heather management practices: under-burning and under-grazing have led to heather becoming senescent over wide areas. When attacked by the beetle, it dies. With on-ground conditions inimical to seedling germination, the grasses take over and the heath land is lost. A contributory cause is that the traditional process of screefing to obtain litter for bedding farm animals (plaggan) has all but ceased. It can still be seen at the Lyngheisenteret on the island of Lynga, near Bergen where manual removal of the litter layer

Heather beetle larvum, x2 natural size. (PHOTO: A. LANCE)

consisting of the roots and stools of the invasive grasses results in good regeneration of heather from seed on the bared and consolidated soil.[79] The restoration of a heather-dominant sward which can withstand beetle attack depends on a well-stocked seed bank.

The life cycle of the beetle is well understood. Over-wintering adults emerge from the litter in spring when the mean daily temperature rises above nine degrees C.[80] On calm sunny days in April, they launch into flight from the heather – occasionally in immense swarms numbering millions of individuals – and are carried by the wind, sometimes for several kilometres.[81] They fly low over the vegetation with little control over their direction of flight and tend to crash land when they encounter a change in vegetation height, which is why new outbreaks tend to be in heather growing at the edge of a fire or along a burn side. The beetles mate during the six or so weeks from the end of April and typically choose damp mossy ground dominated by sphagnum

Successful seeding of sprayed purple moor grass. (PHOTO: G. EYRE)

Stripped by Hymac (plaggan) and sown with seed capsules in 1991. (NN913355). *Closed-canopy heather sward – 2011.*

mosses on which to lay their eggs – the peak hatch is in June. The larvae climb up the younger shoots of heather plants nearby and begin to feed on the growing leaflets and bark. They are a dirty grey-brown in colour at this feeding stage. They are easy to overlook as they are sensitive to vibration and movement – if disturbed, they drop onto the litter where they can be hard to see. The distinctive foxy-red heather foliage starts to draw attention to an outbreak towards the end of June and into July. Larvae pass through three instars between hatching and full size before they burrow into the litter where they pupate. Adult beetles emerge in August and wander about, continuing to feed on heather. Densities as high as 2000 per sq m have been recorded in the Netherlands.[82] When the mean temperature falls back below nine degrees, the beetles crawl into the litter and spend the winter in hibernation. Warm and dry weather in May favours beetles; cold, wet and windy weather reduces activity and breeding success. A run of good springs can lead to a rapid build-up and serious heather damage.

The heather beetle is controlled naturally by a parasitic ichneumon wasp *Asecodes mento*. The tiny wasps seek out the larvae and inject their eggs through their ovipositors into the grubs. The wasp larvae grow and mature at the expense of the grubs, a classic predator/prey relationship with the wasp and beetle populations being roughly in equilibrium most years. However, when beetle numbers irrupt for whatever reason, they achieve plague proportions before the wasp population catches up. When it does and beetles encounter unfavourable breeding conditions, the wasps gain the upper hand and the beetle population collapses. A run of years with beetle damage being rarely seen can be followed by a build-up over a year or two which results in vast areas of heather being affected.[83]

Swards of productive heather are important as winter browse for sheep and deer and provide grouse range. If the beetle destroys a season's growth, the carrying capacity of the moor is reduced. In extensive and severe outbreaks, the heather may take two or three years to recover. If it is degenerate, it may not recover at all. Over large areas of Scotland, heather has been killed by beetle and the ground taken over by cross-leaved heath, grasses and sedges of little value to either ruminants or grouse. In these places, the manager has two options. The first is to exclude animals, break the turf with a scarifier and, if the seed bank is depleted, apply cut trash or seed capsules harvested elsewhere. The second is to spray with glyphosate in July and burn with as hot a fire as possible once the vegetation has dried out. If the litter is removed or destroyed in this way, the surface is ideal for heather seedlings and the moor can usually be restored cheaply.

It is disappointing to note that, on the flagship hen harrier study site at Langholm, Dumfries & Galloway, dead heather which had been growing on wet heath but had been killed by beetle was swiped instead of burned. The regrowth of a vigorous sward after a beetle outbreak is essential if expensive attempts by SNH, RSPB and others to reconcile grouse with hen harriers on Middlemoss Head (NY394865) are to yield data of any general validity. As it is, the mulch of cut material lying on the surface of extensive areas of the wet heath guarantees that it will be many years before ling heather can regenerate, if indeed it ever does. The conditions created by the beetle outbreak and the subsequent treatment are inimical to heather growth – only cotton sedge, cross-leaved heath and possibly blaeberry will manage to fight their way through the mat. Had the site been sprayed with a graminicide and burned under the terms of the project's out-of-season licence granted by SNH under WANE, ling heather would have come back quickly.

Sedges mixed with heather killed by beetle that has been swiped as part of an SNH project. Because the trash has been left on the surface, there is no likelihood of heather regeneration for several years, if at all. (NY394865)

Under-managed heather sward killed by beetle now containing a ten-year-old, self-sown Sitka spruce tree. (NY394866)

We know that heather is more likely to die following an attack if it is in the mature or degenerate phase than in the pioneer or building-growth phase. We also know that good management ensures that swards are made up predominantly of morphologically young plants. A rotation long enough to allow the establishment of self-sown trees predisposes an area to lose its heather but a rotation which keeps the heather young, vigorous and forming a tight canopy checks competitors and keeps the grasses and sedges suppressed, making it more likely to shrug off the effects of a beetle attack and return to full productivity in a year or so. For the manager, there is no plausible alternative.

Direct control of the insect is not an option; even if it were practicable to kill beetles over extensive areas, the cost would be prohibitive. The agronomic solution of applying insecticide to kill adults, larvae or both is unacceptable on moorland as it would have a multitude of harmful side effects. The only way to exercise control is to manipulate the heather, keeping it young and productive and rely on *Asecodes* to hold the beetle population in some sort of check.

Winter moth

Winter moth was first recorded as damaging heather extensively in 1980.[84] Outbreaks in the 1990s attracted research effort from the Institute of Terrestrial Ecology (ITE).[85] Since then, episodes have been sporadic and of limited economic importance but it is necessary to identify them correctly when they occur. As its name suggests, the winter moth is active from October to February. When fully grown, they are about 2 cm (0.8 in) long. In late June and early July, they pupate in the heather litter to emerge as adults in mid October. The male flies but the wingless female disperses only as far as she can walk. After mating, eggs are laid on heather and blaeberry and hatch in late April and early May. The caterpillars are variable in colour; basically pale or dark green, they have a darker stripe down the back flanked on each side by a pale yellow stripe. The heads are 91 green or brown, the bodies hairless. They disperse by spinning a long silken thread which acts like a paraglider; on fine spring days with warm up draughts, they can be carried several hundred metres down wind. A landing on heather or blaeberry is a matter of chance but, if luck is with them, their arrival coincides with the opening of fresh buds and the onset of spring growth in the food plants. They feed until midsummer, sometimes stripping all the foliage off the shrubs and leaving only bare stalks.

Winter moth damage is quite different from heather beetle. To start with, the foxy-red colour on the heather occurs much earlier in the year; the damage is complete by the end of June, the time at which beetle damage is usually just getting into its stride. The damage is usually found on dry faces and ridges whereas the beetle tends to be associated with low-lying, wetter areas and sphagnum mosses. Unlike the beetle, the moth's damage is often circular in shape and spreads down the line of the prevailing wind. Winter moth larvae consume blaeberry as well as heather – everything, in fact, but the grasses – whereas the heather beetle confines its attention to heather. Often the first sign of an outbreak is black-headed and common gulls in loose flocks 'hawking' along a hill face, flying slowly into the wind just above the heather and dipping down frequently to pick up caterpillars.[86] Outbreaks are notoriously difficult to control; the manager can only stand by and prepare to mitigate the damage. The population peaks naturally and dies back to low levels as a result of a shortage of food, unfavourable weather and predation. The danger with winter moth is that the shrub zone may be completely destroyed and the affected area reverts to grasses. This is much less likely if the affected heather is in the building/mature stage as it usually maintains enough vigour to shrug off even devastating outbreaks – yet another reason for ensuring short burning rotations.

Winter moth can also live on Sitka spruce and oak. It is not known if these populations are discrete and food specific. Its performance is inhibited by hard winters with early snow lie. In Orkney, over twenty-five per cent of caterpillars were attacked by parasitic wasps but none were affected in northeast Scotland.[87] Were the incidence of parasitism to become more widespread, winter moth might be controlled biologically in a similar way to heather beetle.

Vapourer moth

Vapourer moth outbreaks are much less common although, when they do occur, they are every bit as damaging to heather moorland as outbreaks of winter moth and heather beetle.[88] Vapourer is a summer moth; as the caterpillars pupate in August, the damage can overlap with that done by winter moth and heather beetle. The caterpillars are roughly the same size as those of the winter moth but are hairy and brightly coloured (dark grey, white lines and red dots down the side of the body, shiny black heads and four yellow tufts of hair on their backs). They are poisonous to birds. Vapourer outbreaks tend to occur on the same type of ground as winter moth ones and can be confused with them except that birds do not eat them so there are no gulls in attendance. As with winter moth, vapourer damage cannot be confused with heather beetle outbreaks.

Snow mould

When snow lies for long periods in winter, the heather underneath can be affected by a fungus which feeds on the emerging plants as the snow thaws. It is more a curiosity than a matter of economic concern as the worst affected heather beds are in gullies, steep-sided burns or places where snow accumulates on the lee side of a ridge and forms a deep drift or cornice. Though the heather is affected most years in such places, they are by definition mostly of local extent so the impact on the average sheep walk or grouse moor is small. Places regularly affected usually become dominated by blaeberry which is unaffected by the fungus. This partly explains the frequent occurrence of back-lying slopes covered with dense swards of blaeberry.

Winter browning

Winter browning caused by desiccation during severe frost. The green plant on right was covered by snow.

If a hard 'black' frost sets in and freezes the water in the soil to a depth of 20 cm (10 in) with no snow cover, green heather shoots can turn red and die.[89] Though often referred to as 'frosting', the morbidity of the green material is caused by desiccation rather than low temperatures. Plants lose water to dry air surrounding them but cannot make good the transpiration loss because moisture in the soil is frozen. It typically occurs in a spell of anti-cyclonic weather in January and February when the sun is gathering strength, humidity is low and day temperatures are several degrees above freezing. A strong, drying wind makes matters worse. The damage is more intense where hares, sheep or deer have been browsing or when sheep and deer have trampled on the heather. If a vehicle runs across frozen heather, the tracks soon turn a tell-tale red. 'Frosting' does not occur when the plants are covered with rime or snow as transpiration loss is lower.

Browned heather has no nutritive value so territorial grouse forced to feed on it suffer a check which is only made good when cotton sedge flowers or blaeberry buds and leaves start to emerge. There is some evidence that the larvae of strongyle worms actively select green shoots to climb up prior to encysting on the shoot tips because green shoots hold a film of moisture more readily than dead, brown ones. The moisture facilitates the ascent of the larvae. Severe winter browning can result in grouse mortality in spring. Though unproven experimentally, a plausible explanation is that birds tend to collect more larvae than if the heather had been undamaged over a wide area and the larvae climbed plants at random.[90]

Salt spray

Winter gales from the Atlantic ocean can bring sea water on shore in the form of a fine mist or aerosol. Although the quantity of salt is normally small, it can be enough to scorch plants on the western seaboard. After a severe gale, damage may be sustained up to 10 km from the coast.[91] As the heather goes red-brown; it may seem that the damage was caused by insects but the real cause is readily distinguished because there is no physical damage to the plant (the leaflets and bark are intact), the red colour typically appears in the winter and early spring and is often associated with a memorable gale. The plants recover completely as soon as growth restarts. Salt damage can also occur in a narrow band on either side of main trunk routes which are heavily salted during cold weather in winter. Again, it is easily identified as it occurs only on the road sides, typically on the lee or downwind side, and is associated with periods of cold weather.

The distinction between insect, sheep, drought, salt and frost damage

February to April: Heather shoots intact; bright foxy-red in colour – Frost/desiccation.

April to June: Heather/blaeberry shoots tattered and torn, often completely stripped and red or grey in colour; affected area often circular with a 'hard' edge; damaged plants typically sited on a ridge or hill face covered in dry heath often with over-flying gulls; fat, smooth greenish caterpillars with yellow stripes feeding on heather – Winter moth.

May to August: Heather shoots tattered and torn, red or grey in colour; affected areas usually of limited extent, typically on dry heath; no gulls; brightly coloured hairy caterpillars feeding on heather – Vapourer moth.

Late June to August: Heather leaflets ragged, tattered and torn, cambium chewed; red or grey heather often alongside streams or a division between age classes; no blaeberry damage, no gulls; dirty grey or brownish spotted caterpillars drop to the ground in response to vibration – Heather beetle.

July to August: Foxy-red, two-to-five-year-old seedling heather plants lying on soil surface, often intact – Sheep.

May to August: Heather shoots red but complete; damage confined to young plants or mature plants on thin soils and among rocks – Drought.

End notes

1   Meikle *et al*, 1999.
2   Moorland Association, 2010.
3   Cunningham *et al*, 1978; Phillips, 1990d.
4   Eaton *et al*, 2009.
5   Fletcher *et al*, 2010.
6   Rodwell, 1991.
7   *Country Life*, vol. CCIV, no. 45 (10 November 2010).
8   Baynes, 2011.
9   Birse & Robertson, 1976 named this type Atlantic heather moor.

10 Birse & Robertson, 1976 named this type Boreal heather moor.
11 Cowberry is frequently misnamed cranberry.
12 Yalden, 1972; Anderson & Yalden, 1981.
13 McGilvray & Perman, 1991; McGilvray, 1996; Cobham, 1997; PACEC, 2006.
14 Watson & Miller, 1976.
15 McIntyre 1926. Writing of Argyllshire in 1926, McIntyre said that 'in the Great War, graziers were given full and free liberty to burn heather as they pleased'. Tucker 2003.
16 Watson Lyall, 1912; Usher & Thompson, 1993.
17 Nelson, 2010.
18 Hunt, 2003.
19 Scrope, 1838.
20 St John, 1845; Hare, 1939.
21 Trees for Life, 2011, *Wildlife Scotland* (www.visitscotland.com).
22 Lovat, 1911.
23 Savory, 1978; Lance, 1983a.
24 Hill Farming Act, 1946.
25 Wildlife & Natural Environment (Scotland) Act, 2011.
26 Lance & Phillips, 1975; Lance, 1983b; Steen, 1989.
27 Phillips *et al*, 1992.
28 The Heather and Grass Burning Code, 2007 (England & Wales); Muirburn Code, 2011 (Scotland); Scotland's Moorland Forum, 2003.
29 Rodwell, 1991.
30 BS2092.
31 Scott, 1937.
32 http://www.metoffice.gov.uk/public/pws/invent/ provides access to detailed Met Office forecasts for anywhere in the UK that are updated every three hours.
33 Muirburn Code, 2011.
34 Cazenove, 1936.
35 Lovat, 1911.
36 Watson Lyall, 1912.
37 Malcolm & Maxwell, 1910; Wallace, 1917; Mackie, 1924.
38 Hutchinson, 1903.
39 Watson & Miller, 1976.
40 Phillips, 1991a; Straker-Smith & Phillips, 1994.
41 Hutchinson, 1903.
42 Hester & Sydes, 1992.
43 Wallace, 1917; Hobbs & Gimingham, 1987.
44 Jenkins *et al*, 1970; Miller *et al*, 1970.
45 Quadrat is used here in the sense of a sampling method.
46 Miller & Miles, 1970.
47 Lance, 1983b.
48 Phillips, 1990a.
49 Welch *et al*, 1994; Welch, 1998; Phillips, 1990c.
50 Richards, 1973.
51 Brys *et al*, 2005; Jacquemyn *et al*, 2005.
52 Walters, 2011a; Todd *et al*, 2000.
53 Scott, 1940.
54 Milligan *et al*, 1997; Milligan *et al*, 2003; Marrs *et al*, 2004.
55 Scott, 1937.
56 Pakeman *et al*, 1998.
57 Brown, 1997.
58 Mabett, 2011.
59 Walker, 1808.
60 Mitchell *et al*, 1999; Le Duc, 2000; Marrs *et al*, 2000.
61 Anderson, 2001.
62 G. Eyre, pers comm.
63 Sutherland, 1997.
64 G. Eyre, pers comm.
65 Todd *et al*, 2000; de Smidt, 1995.
66 Phillips, 1991c.
67 Keatinge *et al*, 1995.
68 Savory, 1978.
69 Phillips & Moss, 1977; Miles, 1988.
70 Hudson, 1986; Lovat, 1911.
71 Moorland Association, 2010.
72 Nelson, 2010.
73 Rosenberg & Marrs, 2010.
74 Wallace, 1917.
75 Webb, 1989.
76 Lee *et al*, 1992; Bobbink & Heil, 1993; de Smidt, 1995.
77 Brunsting & Heil, 1985.
78 Berdowski, 1987; Scandrett & Gimingham, 1991.
79 Kaland, 1986.
80 Rosenberg & Marrs, 2010.
81 Brunsting, 1982.
82 Brunsting, 1982.
83 Phillips & Wilson, 1987.
84 Picozzi, 1981.
85 Hartley, 1993; Hartley *et al*, 1994; Hartley & Kerslake, 1995.
86 Picozzi, 1981; Hartley *et al*, 1994.
87 Hartley & Kerslake, 1995.
88 Pinder & Hayes, 1986.
89 Watson *et al*, 1966.
90 Delahaye, 1992; Delahaye & Moss, 1994.
91 A severe westerly gale which battered west Scotland on 23 May 2011 scorched vegetation many miles inland.

Chapter 2
The Management of hill grazings

Cattle on hill. (NN962096)

South-country Cheviots on white bent in winter. (NS949132)

The utilisation of hill vegetation

For millennia, mankind has used herbivorous animals to gather an annual crop of solar energy fixed by photosynthesis and stored in plant tissue and convert it into more portable and palatable products. Early man moved from being a hunter-gatherer living opportunistically to a more settled existence with an economy based on domestic animals. Across the world, milk products and flesh meat became staple foods and provided the basis of trade. The process reflected the principles of the animal's own bank account: saving and storing the riches of summer and drawing down stored nutrients through the winter to sustain life with sufficient surplus to allow reproduction in the form of a calf, kid or lamb. The breeding animal was a bit like a balloon: it blew up on summer plenty and gradually deflated into a flaccid and wrinkled thing by the end of winter. In extreme cases, the balloon had collapsed completely by the spring and the animal died.

The principle of hill sheep farming has always been that 'you keep as many animals on the farm as you feel you can safely starve through the winter'. The build-up of fat depends on the quality of the summer grazing and its decline on the rate of draw-down of the reserve in winter. This in turn depends on the quality of available forage during both seasons.[1] On high-quality winter range, the loss is slow and the animals come through times of privation in good fettle with no need for supplementary foods. The opposite happens if quality forage is scarce. Because of the disparity of palatable food between seasons, grazings tend to be under-used in summer and over-used, sometimes grossly so, in winter. Many grasses and forbs are at their most palatable and digestible for brief periods only in their growth cycle; there may not be enough mouths to crop them at the optimum moment so they become overripe and thus wasted. From autumn onwards, animals choose to graze the evergreen fescues and wavy hair grass until they are exhausted. They are then forced to turn to browsing shrubs. If sheep numbers are too high, it is common to find extensive swards of heather with no current year's growth left by the end of December. Whin bushes can be browsed into prickly domes and clumps of rushes can look as if they have been cut mechanically so desperate can sheep become when closely confined on poor range. The same symptoms of over-use can also be seen on land carrying too many deer.

Building heather bitten to the quick over winter.

Heather nibbled to ground level on over-stocked land. (NC360617) (PHOTO: D. PHILLIPS)

Over-browsing the better plants progressively reduces the productivity and dominance of dwarf shrubs and leads to their replacement by poor quality grasses of low palatability and digestibility, particularly purple moor grass and white bent although cotton sedge also increases. On peaty flats and slopes, trampling puddles the top layer of soil, inhibits natural drainage and results in the decline of ling heather, often to extinction. Sometimes, cross-leaved heath, which has no food value, is the only ericaceous plant left and the dominant plant cover may be sphagnum moss and deer sedge. Unsettled, hungry animals spend more time in search of food and sheep paths on soft ground become grooved and progressively deeper. So starts the process of peat hag formation.

Erosion triggered by sheep traffic. (NC360617)

Over decades, changes in plant associations have reduced the productivity of very large areas of sheep and deer range. Over successive years, animals go into winter carrying less and less condition and, unless the issue is addressed, are flaccid balloons by early March. Farmers then (often rather grudgingly) spend money on supplementary food to get the animals through to spring. Though rarely recognised as such, money spent on imported feeding reflects the failure to equate stock numbers to food availability and is a measure of inadequate range management and over-stocking. Without supplements, lambing percentages on marginal sites in the north and west fall to between fifty and seventy per cent. As a result, the farmer has a maximum of twenty-five to thirty-five wether lambs per hundred ewes to sell in autumn; most, if not all, the ewe lambs have to be retained as breeding replacements so no selection for recruits of good quality takes place.

Table 6.

Gross income according to flock size				
Lambing (%)	100	300	600	1000
70	£5,600	£16,800	**£33,600**	**£56,000**
90	£7,200	£21,600	**£43,200**	**£72,000**
110	£8,800	**£26,400**	**£52,800**	**£88,000**

Lambs weaned as % of ewes tupped and financial performance: productivity is the key to the profitability of a sheep farm.

***Bold** equals a potential profit situation.*

Because productivity and profitability are poor, either breeding numbers decline or the stock is cleared altogether. This sorry state occurs nowadays on many farms on which eighty to a hundred per cent lambings were recorded year after year a century ago. The conclusion is inescapable that the range available to many sheep in highland Scotland as well as in much of Wales and Ireland has been severely degraded in recent decades.

Hill grazing

Sustainability

A hillside where the sheep are in balance with the heather, 1992. (NS920164)

The same place in 2011: the heather sward has been largely destroyed by sheep.

Destruction of moorland can take place insidiously over years or be dramatic and brutal. The former has occurred over most of unenclosed uplands whereas the latter is often the pattern on more accessible sites, some of which have been enclosed for decades. Careful examination helps to explain why farmers carry out certain works. The methodology, the medium- to long-term effect and the likelihood of reversion, whether accidental or intentional, have substantial bearing on the future of moorland.

Moorlands have been declining for years over the whole of western Europe.[2] Deterioration may even have accelerated but few reliable data are available. Such as there are show regional losses of up to forty per cent of heather moorland since 1913.[3] The anecdotal evidence from many districts in Scotland and the north of

Progressive degradation caused by over-browsing between (LEFT) 1968 and (RIGHT) 1995.
(NO711913)

England is of the regression of heather and reversion to white grasses, the afforestation of substantial areas or up-grading by fencing and cultivation of ground initially dominated by heath plants. In Dumfries & Galloway in south-west Scotland, which has seen the largest proportional increase in afforestation, tree cover has increased from four per cent of the county in 1945 to twenty-eight per cent in 1990 though even that figure masks the effect on hill land because it relates to total land area whereas afforestation is confined largely to uplands. Elsewhere in Scotland, the pattern is much the same though less heather moor-land has been lost to trees in the east. There are plentiful examples of heavy and persistent winter grazing, sometimes allied to artificial feeding, which have caused heather to be replaced by inferior grasses. They are often a precursor to planting trees. Parts of Perthshire over which good bags of grouse were obtained in the 1960s can no longer be classed as grouse ground because set-stocked hill ewes at a density of more than one per hectare (2.5 ac) together with ponies and cattle in winter have made such demands on the dwarf shrub heath that former heather land has reverted to grass. On one site, the outlines of fires burned in 1970 are still recognisable as burned sites forty years later – the regrowth consists of a scattering of prostrate heather plants which have never been given a chance to form a closed sward because of over-browsing. See p.102.

In the east, where conditions are more favourable and peat depths usually shallower than in the wetter west, a higher proportion of heather has been reclaimed to grassland. Recent increases in red deer numbers and their patterns of winter movement have intensified the pressure on heather. Many former grouse moors formerly of high quality in the central Highlands have succumbed to the siren voices of red deer. Year-round densities of a hind per 4 ha (10 ac), though common, are incompatible with grouse on two counts. First, the grazing pressure on the heather is unsustainable – as the habitat declines in height and

Regressing heather, west Perthshire.

Heather burned in 1970 is unable to re-establish dominance forty years on because of browsing pressure. 2011. (NN805515)

---

feed availability, the grouse are displaced. Second, if the deer access moorland below 450 m (1500 ft), they host many ticks and so affect grouse breeding. The problem has been exacerbated by a lack of resolve on the part of many owners and keepers to address the issue vigorously.

The harmful effects of concentrated winter use of moorland by sheep have long been recognised. In the mid-1960s, one owner took steps to rectify browsing damage which was local in extent but important as it affected an important grouse drive. Five 0.4 ha (1 ac) areas were fenced against sheep, initially for the whole year but latterly in winter only in what was a forerunner of the Nickerson System of Sheep Grazing (see Chapter 1). The heather recovered so well due to the change in the sheep system that the fences were dismantled in 1991 (NN815398).

What brought about over-use on such a scale? Changes over the years in the ratio of the value of the product to the wages of the man producing it offer a clue. By the end of the war in 1945, food was scarce and the country virtually bankrupt with a large, adverse balance of payments. The thrust of government policy was to produce as much food as possible from home sources almost regardless of cost. High-quality farm labour was plentiful and relatively cheap but hill farming techniques had changed little over the previous twenty years. The Hill Farming Research Organisation (HFRO) was set up in 1954 and charged with improving the agricultural productivity of hill land. At the time, a one-man sheep unit was regarded as twenty score (four hundred) breeding ewes and one hundred and twenty hoggs (female replacements) but many full-time shepherds looked after fewer; twenty-five score was regarded as a big handling for one man. In 1952, a top-quality hill lamb fat off its mother in August and weighing 18 kg (48 lbs) on the hook was worth £4 10s 0d to £5 0s 0d (£4.50 to £5.00) and a single-handed shepherd had a cash wage of £7 to £8 a week plus perquisites of house, fuel, keep of cow, oatmeal, potatoes, dog allowance and so on – worth perhaps another £3. The ratio of wages to fat lamb price was two to one. By 1987, when change in the hills was at its height, good store lambs made £35 and the shepherd got £120. The wages to fat lamb price was 3.4 to one. By 1992, the gap had widened substantially with store lambs making less than £20 in August while shepherds were being paid £170 a week (8.5 to one). By 2002, lambs were worth £18 and shepherds paid £250 (13.9 to one). The gap has narrowed slightly since. In 2010, lambs were fetching £40 and shepherds earning £400 (10 to one); in 2011, when fat lambs were making £50, it took eight lambs to pay a shepherd for a week. Oats, in contrast, were worth £20 per ton in 1952, the price of four or five lambs; in 1993 they cost the equivalent of six lambs. By 2002, they were back to £60 per ton, a little over three lambs. The ratio is unchanged in 2011.

In short, wages have generally risen faster than lamb prices over the last fifty years while feeding-stuff prices have remained relatively stable; rises in wage costs have driven the move towards extensification and encouraged the development of ranch and easy-care systems of keeping hill sheep. It became more profitable for farmers to buy supplementary food to provide premium nutrients at key times of year (tupping and late winter/lambing) than to employ labour to ensure that stock 'raked' the hill properly and harvested 'wild' nutrients from the native vegetation in a sustainable way. Farmers seeking to improve the profitability of a hill farm could either cut labour costs by asking staff to look after more animals or increase the animals' productivity.[4] By the 1960s, many shepherds were looking after forty score of ewes; by the early 1990s, some were looking after sixty score under easy-care systems imported from New Zealand. In 2011, there are contract shepherds looking after a hundred score of breeding ewes. Animal husbandry is not of the standard of forty years ago but, due to substantial investment in fencing, better nutrition, better prophylactic medicine and (not least) higher levels of grant aid via the European Union (EU), profitability is higher. But it comes at a price. In former times, a shepherd looking after four hundred ewes spent a lot of his time looking at his charges (as all good stockmen should) and knew every animal individually. As numbers per man rose, the shepherd spent less time looking *at*, and more time looking *for*, his stock. Farmers, begrudging what they saw as poor use of expensive time, invested in fences to shift the effort back to looking *at* the animals. Many abandoned the old 'hefting' method where each ewe was part of a kinship group kept on her home range.[5] The result was an imbalance in grazing pressure whereby, for convenience sake (and because it was probably better land), the lower slopes of the 'out-by' hill became the year-round home range of most, if not all, the ewe stock.[6] The heather there became over-used while the top ground was virtually unused except as a summering ground for stock hoggs and a few dry ewes. Over-use low down and under-use higher up replaced the more even grazing epitomised by the traditional transhumance system. Though easy to manage, the retreat from hefting to the two-pasture system overlooked the crucial points that grazings on the lower slopes were not inexhaustible and that substantial imports of foodstuffs quickly became essential if ewes occupying a reduced area were to perform profitably.

The result was the evolution of wintering systems that depended on feed blocks from which ewes can self-feed at will, supplementing their protein intake and bolstering the performance of the gut organisms that enable them to utilise a higher percentage of nutrients from increasingly poor forage on the open range. The satisfactory (if, perhaps, not outstanding) performance arising from natural use of the whole range under hefting has been replaced by one which is attainable only if the regime is supplemented by imported nutrients, expensive

Feed blocks concentrate sheep in winter and destroy heather swards. (NS531847)

medications, pour-on sprays, anthelmintics, clostridial inoculations and so on. The sheep perform as well as they did under the hefting system but do so under conditions of semi-confinement. A quasi-natural system that had stood the test of centuries has been replaced by one whose chief merit is lower labour costs. They were achieved by sacrificing the associations of moorland plants: on farm after farm, suppressed heather with sedges, poor-quality upland grasses and bracken have come to dominate the lower slopes before giving way to a narrow band of well maintained and productive dwarf shrub heath in the 'mid hill' which in turn fades into an extensive higher area which barely sees the hoof of sheep or foot of shepherd from one year's end to another.[7] Because the higher area is ignored agriculturally, it does not get burned and its productivity continues to decline.[8] Today's shepherd is no longer expected to walk the hill in the way his father and grandfather did; the ubiquitous 'red bike' has seen to that. Sadly, instead of restoring the ancient craft of running hefted flocks to the benefit of wildlife and sheep alike, the bike has contributed to its destruction.

The problem of nutrient shortage limiting sheep production led the HFRO in the 1960s to develop the two-pasture system whereby, on the typical farm, a tenth of the moorland was fenced into enclosures and reclaimed to improved grassland. The scheme relied on Aberystwyth strains of 'bred' grasses which, developed after the war, were well adapted to hill conditions. It created a halfway house between total reclamation of extensive moorland areas, which was often impracticable and of doubtful economic value on a large scale, and the unimproved wild land beyond.[9] The improved area addressed two weak spots in a hill ewe's nutrient cycle, the first between late October and mid-December, when a high plane of protein-rich food improves the incidence of multiple ovulations, implantation of fertilised ova and hence the number of lambs born the subsequent spring, and the second, from mid-March to end April covering the three or four weeks immediately prior to the onset of lambing and the first two or three of lambing itself. Increased grass-based protein intake at that time of year improves milk output and so influences lamb survival directly as well as indirectly by improving the mother's maternal behaviour. (Thriving ewes tend to mother lambs with more persistence than those debilitated by lack of nutrients so lamb survival is enhanced.[10])

Widespread adoption of the two-pasture system substantially increased output but did nothing to raise awareness of the value of unenclosed upland as a grazing resource. It continued to be managed in a haphazard and casual way if indeed it was managed at all. An unforeseen effect of the system has been that, as fewer hill acres are needed to produce the shepherd's wage and the farmer's profit, few farmers today see heather moorland as a valuable resource. In the eyes of many, it would be better used to grow something more useful like 'grass' (of unspecified type). Though stocking densities have generally increased on the areas that are used, hundreds of acres of 'top' land have become farm-redundant. Farmers and owners generally have not been slow to cash in on this otherwise unused asset at inflated forestry prices. Unlocking the forestry money box has often relieved the crippling pressure of bank borrowing (much of which was incurred to pay for 'improvements' in the first place, especially in the heady days of the 1970s) and provided the farmer with a cushion of unearned income, something he rarely had in former times. Before the rude awakening of the late 1980s, farmers could be forgiven for thinking that every day was Christmas Day – more and better sheep, fewer staff, lower costs and money in the bank as well. The changes were driven largely by out-of-control EU spending on the Common Agricultural Policy (CAP) which led to over-production of farm goods generally.

Differential grazing: heather in the foreground is growing on blanket bog/mor humus and is neglected by sheep. In the background, the heather is on mull humus and has been heavily browsed. (NO958369)

105

These were often stockpiled, destroyed or sold at below the cost of production on the world market.

Because of past land management practices, the options available to the average farmer by the early 1990s were fewer than they had been even fifteen years before. He was now saddled with the unenviable task of making sheep farming pay in a difficult hill environment and a different economic climate in which many of his lowland cousins were ready and anxious to produce sheep rather than surplus cereals or oilseed rape – at a lower support cost to consumers into the bargain. By the turn of the century, hill farming was starting to take yet more knocks as a farmer-bashing conservation movement became increasingly vociferous amid a collapse in profitability allied to bovine spongiform encephalitis (BSE) and chaotic government handling of the 2001 foot and mouth epidemic. Ten years on, official emphasis is now on 'biodiversity' and 'sustainability' with hill farmers being cajoled if not forced into participating in cross-compliant agri-environment schemes where subsidy is paid only if certain perceived environmental advantages are delivered. As a result, ewe numbers in Scotland have declined from 3.88 million in 1999 to 2.64 million in 2010, a loss of thirty-two per cent in eleven years.[11] The brunt of the loss has been borne by the hill flock which has traditionally provided foundation stock for low-ground farmers. With hill farmers looking at alternative ways of managing stock and earning a living, the wheel is turning full circle as, in order to satisfy conditions imposed by cross-compliance, they begin to relearn methods of range management that were second nature to their forebears. The highly developed sheep management skills of former days are now probably redundant but, if policymakers could only come to understand the conservation value of the old system, farmers might be retrained and encouraged to appreciate what native vegetation can do for live-stock farming and wildlife conservation alike if the two are reconciled and properly managed.

The hill farmer is on the horns of a dilemma. He is exhorted to do this and that by factions of the 'green' lobby and others, particularly scientists, who were rarely raised on farms, have limited hands-on skills and whose knowledge of hill farming is sketchy at best. The information they peddle has typically been garnered from college courses in environmental management that have limited relevance to practical farming and livestock keeping. The farmer is expected to conform to a rash of regulations stipulating when he can and cannot carry out basic operations because perceived wisdom has it that they affect wildlife conservation. Indeed they may but populations of foxes, badgers and crows that do equally severe damage to the same conservation interests go uncontrolled because they constitute a politically sensitive problem. Quangos extol well-managed heather moors but hamper farmers (a group with a key role in delivering how moors look and perform) with a raft of petty regulations and subject them to constant criticism. Policymakers forget that, were it not for farmers grazing our uplands, there would be less open space and substantially more plantation forest. Of course, poorly managed domestic livestock causes damage here and there but the solution to that lies in rectifying malpractices, not in discarding a land-use system which has served the country well for many decades.

Value-added diversification into the provision of leisure facilities works for some farmers; the production of farm-assured and organic produce assists others but the initiatives are not a panacea. Given a bit of well-targeted support, farmers could and would encourage game on their land if it created a new income stream.

Many of the points awarded under the Higher Level Stewardship Scheme (HLS) (England and Wales) and the Scottish Rural Development Programme (SRDP) are already the norm for shoot managers primarily because they benefit game production and wildlife generally.

The reversion to other plants

To understand how different grazing patterns and agriculture-based surface treatments produce different results in different districts, it is helpful to describe typical sites, their flora and the effect of different treatments. Change can be brought about by the manipulation of grazing and of species whether singly or in combination by season, or by various combinations of surface treatments including burning, scarifying and sowing seeds.

Transhumance

Where a large rural population depends on subsistence agriculture and lives permanently close to starvation, every scrap of vegetation has to be utilised. Up to two hundred years ago, transhumance – the seasonal movement of domestic stock to upland summer grazing when the snow clears and growth begins – was universal in our uplands, as it is today in the Alps and Pyrenees where it has great conservation worth. In Scotland, 'greens' in the remotest of glens would have been visited by a mob of cattle three or four times in the summer with the animals being folded on these better sites. The stock were not held in by fences or pens but confined by hunger reinforced by the attentions of the women and children who lived with and tended them daily. Animals using the high pastures were mainly cattle which had wintered in sheds or sheltered pastures on low ground until spring arrived, the snow cover broke at altitude and there was a bite of grass. They were then moved to mountain shielings, herded onto suitable pastures and minded and milked daily by the women and girls. The menfolk stayed in the valleys where they earned a wage, fished or tended growing crops. Butter and cheese were made from the summer abundance. The cows' milk output was commonly shared with their calves by penning them separately at night and not allowing the calves to access their mothers until the cows had been milked in the morning: the cows served two masters. Of course, the system was labour intensive but there were few other work opportunities for country people at that time.

Transhumance fell out of use in Scotland at a time of agrarian change in the first half of the nineteenth century with the breakdown of the clan system and the arrival of Border flock masters in large numbers.[12] It had been in use all over upland Britain but had declined earlier in England and Wales. Isolated examples persisted in the remote parts of the Highlands into the last quarter of the twentieth century.

Walking in the hills today, one frequently comes across pockets of better land supporting short, clipped grass dominated by clover and daisies. Generally a pile of stones is all that remains of the shieling which once sheltered those who attended the cattle all those years ago. Look closely at the immediate surroundings: small pockets of better mineral soil, often partially hidden by rank grass or bracken, are visible among the rocks and on burn sides. In the time of the cattle, the daisy turf would have been a mass of prostrate rye grass, wavy hair grass, fescues and clover. Bracken would have been scarce and the rank wild grasses

Young Simmental cattle on mountain pasture in the European Alps. (PHOTO: R. CLARKE)

Probably how the typical Highland township looked in the late eighteenth century.
(COURTESY MIKE TAYLOR/WAYLAND PUBLISHERS LTD BASED ON ARTWORK BY RICHARD PROUDFOOT, NOW OWNED BY 'LANDMARKS')

cropped bare and suppressed. The small pockets would have had the appearance of miniature fertile fields interspersed with and confined by rocks, turf dykes or natural features. The rougher vegetation beyond, including the heather-dominated dwarf-shrub heath, would have been hard cropped in some places and trampled in others, opening up the lightly poached sward. Patches would have been flattened and dunged upon where animals lay at night. Transhumance created a complex and intimate vegetation diversity.

A fermenting vat on legs; the solution to many upland problems. (PHOTO: D. PHILLIPS)

Just as it is closely related to the summer grazing of marshlands and water meadows which can be used only in the drier months of the year, so is it akin to the widespread practice of letting out or acquiring lowland fields of grass on short lease for summer grazing, typically from 1 May to 30 September. Fields so let develop a dense, productive turf full of forbs and clover that withstands summer drought and trampling damage. Even without chemical fertiliser, they often support and fatten two bullocks to the hectare (1 per ac) with permanent grass holding reserves of fertility that succours many rare plants and insects.

The significance of transhumance lies partly in the way that different species forage. Adult cattle tend to graze unselectively; the width of their mouths sees to that. They have prehensile tongues which they curl round a bunch of grass before, with a combination of 'pull' and 'slice' from the lower incisors, gripping and tearing off a mouthful. Because the action takes place over a width of 10 cm (4 in) or more, coarse as well as fine grass, weeds as well as clovers, are torn off. The effect is to free smaller plants from competition and enhance plant diversity. At the end of the digestive process, large volumes of dung are voided after passing through the animal. Tough plant cell walls are broken down in the gut during mastication and fermentation and nutrients are made available for absorption by the animal or recycling as soon as they land on the ground. Mineral extraction from the food intake is relatively low. The majority of inorganic chemicals contained in the plants are returned to the soil – the process is good for the land because it accumulates fertility and helps to build up an enriching mull humus. In contrast, sheep are highly selective 'pickers' which allows coarser, less palatable and digestible material to gain the upper hand.

Grasses removed selectively from a diverse sward grazed by cattle in summer. Note the untouched heather.

109

The shepherds and their sheep

When the Border flock masters took over the north Highlands, they brought with them the white-faced Cheviot. It displaced the small dun-faced native sheep which looked like the primitive Soay and Hebridean sheep of today and yielded a meagre fleece weighing under a kilo (2 lb). Reckoned delicate in constitution; they were customarily housed in winter. They were not kept in large numbers, their meat was not prized and their wool used only to satisfy local needs. The incomers found extensive clean grazing dominated by good grasses, forbs and shrubs. Their sheep throve and the flock masters became wealthy. Within two or three decades, selective breeding had split the Cheviot into distinct south-country and north-country types. The latter were substantially larger and longer in the leg and so better suited to ranging widely over rough country but their management regime did not differ from the system of hefted sheep in their original home in southeast Scotland.

Hefting, which is still remembered by many living today, started to break down under the pressure of rising costs in the 1950s.[13] Until that time, the method used to run the sheep, though not as simple or as valuable in maintaining diversity as transhumance, worked reasonably well; if the fertility and conservation value of the hill was declining, it was doing so at an imperceptible rate. Shepherds looked after between three hundred and fifty and six hundred breeding ewes run at about a ewe to four hectares (10 ac). In the north Highlands, the range of the ewes and hence of the shepherd's handling was around 1600 to 2400 ha (4000 to 6000 ac). The land grab (for that is what it was) saw sheep and their shepherds colonising the remotest corners of wild Scotland and everywhere between and imposing a different system of husbandry to that of the crofters they displaced.

The sheep year began in late November when the tups were put out. Each male was given a 'cut' of forty to sixty ewes on their home range (there were no fences until the mid nineteenth century) with the male and ewes 'shepherded' together until the end of December when the mating season was over. The spent tups then came home to sheltered pasture and the stock ewes were allowed to rake at will over the areas to which they were hefted. If a snowstorm threatened, the shepherd would gather the flock and confine it in circular pens built of dry stone 'stells' or shelters. The stell held a 'cut' where the sheep picked at a meagre supply of bog hay supplied daily by the diligent shepherd until the storm was over and they could be turned loose again. No systematic winter feeding took place and there were no cobs, blocks or cereal mixes to paper over nutritional cracks. Ewes had to survive on what they could find by foraging on the open hill supplemented by metabolising fat reserves to maintain themselves and the unborn but growing lamb. At the end of March, they were brought down to low ground or sheltered parks surrounded by stone dykes (if such existed) for lambing in April. As soon as the lamb had bonded with its mother and was strong enough to follow her, the pair would be turned away to the hill and the shepherd's summer work would begin.

To get sheep to utilise the open hill and graze it in a manner as close as possible to the tried and tested transhumance system, the shepherd would rise before dawn, go round the high ground and cast out his dogs to rouse the sheep and turn them down the hill. The flock would make its way down at the pace of the slowest lambs feeding as it went and spend most of the middle hours of the day on the better grasslands which made up the low ground, food-gathering and chewing the cud. The shepherd would go out again about five o'clock and,

A Y-shelter in north of England.

turning the flock off the better pastures, point it back up the hill, reversing the morning process. Of their own volition, the sheep would string out to the high ground where they would settle for the night away from flies, clegs and midges. There they would cud and rest, leaving behind significant quantities of fertility brought up from the productive lands below.

This process went on daily at least until the milk ewes were clipped in July and usually until the lambs were 'speaned' (weaned) in August. This four-month period coincided with the maximum number of mouths on every corner of the range where there was newly grown herbage. Thus far, the sheep grazing system mirrored transhumance fairly closely in that increased numbers of sheep-mouths could be brought to bear on the vegetation during the growth months of summer when it was green and palatable (May to October). The trouble was that whereas cattle would be taken off the hill and housed in winter – giving the hill range the rest essential for its stability – sheep had to stay out year-round. From autumn to April, each animal was harvesting a daily crop: the dwarf shrub heath was being over-browsed. Progressively, unpalatable and unutilised vegetation gained the upper hand over large areas. Something had to give.

Palatability of vegetation and its ability to withstand grazing

As a general rule, the earlier in the growth cycle a plant is eaten, the more palatable and digestible it is and the more value it has to the animal that eats it. For example, short wavy hair grass in May is better for beasts than dead yellow purple moor grass in October. Blaeberry shoots in October and ling heather shoots in February are more palatable and more sought-after than dead white bent. However, if defoliation bears too hard on palatable plants, they decline and others, hitherto under-utilised because they are disliked by grazing animals, come increasingly to dominate the sward.

111

On a typical hillside supporting a mixed sward of grasses and heath, heavy summer grazing maintains grass productivity and frees shrubs and forbs from excessive competition so that they too can thrive.[14] Transhumance not only suited farm stock but was excellent for wildlife and game. If summer grazing is light or non-existent, coarse grasses over-top shrubs, fine-leaved grasses and forbs by the beginning of October. As the dominant plants die back and fall over, they create a soggy mat which can stifle and kill shorter, less vigorous constituents in the sward. Enter the herbivores – be they red deer, sheep or mountain hares. Their first choice in autumn is the evergreen fine-leaved grasses like wavy hair grass, fescues and bents. As these are eaten, the choice falls on the next most palatable plant, probably blaeberry. When this too becomes exhausted, browsing pressure falls onto young heather before moving onto older plants as scarcity bites. When most of the current year's growth has been eaten and every plant defoliated by fifty per cent or more, the animal goes into nutrient deficit as the purple moor grass and/or white bent (according to site) are all that is left. These have little or nothing to offer once the top growth becomes senescent and unpalatable. Poor quality forage can cost the animal more to harvest and digest than it gains in nutrients thus accelerating the draw-down of its reserves.

The problem does not end there. When growth begins in spring, the coarse grasses get a head start because their meristems are sheltered and undamaged and their over-winter reserves are intact. The shrubs have had a hammering from teeth and hooves, have lost a high proportion of their green material and are slow to start growing; their photosynthesis is impaired and they tend to perform poorly. If the scenario is repeated year on year, heather vigour declines and grazing-induced changes become inevitable. The situation is worse if the dead grass is burned in spring, almost an annual operation in many parts of Scotland. The only effect is to increase the dominance of poor-quality grasses at the expense of the shrubs.[15]

The process may go unnoticed on a five-year timescale but it is significant over a twenty-year period and can be dramatic over a century. One problem for those seeking to measure the speed of change is that, for example, back in 1911, when Lovat was writing, nobody was recording the state of hill swards. All we have are occasional photographs with an identifiable background that allow us to take another picture from the same vantage point and compare the two. Even this only identifies gross changes such as what was formerly forest land, regression from heather to bracken or places where 'white' grasses have excluded dwarf shrub heath.

When sheep started to arrive in large numbers at the end of the eighteenth and the beginning of the nineteenth centuries, we have few accurate records of the quantities of winged game, deer and other wildlife present over large areas at that time although we can glean some information from the writings of contemporary residents and travellers.[16] Osgood Mackenzie, founder of the famous gardens at Inverewe in Wester Ross, described shooting days in the mid-nineteenth century when large numbers of grouse were shot but red deer were hard to come by.[17] Observers wrote of events over widely different regions so what they say is probably representative of the country as a whole. William Scrope stayed at Forest Lodge in Glen Tilt and described many stalking incidents in the district. Deer were so scarce that days at a time would often pass without anyone setting eyes on a beast of any sort. A hundred and seventy years on, the beat, which is still known as Forest Lodge, yields an average of a hundred stags a year – a change typical of much of the central Highlands today.

Cattle on moorland in the Highlands.

Cattle in the Highlands

Large numbers of cattle were exported annually from the Highlands in the nineteenth century. Many were raised in the islands while others swam in ordered droves over to the mainland from Skye.[18] Fifty thousand black cattle went to the trysts at Crieff and Falkirk each year 'from the counties of Inverness, Ross etc'.[19] As they were not fit to trek the long distances to the main trysts in central Scotland until they were at least three and more usually four years old, the crop exported annually represented the oldest age-group of a base population of about 200,000 breeding cattle living in the region. The foundation stock of resident adult cows was always stressed nutritionally (crofters kept as many as possible over winter) making it unlikely that they raised a calf every year, more likely a case of 'one year off and one year on'. That would suggest an additional 100,000 adults, making about 300,000 adult breeding cattle along with 240,000 stirks on all the farms north of Inverness. The annual crop available for sale would have been around ten per cent of the total cattle stock. Today, there is just under half that number of cows (154,000) in the whole of the northwest, the majority of them never on the hill. Nor is this the full story because modern-day cattle mature much earlier than they did in the nineteenth century. Few calves spend their first winter at home, being normally sold in autumn; far fewer followers (young cattle) spend the first three years of their lives independent of their mothers, eating large amounts of native vegetation and growing slowly. It is a fair estimate that the beneficial effect of cattle on the hill today is probably a tenth or even less than it was two hundred years ago. Cattle numbers overall are declining currently at around five per cent a year but the loss in the northwest was around twenty per cent between 2005 and 2009.[20] This is a matter of official concern.[21]

Even if the cattle of the time were small in both size and appetite in comparison with today's animals, the gross figures give some indication of the grazing pressures exerted on the lands north of Inverness in the half-century 1800–1850. Each district north of the Highland Boundary Fault would have had its annual export of cattle to swell the torrent of animals passing south, most of which would end up being fattened in England.[22] Cattle droving was at its height in the mid-nineteenth century when 150,000 cattle would pass through the autumn Falkirk trysts. Coming from all over the north, they represented the annual cattle-crop, suggesting a total population of well over a million animals.

When sheep took the place of the cattle, they initially had access to wonderful, fine-leaved pastures on which they throve mightily. But, after decades of mono-grazing in the nineteenth century, the pastures became rougher, the fine-leaved grasses over-grazed and choked out by coarse grasses. Shepherds increasingly resorted to fire to dispose of uneaten plant residues. Native people were appalled at the change and criticised the shepherds for 'burning the hill'. The frequent fires changed the make-up of the swards and started the steady run-down of fertility to the level we see today.

The moorland problem is easier described than solved. Revising the way we manage it by reinstating transhumance in some form or other would deliver many benefits. Problems arose because of government agricultural policies and subsidy regimes; farmers have been encouraged to graze the hills with sheep and given little or no inducement to run cattle either alongside sheep or on their own or to do anything about the plague of red deer, now just about everywhere. If we want to achieve stated conservation objectives and deliver biodiversity, sheep numbers must be reduced in locations where they are causing identifiable damage (though maintaining a substantial sheep presence; the clearance of individual

A cattle drove passing through Strathyre, 1926. During the General Strike, there were no trains; this mob of two hundred cattle was ferried from Mull and walked the seventy-eight miles (125 k) from Oban to the market in Stirling. It was possibly the last drove ever to come from the west coast to central Scotland. (PHOTO: J. BLACKWOOD)

farms solves nothing) and mobs of cattle of suitable provenance must be brought back to summer on the hills. It remains a matter of concern that, while SNH strongly advocates the reduction of sheep and deer numbers on the basis of their perceived negative effect on conservation objectives and commissions reports on the problem, the one proactive initiative that would deliver most conservation objectives is never discussed. Re-creation of a suitable cattle-type, a modern-day kyloe, which would carry out stated grazing tasks would go a long way to delivering the objectives.[23] Better by far to invest grant money in remote areas to encourage farmers and crofters to breed up mobs of cattle to summer-graze and enrich the uplands than to endorse the widespread but nevertheless negative proposal of fencing areas to grow scrub.

More controversially, to deliver conservation objectives, deer numbers must be significantly reduced – perhaps to a quarter of the present 350,000. But who has the political will to authorise a cull of such magnitude? What would the average deer forest owner say if he is told that his hind stock has to come down to twenty per cent of its present level and that his stag cull – the foundation of the capital value of his estate – has to be reduced by the same amount? What would the owner of a modest-sized estate on the edge of traditional deer forest ground say if an edict made it an offence to have *any* red deer breeding on his ground? Within the last two decades, deer clearances have taken place at Mar Lodge in Aberdeenshire (NTS), at Glenfeshie, Inverness-shire (a National Nature Reserve (NNR) in private ownership), at Abernethy, Inverness-shire (RSPB) and areas on the western seaboard owned by the JMT. Every case provoked strong opposition from parties within and without the estates who saw their interests affected. If policy changes involving either large-scale killing or local extermination are going to be imposed with the current lack of consultation, the squeals of anguish will be deafening. If changes are to occur, the relationship between politicians, sectional conservation interests and landowners must improve otherwise it will sink into ever more open hostility.

Grazing manipulation

The wetter the climate, the more liable heather becomes to destruction from grazing and browsing animals. The plant prefers dry ground; its ability to maintain a sward is in inverse proportion to the dampness of the soil in which it grows. Wet soil inhibits its growth and tends to produce conditions which favour competitors such the 'weed' grasses, sedges and rushes. Under wet conditions, the plant is less tolerant of stress from whatever cause. One of the worst things that can happen to heather is for it to be over-defoliated by grazing, trampling, heather beetle or frost-induced die-back. As a shrub, damage in year one means reduced performance in year two. The converse applies. If heather is released from pressure in year one, year two will usually yield a limited recovery in ground cover and an increase in shoot length but year three often produces dramatic growth. It can be misleading to lay down general levels of use which suit heather in all field situations (rules of thumb cannot take account of off-take by red deer, sheep, mountain hares and rabbits) but, as a broad guide in a sheep-only situation, a ewe set-stocked all year on 1.4 ha (3.5 ac) on most terrain in the west of the country allows the maintenance of heather as a dominant plant and a ewe on 0.8 to 1 ha (2 to 2.5 ac) in the drier east is compatible with a productive heather mosaic. Unfortunately, the opposite is more common: many hill farms in the west carry a ewe to 0.8 ha (2 ac) on grassy heather moors while the heathery moors of the north and east where the winters are long and cold and grasses make up a smaller proportion of the sward may support one ewe to 2 to 4 ha (5 to 10 ac) or more. Under normal farm practice, the moorland least able to cope with heavy grazing is expected to carry the most animals.

Pure stands of heather such as occur over extensive areas in Scotland and northeastern England generally provide poor range conditions for grazing animals. Sheep cannot live on a diet consisting solely of heather; they need access to more readily digested nutrients supplied by grass species or to have their diet supplemented by bought-in food. Moorlands featuring a mosaic of even quite small areas of fine-leaved grasses can carry higher densities of farm animals than those without while others, such as those in parts of Aberdeenshire, Moray, Angus and Inverness-shire where hillsides are covered in an unrelieved carpet of heather, are poor prospects for sheep farming. Others, with out-crops of base-rich rocks and flushes which result in streaks and patches of fine-leaved grasses and herbs, supply a vegetation diversity which can provide excellent sheep-walks.

The influence of even modest areas of grassland can be surprisingly high. A moor capable of carrying an extensive breeding flock (and often a fairly unproductive moor at that) at the rate of a ewe to 4 ha (10 ac) can be transformed by the addition of five to ten per cent of 'greens' to a point where it can provide the nutritional needs of a ewe to the hectare (2.5 ac) that raises a lamb a year. Land improvements of this kind are commonplace in east Scotland where ewes are set-stocked. In the west, not only is the climate less harsh but the mineral substrate is often quite 'good' and capable of supporting a natural sward of bent or fescue grasses in patches and strips. The higher the proportion of the hill with these grasses, the higher can be the stocking density. All is well with a stock of ewes at one to the hectare in summer and autumn but, as soon as the fine-leaved grasslands are exhausted in September or October, the sheep have to move down a notch in terms of plant selection for palatability and digestibility. West coast moors with a predominance of white bent and purple moor grass in association with heather are poor subjects for sheep to graze in winter as the grasses are both unpalatable and indigestible. Heather is also indigestible but it is at least evergreen; sheep can gather a proportion of their daily dry-matter requirements from this source while it lasts. On range dominated by dead grass, a sheep is in 'negative energy balance' with its food resource: it expends more energy gathering and digesting the poor-quality food than it gains from it. However, the energy deficit is less when animals have access to good heather than if they are confined solely to dead grasses. In practice, sheep (and deer) lose weight more slowly if they have access to a good diverse heather sward than when forced to gather food on dead and bleached moor grass and white bent. A good heather sward is made up of a variety of age classes with the oldest plants showing densely packed stems, finely branched and about 30 cm (12 in) long (see Chapter 1, *Heather and its management*). This is more likely to occur on well-drained, well-burned ground in the east than in the west where heather plants often look like miniature umbrellas as they try to force their heads above the grass competition. Normally, the standing crop of nutritious and useable shoots on western heather swards can be as low as five per cent of that in the east. It is not hard to understand how readily western heather can be abused, over-browsed and driven into terminal decline.

Sheep browse heather more heavily on the edges of patches. They seem to like to stand on short grass or a bare path and browse with their heads raised off the ground. Hills which consist of grass interspersed with heather have a high proportion of this interface. Through time, the heather regresses and its place is taken by grasses, sedges and bracken. The process is slow, insidious and rarely recognised by the manager because the changes take place slowly and are not obvious year to year.

Though attention has focussed on the broad effects of vegetation change brought about by sheep grazing, sheep are not the only grazing animals to use the range in significant numbers: cattle, red deer and hares can also be important locally. That the three species have different patterns of vegetation use is obvious to anyone who has ever looked into their mouths. The width of an adult cow's lower incisors is approximately 100 mm (4 in), of a hind about 40 mm (1.5 in)[24] and of a sheep 25 mm (1 in). A mountain hare has a bite width of 5 mm (0.2 in). The daily dry-matter intake of a cow is about 12 kg (30lbs),[25] a hind about 1.7 kg (3.75 lbs)[26] and a hill ewe roughly 1.25 kg (2.75 lbs).[27] It follows that a cow is likely to be much less selective when food-gathering compared to a sheep as the latter can nibble out the most palatable and nutritious material in a way a cow cannot. A hind grazes in a manner intermediate between the two.

Sheep grazing is often held to be detrimental because it is so selective.[28] Grazing the same weight of material by a combination of sheep and deer benefits the sward by removing some of the less valuable plants; a mix of sheep and cattle is even better. Mixed grazing can have conservation benefits.[29] Cattle grazing, either on its own or in combination with sheep and deer grazing, is best because the cow's offtake removes grasses indiscriminately thereby reducing competition with adjacent heather and depositing processed nutrients locally as dung. The effect on adjacent vegetation is very different from the sprinkling of small pieces of faecal matter over a wide area as happens with deer, sheep and hares.

Mixed grazing offers the best range management for the maintenance of dwarf shrub heath but is not always easy to implement over a wide area of upland. Both the economics of hill cattle and our systems of husbandry have changed; the great days of the Highland cattle prior to the end of the eighteenth century have long gone and, with them, a breed peculiarly adapted to local conditions. Alas, there was no Rare Breeds Survival Trust in the days of the kyloe's decline. Would there had been because we could have used the genes today. The breed – the main wealth of many clansmen and the subject of bitter feuding – was small by today's standards, probably weighing a maximum of about 300 kg (6 cwt) live-weight compared with the 500 to 600 kg (10 to 12 cwt) of the Simmental/Luing cross or blue-grey hill cow of today. They existed almost wholly on native vegetation for ten months of the year and on small allowances of straw and hay through the worst of the winter. Many were so debilitated by spring that they could not stand and had to be carried to the grass, their very existence on a knife-edge. Winter mortality would have been high.

At the end of the summer, the drafts for sale were walked to the trysts in autumn (of which Crieff and Falkirk were the largest and best known) but many cattle went as far as Norfolk before finding a fattener and thereafter some were shod before being walked to London for killing and carcase sale at Smithfield market.[30] Unproductive, slow-growing, of poor conformation and uneconomic by today's standards, they were of great environmental benefit in their home-lands, recycling large quantities of plant nutrients and taking little away when the time came to walk to England. Compared with sheep, they were benign grazers who maintained productive moorlands of high fertility which generations of hill farmers have run down over the last two hundred years. Early last century, a writer described the agricultural economy of the parish of Gairloch in the second half of the nineteenth century in some detail.[31] The emphasis was on cattle; sixty cows lived all year round in the immediate district but even as late as 1860 there was little mention of sheep. The richness of the area is well illustrated by anecdotes covering everything from flower meadows to wild bees, blackgame to wild geese. A hundred and fifty years later, it is hard to relate his word picture to the depopulated, sheep-sick parish we see today, devoid as it now is of species common a century ago and where hardly a cattle beast is to be seen.

Despite their wide bite, any heather that cattle eat is usually incidental to their choice of grass for food. The selective removal of grass from among and within a heather sward creates areas of short vegetation which frees flowering plants from competition, allowing them to thrive, flower, seed and recolonise bare spots. Cattle only eat heather when they are very short of grass. By the time that happens, they should be off the hill anyway. The presence of cattle on the open moor – from the time of year when there is a fair mouthful of green grass on offer until it starts to die back in autumn, normally mid-June to mid-October – is highly advantageous. They must under no circumstances be given supplementary foods on the hill as the trampling and excessive dunging can destroy the sward, tends to reduce heather dominance and may even exterminate the plant. There is also a risk of creating a morass around feeding points if the weather is wet and the ground peaty. Cattle have a large gut and can process substantial quantities of vegetation. As the diet of hill beasts (as opposed to low-ground cattle) is high in roughage, they drop partially digested remains in the form of relatively hard, solid cow-pats very different from the semi-liquid faeces deposited by cattle grazing rich, heavily fertilised lowland pastures. They break down gradually through the action of scavenging insects, fungi and bacteria and release nutrients slowly into the soil round about that are taken up by the surrounding plants. These then have a higher nutrient content, making them much sought after by other herbivores. Provided the cattle have not been treated with Ivermectin (a powerful anthelmintic and systemic treatment against parasites), the spectrum of insects supported by the dung is wide. Insect-eating birds in general benefit but especially local rarities such as choughs. Cow-pats are also probed for insects by wading birds such as woodcock, snipe and curlew. Though a cow-pat may kill the piece of heather on which it falls, the open site so created is often colonised by forbs such as pansies, violets and heath bedstraw as well as fine-leaved grasses like sweet vernal, red fescue and crested dogs-tail. Floral diversification resulting from extensive grazing by free-ranging cattle in summer can have a dramatic effect on the carrying capacity of moorland for red grouse as well as markedly increasing biodiversity. The right amount of cattle running on heather can create conditions which support a high density of breeding grouse without intervention by fire.

A downside of cattle can be local trampling damage on soft ground. It can be severe with the heavy cross-continental cows now almost universal in hill-cow herds. Weighing perhaps thirty per cent more than smaller, native breeds such as Highlanders, Galloways and pure-bred Luings, they can make a serious mess in flushes and wet peat because their ground pressure is so much greater. The typical load from the hoof of a 500 kg (12 cwt) cow is 2.5 kg per sq cm (35 lbs per sq in) compared with 0.25 kg (3.5 lbs per sq in) from the foot of an 85 kg (13 stone) man and about 150 gm per sq cm (2 lbs per sq in) from an Argocat. But there is a plus point in that the creation and maintenance of small morasses and the dung associated with them can benefit moorland waders, particularly snipe and curlew. Provided it is not extensive, 'damage' of this sort is an important component of biodiversity. The creation of enriched wet flushes is analogous to the current fad for blocking up sheep drains (grips) to improve insect abundance, particularly on wet heaths. Cattle of whatever breed are often criticised for rubbing on grouse butts and damaging them, particularly those made of turf within a wooden frame. Keepers dislike them because repairing butts is just one more task to be tackled in the run-up to shooting. The solution is simple: fence the butts.

Red deer create green patches within a heather sward through their habit of lying in traditional resting-beds that become enriched with dung which encourages grass at the expense of shrubs.[32] The pressure from a large animal body helps to create a flat surface. When the ground is hard in times of frost, the soil thaws out from body-heat and becomes plastic. The subsequent solifluction creates a small shelf that alters the soil hydrology, making the site better for grass colonisation. It is a common sight to see green flecks dotted about in a heather sward frequented by deer. These small green oases attract other herbivores, including hares, which browse the heather edges and may cause heather regression over the years.

Intensive farming and over-grazing

It is neither realistic nor desirable to suggest that managers looking after moorlands should do so without grazing animals but it is hard to anticipate what level of pressure is going to damage heather and cause it to decline. Workers at the HFRO in the 1970s compared the effects of defoliation of heather in autumn and spring and measured the effect on subsequent performance and survival. They concluded that heather in the building and mature phases could sustain removal up to a level of forty per cent but that off-take in autumn was more damaging to the plant's long-term performance than browsing in spring.[33] Many moor managers at the time were uncomfortable with the finding as it was apparent to them that heavy spring use damaged heather more severely than the same level of browsing in autumn. The explanation is simple. The trials were performed by clipping heather to simulate grazing. When heather is clipped, the only damage the plant sustains is the surgical removal of a proportion of top growth. If browsing occurs in late winter, the plants are dry and brittle and animal hooves do as much damage as teeth; truly it can be said that the animals have five mouths. Attempts were made to refine the assessment of the browsing impact of sheep on heather by modelling.[34] More recent work has suggested that a sustainable level of off-take is a mere sixteen per cent in summer and twenty-four in winter, calling into question long-standing perceived wisdom on the sort of stocking densities that heather can sustain.[35] The topic clearly merits further study. How can the damage that sheep can do to dwarf shrub heath at certain times of year be minimised while maintaining viable sheep enterprises on the hill? If it is accepted that more than a ewe to the hectare of heather year-round is liable to put the long-term future of the sward at risk, what grazing system can accommodate sheep and maintain a vigorous and diverse dwarf shrub heath? If an answer could be found, many of the conflicts between farmer and wildlife manager would disappear. The following 'worked examples' assume that the site is well-burned in a fine mosaic with about five per cent by area of fine-leaved grassland intimately mixed with the heather. By and large, at one per hectare, sheep do little if any damage to heather outwith the winter period. As long as

Table 7.

How flushing breeding ewes and the use of the two-pasture system relieves grazing pressure on moorland over the winter.

	Days on the hill	Days off the hill	
1 October to 7 November (1st visit)	38		All ewes on the hill
7 November to 30 December (tupping)		55	All ewes off the hill (tupping)
31 December to 30 March (2nd visit)	89		All ewes back on the hill
7 February to 20 March		42	Ewes with twins off the hill
Total equivalent days on the hill			
If 5% of ewes have twins	125		
If 20% of ewes have twins	119		
If 40% of ewes have twins	110		
Example 1			
Hill size (hectares)	1600		
No. of ewes	1000		
Ewes per hectare	0.6		Hill size ÷ No. of ewes
Ewes with twins (%)	40		
Thus, sheep-days per hectare	69		38 + 89 (1st & 2nd visits) – 40% of 42 (days twin ewes in-bye) x ewes per ha
Example 2			
Hill size (hectares)	400		
No. of ewes	720		
Ewes per hectare	1.8		
Ewes with twins (%)	40		
Thus, sheep-days per hectare	198		

there is productive grass to be had, the sheep concentrate on it. If there is blaeberry on the hill, they may graze it heavily in autumn.[36] The problems begin when blaeberry and heather are the best foods available, normally in early October. From then on, grazing animals concentrate increasingly on the dwarf shrub heath. When the blaeberry is finished, the sheep turn to ling heather, which is less palatable and digestible. Though the date varies according to the severity of the winter, the onset of growth in cotton sedge in mid-February generally heralds the arrival of spring and early-growing, fine-leaved grasses such as wavy hair grass, and blaeberry provide an early bite in the second half of March. Pressure on the heather is then relieved.

If the six months from the beginning of October to the end of March are assumed to be one hundred and eighty days and ewes are running on a reasonably diverse heathery hill at a density of one per ha for the whole winter, the 'grazing pressure' is a tolerable 180 sheep-days per ha (70 per ac). Where a farmer runs two ewes to the hectare (which is commonplace), the grazing pressure is 360 days per ha (140 per ac). Heather cannot tolerate such high levels of use and declines rapidly but many farmers on limited acreages often have to keep sheep in this way to make a living. Expecting them to reduce their stock is impracticable so a compromise must be sought. If sheep numbers cannot be reduced, then the number of days they use the heather must be cut back. Somehow or other, winter grazing pressure has to be brought down to 180 days per ha or below.

The two-pasture system and the flushing of breeding sheep

Normal practice with the two-pasture system sees ewes brought in from native vegetation and put onto sown grass pastures for 'flushing' prior to tupping. Ideally, they should be on the better forage for three weeks before the tups are put out so they are on a rising plane of nutrition. The traditional date to start tupping hill flocks in Scotland is 20 November (earlier in the north of England) which brings the first lambs on 15 April. The tups are left with the ewes for about forty days to cover two oestrus cycles; shepherds like to have the tups off the hill and back home before the end of December. The ewes are thus off the hill for the whole of November and December, about sixty of the one hundred and eighty-day winter total. They should be back on the hill by the end of December where they remain until the flock is brought down onto in-bye ground about 20 March in the run-up to lambing. The period from 20 to 31 March is another ten winter days when no sheep are on the hill.

Flushing ewes increases the number of multiple pregnancies. Hill farmers do not generally want the high lambing percentages sought by their lowland cousins because the ewe that has twins this year is often the yeld (barren) one the year after. In true hill flocks, the aim is to have just enough twins to provide spares for 'twinning' onto ewes that lose their lamb at birth. It is increasingly common to examine ewes with an ultrasonic scanner in February to identify those carrying twins. They can be drawn off, put on in-bye fields or housed in purpose-built sheds and fed supplementary food to keep them in good condition. Between five and twenty per cent of a hill flock commonly falls into this category though it can be as much as forty per cent on the best places. Moving the ewes off the hill in early February reduces grazing pressure by between two and seventeen days per ha. The nominal 180-day winter period has thus been cut to somewhere between 110 and 125 days.

A common problem is that the hill farmer does not have enough in-bye land on which to flush his ewes. He may be home-wintering ewe hoggs as an economy measure (away-wintering was costing £20 to £25 in 2010), compelling him to keep the ewes on the hill the whole time (and perhaps even flush them with supplementary food). If ewes are fed on the hill, they concentrate at feeding points, cause significant local damage and lower the level of utilisation in the long term. Inevitably, the productivity of the heather declines. The remedy is straightforward but, contrary to what many suggest, does *not* inevitably mean reduced sheep numbers.

Consider first a lightly stocked hill farm with 1600 ha (4000 ac) of heather situated in the eastern UK and carrying a thousand ewes. They are on the hill on 1 October (the first day of 'winter'), brought in for flushing on 7 November, put with the tups on 20 November and turned out again on 31 December. At scanning in mid-February, up to forty per cent are found to be carrying two embryos and are taken to in-bye ground for the run-up to lambing; the rest are left on the hill until gathered for lambing at the end of March. To accommodate the system, the sixteen hundred hectares of heather have to provide 110 sheep-days per ha (28 per ac), which is well within sustainable limits.

Now consider a farm with 400 ha (1000 ac) of heather hill with seven hundred and twenty ewes running on it which is 2 per ha (1 per 1.33 ac). The calendar is the same and, again, forty per cent are scanned with twins. The sheep-days total 198 per ha over the one hundred and eighty-day winter period (83 per ac). In the first example, the heather cover is maintained; in the second, it goes into decline, probably steeply.

The average farmer who wants to put the matter right can accommodate approximately 1.25 ewes to the hectare without damaging his heather if he acts on these simple calculations. Options include putting a proportion of the flock to graze in fields, particularly during the second half of the winter, feeding some on 'sacrificial' areas where the vegetation is unproductive, housing gimmers and/or old ewes, perhaps for a few weeks only, and way-wintering some stock and discarding ewes not in lamb at scanning-time. A mix of these options can reduce upland sheep-days by enough to avoid over-use of heather moor in winter.

The rule-of-thumb stocking densities outlined here take no account of the size of the animals making up the ewe flock. They vary greatly between districts. The liveweight of an individual has a bearing on its daily dry matter intake (its appetite) and so its impact on the vegetation. Body size varies significantly between different hill breeds and depends both on inherited characteristics (the genotype) and living conditions (the phenotype).[37]

Red deer and associated complications

These examples are only a guide. Few places are *that* simple to assess; some farms may have plenty of good upland grass-pasture while others may have extensive swards of grass in in-bye fields. The increasing number of farmers plagued by red deer can have their careful sheep calculations upset. In the many cases where red deer are a problem, they divide their time between several farms; so coordinated action between neighbours is needed to address the problem (see Chapter 4, *Grouse groups*). The situation is often exacerbated by well-meaning proprietors feeding deer, particularly stags, in winter.[38] Sometimes, large concentrations of animals gather in one location, trample the ground and weaken or destroy the heather, often not dispersing until well into June.

Box-feeding stags in winter improves body size and next year's antler growth, keeps the animals at home and may reduce natural mortality. The result is often that a higher number of animals is carried than the range can support.

Both stags and hinds eat grouse eggs; how often this happens is unknown but often enough to merit mention in the literature of the 1930s.[39] The author was given an eyewitness account of a stag eating a nest of eggs by Johnnie Christie, head keeper at Cuaich, Inverness-shire, in 1973.[40]

The effects of feeding livestock on heather moor in winter

Concentrations of sheep eating hay fed on heather in winter lead to its destruction. (NO375415)

Feed and mineral blocks

Putting feed blocks on heather is so obviously damaging that it is surprising to see it still so widely practised. The majority of hill farmers have come to depend on feed blocks in winter. That they are now a vital component of the hill-farm economy does not mean that there is no way of minimising the damage they do to heather. The impact of localised grazing and trampling for, say, the ninety days after Christmas greatly exceeds the capacity of the heather to tolerate it but, if blocks are distributed at random over the hill and put out on robust vegetation such as building heather or grass (both of which can withstand a short, sharp burst of heavy use), damage can be minimised. When the blocks are replaced, the new ones should never be put on the old sites. In practice, an energetic shepherd can reinforce the home-range instincts of the ewes, effectively folding the flock so that every corner of the hill is cropped. Owners and keepers are often quick to accuse farmers of damaging the hill with feed blocks but few offer to help to spread them far and wide. Considering that most hill keepers have access to an ATV of some sort, it is odd that this conflict of interest is so widespread.

Hay and hay hecks

Shepherds prefer to feed hay on swards of heather as it is less likely to blow away. Fine, well-made, herb-rich meadow hay – the best for hill ewes – is light and fluffy and blows about even in a modest breeze so wastage can be high. If a simple heck is made out of a length of sheep net suspended like a hammock on a double row of fence-posts, the sheep feed from underneath and the hay stays put. However, such a piece of fixed equipment – like a ring-feeder – is unlikely to be moved during the course of winter. If it is located on a dry ridge, the heather sward will be destroyed by trampling and the seed-fall coupled with heavy dunging will result in the area being colonised by grasses within a couple of years. The answer is to feed sheep on grass or sacrificial areas wherever possible.

Cobs

In the latter part of winter when ewes are heavy in lamb, processed food cobs and pellets are often scattered straight onto the ground. Ewes quickly get the hang of this and feed in a long and competitive line. A feeding-station in constant use can become trampled into a morass. If the feeding-point is changed daily, no harm results, but very often this is not done.

Hefted sheep flocks

Given the chance, hill sheep behave in a semi-wild manner. For many, 'silly sheep' sums them up but they have evolved a method of range use which is far from silly. If they had not, they would never survive the harsh conditions in which they are expected to live year in, year out while producing a lamb a year. They do it using acquired knowledge of their home range, part of the acclimatisation value placed on hill ewes that inhabit a particular piece of ground. They know to move to sheltered places when storms threaten, which place is best in a given wind, which path to take to get there quickly and safely, where and when to go, how to look for the first stems of 'draw-moss' in the hungry days of spring and how to find a secure corner in which to drop their lambs safe from eagle or fox. The knowledge is passed from mother to daughter by imitation. Though the daughter usually spends her first winter away from the home range on a low-ground grazing, she bonds with her mother, sisters and aunts as soon as she returns to the hill in April. The kinship group spends the whole year within a home range. Ranges overlap but each group in a properly shepherded hefted flock rakes its own area where it can be found at all times. Hefting, one of the most valuable attributes of hill sheep, has been appreciated for centuries and was (in rare cases, still is) crucial to low-input hill sheep farming before the development of more intensive methods in the last six decades. It was the only way to keep sheep profitably before the advent of doses and inoculations designed to combat diseases and nutritional deficiencies. Even today, shepherds voice the old adage that 'a sheep's worst enemy is another sheep' knowing as they do that parasites are related to grazing density. If sheep concentrate on particular areas of choice grazing, nematode and tapeworm infestations surely follow.

Hefting is a simple kind of management based on the inborn and the learned habits of sheep.[41] They know to pull out to the higher ground at night and drop down to the more fertile flats during the day, a diurnal movement pattern that, curiously, is the opposite of what red deer do naturally. Because sheep gather food and defecate on the move, they not only graze the hill evenly but also spread the excrement (including any worm burden) widely, thus avoiding concentrations which might create pockets of contaminated land. Each heft occupies a discrete area such as the catchment of a burn, a hill face or even a whole hill; every piece of the ground is grazed evenly. Conscientious shepherding reinforces this natural raking behaviour, ensuring that no piece of ground becomes sheep-sick; the animals do not conflict with their environment or the wildlife that shares it.

Until sixty years ago, once lambing was over and the sheep had been turned out to the hill, the hill shepherd went round his ground with his dogs every day before breakfast to turn the sheep down off the heights where they had spent the night. He would be checking for mis-mothering and 'coupies'. In high summer and before clipping (shearing), he would also be looking out for blowfly-strike particularly in bracken areas. After breakfast, with several hours walking under his belt, the middle hours of the day would be occupied by light work such as sorting a broken 'slap' in a dyke or repairing a lunkie-hole gate. Around four o'clock, he went out again, this time to turn the sheep off the lower pastures and slopes. They would throw up their heads in response to his piercing whistles and the barking of his dogs, string out at a leisurely pace away from the enclosed pastures and low ground and take station on the heights for the night. The 'herd was like a benign schoolmaster, he and his dogs' presence being enough to ensure the sheep obeyed the rules of hefting.

Sadly, running hefted sheep was a casualty of intensification. It was labour intensive and has for the most part died by default. However, there are places where it continues and where the sheep do well and do less harm to the vegetation than if they were paddock-grazed. Reconciling sheep at economic densities with the uplands they inhabit will become critical to hill management as more extensification becomes the rule. The shepherd of the future may well not spend many hours hunting his charges out to the higher ground of an evening but much the same goal can be achieved by improved standards of heather burning, perhaps allied to paddock fencing. If sheep are presented with a well-burned hill face supporting a fine mosaic of heather age classes, they largely heft themselves. What is all but impossible is to get ewes to heft properly on a hillside covered in large beds of old heather with only a patch or two of young growth here and there.

One aspect of hefting relating to heather burning patterns is often overlooked. Once sheep rouse from rest, they start food gathering. As they move from place to place, they snip and pick as they go, collecting edible material from whatever association of plants they are traversing. Although they do not generally choose to browse mature heather, preferring to concentrate on younger plants, they still manage to collect a significant proportion of their daily intake as they pass through it. However, sheep left to themselves on hillsides burned in a coarse pattern go round and round large patches of young plants and browse them to extinction. Hefted sheep on a skilfully burned moor have less impact on long-term heather survival than the same number of unhefted sheep on a moor burned in a few large fires even if the proportion of young heather is the same in both cases.[42] The widespread decline of hefted sheep has made a major contribution to the problems of over-browsing and has directly caused the loss of many thousands of acres of heather.[43]

Vegetation changes caused by grazing

It is beyond dispute that the grazing animal has altered the British uplands over millennia.[44] However, if grazing is properly managed, it is possible to reconcile

Cheviot ewes on left; Blackfaced ewes on right, both stocked at 1 ewe to 1.4 ha. The white moor was created by shepherds who burned the hill almost every year and the sheep exterminated the heather. (NS934176)

profitable sheep farming, sport shooting and the maintenance of heather moorland.[45] Shepherds are quick to point out that a badly burned or, worse, unburned hill makes the hefting of sheep all but impossible but often do little to help. (They may well be responsible for the bad burning in the first place but loath to admit it.) Animals are reluctant to move through large areas of long, even-aged heather; extensive areas of young heather attracts concentrations of grazing animals to the detriment of both the sward and the health of the animals. If the hill is burned in many small fires, each home range contains a sufficiently diverse mixture of age classes of heather and other vegetation for the sheep to hold to their home range. They are often reluctant to move off when required to do so at a gathering. The burning is good for biodiversity with sheep and deer both having a role in delivering conservation objectives.[46] Good burning for game and wildlife is demonstrably good for sheep. By the same token, the correct level of grazing can yield sustainable benefits *on its own.*[47] In contrast, the infrequent use of large fires upsets individual home ranges and causes most if not all the animals from a heft to congregate on a newly burned site with disastrous effects on the regeneration of hill vegetation as well as on their health. Vegetation changes are induced by the sheep but they are ably assisted by the shepherds. Though not necessarily caused by over-stocking, the line between 'adequate numbers' and 'too many' can be a fine one.

In parts of the north of England, particularly the south Pennines where much of the open hill is unenclosed, a stocking density has evolved which is dictated by the overall stocking rate. A hill on which the grazier proposes to run a ewe to, say, 0.8 per ha (2 ac) – a level compatible with the maintenance of productive heather if she is run under the system described above – is unattainable if the neighbour(s) are running at 1 to 0.6 ha (1.5 ac). Immigration from the more heavily stocked neighbouring ground evens out the grazing pressure and pulls the management of grazing down to the lowest common denominator. This works strongly against improved moor management on a parish and regional basis. An obvious solution is to fence the perimeters of the farms but often it is not viable. Much of the moorland in the Peak District National Park is subject to Open Access Agreements which preclude stock fencing on the grounds that it impedes public access and adversely affects landscape quality. Natural England (NE) and the National Park Authorities wring their hands in frustration as they see progressive declines in the quality of moorland grazing control and management; everyone is a loser. The authorities have sometimes paid farmers to reduce livestock numbers dramatically over a wide area. A sounder solution would have been to encourage them to run stock on the sustainable basis of decades ago.

In 1858, a well-respected estate-owner said that 'overstocking of the moors with sheep over the past thirty to forty years since the opening-up of the Highlands' was seriously disadvantaging the grouse.[48] He went on to say that 'I quite believe that if the heather disappeared in our country – and I think there is every appearance that the green will usurp the purple – that grouse will disappear too'. Around that time, others were correctly identifying the cause of heather loss – a process which started about 1820 in the west of Scotland.[49] The loss was far advanced by the 1930s and virtually complete by 1950. Much the same thing was being said of the central and north Highlands in 1884.[50] The observed difference between the effect of sheep on heather in the wet west and in the dry northeast is still important in heather survival to this day.

In more recent times, whole hillsides have been converted from heather to moorland grasses by burning followed by heavy stocking with sheep and cattle. The section of North Logiealmond moor which makes up Dalreoch farm (NN925336) was dominated by heather in 1940. Following a hot fire and subsequent heavy stocking of regrowth by sheep and cattle on good-quality, brown-earth moorland soil, it went 'white'. Most of the hill remains 'white' to this day.

The lime-loving mountain avens can be a casualty of too much grazing and hard burning. | 127

Blanket afforestation on a former sheep-walk, Eskdalemuir, 1972.

A part of the Roberton moor on the Douglas and Angus estate in Lanarkshire which lies alongside the M74 motorway (NS909275) yielded a day of over four hundred brace immediately after the Great War.[51] During the 1930s, after suffering an extensive and very hot fire, it was grazed heavily by sheep. The heather was extirpated and its place taken by moorland grasses of poor quality – a result that the passing motorist can still see today. There are plenty of examples throughout Scotland which replicate this formula: intense fire + many sheep = total loss of heather cover.

When moorland goes 'white' and becomes dominated by purple moor grass, white bent and rushes, profitable sheep farming is hard to achieve. In the most unfavourable areas, particularly in the western half of Scotland, proprietors often clear the hill and turn the land over to forestry. Explanations are often offered that staff are hard to get or that the farm enterprise is unprofitable. When the sheep go, deer move in and the proprietor consoles himself by encouraging tenants to shoot a few stags, creating a modest income stream from minimal investment. If the keepers of a hundred years ago were to return and see the state of the hills they had looked after, often with care, knowledge and dedication, they would not be impressed. When degradation takes place, it tends to be progressive – the productivity and conservation interest of hill land declines in steps. Few extensive stretches of moorland have, over recent decades, become better for farming and general conservation but many hundreds have got worse. What is tragic is that a multi-disciplined approach to upland management *can* deliver a broadly based result that pleases everyone – there *is* a happy medium but few people have the courage or competence to deliver it.

Vegetation changes caused by surface treatments

Changes in grazing can alter vegetation but surface treatments are more drastic. They have relied largely on ploughing or burning and scarifying, usually accompanied by the addition of plant nutrients and the sowing of non-native seed species. Draining was sometimes also performed. The techniques were known before 1939 but the stimulus to produce more food came from the desperate straits in which Britain found herself during and immediately after the war. Two important developments of the time were the Cuthbertson drainage plough and the crawler tractor to pull it. From the outbreak of war, successive governments sought to expand production from all sectors of the farm economy. Guaranteed prices allied to production subsidies launched a new enclosure movement. With the available equipment, steep ground, rocky terrain and areas covered by scrub and bracken could all be reclaimed. Our forefathers coined the aphorism, 'Gold under bracken, silver under whins (gorse) and copper under heather'.[52] Before subsidies distorted basic farm economics, only the bracken sites had been worth tackling and even then only in stable periods of agricultural prosperity. Suddenly, it became feasible, affordable, desirable and even patriotic to tear into the wild land and 'improve' it. The interventions were large in scale: existing vegetation was burned, ground was ploughed and stones were removed. Tile or open drains were sometimes dug at this stage and over-turned sods treated with heavy doses of ground limestone that were so heavily subsidised that they cost the farmer almost nothing to apply. The heavy doses were soon found to lock up and immobilise trace elements such as manganese, cobalt, copper, selenium and magnesium, vital minerals for healthy animals. Phosphorus, a plant nutrient normally in short supply in moorland soils, was applied in the form of basic slag, a high-calcium by-product of the Bessemer steel-making process. After cultivation and the addition of compound fertilizers, a 'pioneer' crop of brassicas, usually forage rape was grown and eaten off in the autumn by fattening lambs. The following spring, the ground would again be disced or scarified and another crop of rape sown, usually as a nurse for a semi-permanent grass seeds mixture based on perennial ryegrass and wild white clover. Prescriptions tailored for different field situations at the Welsh Plant Breeding Institute and Newcastle University's Cockle Park farm allowed the frontiers of cultivation to be pushed up the hill.[53] However, for every successful reclamation (there were many, the heather moorland on Exmoor overlying the free-draining Old Red Sandstone being a notable example), there was just as often a disaster. Where the substrate was acidic and poorly drained glacial till and the native vegetation, limited in productivity as it may have been, was replaced by sown grasses, it took only a year or two for them to die out due to the poor soil conditions. Trampling by livestock often accentuated the problem with useful plants replaced by solid stands of rushes.

One of the earliest, and probably the classic example, of these misguided developments was in Sutherland on either side of the A838 Lairg to Altnaharra road. In the mid-1950s, the Lairg 'flow' was made to bloom and the peat was 'reclaimed'. Big North Country Cheviot ewes grazed lush swards of ryegrass and wild white clover, producing fat lambs where none had been seen before. But gradually and insidiously, rushes took over and liver fluke became endemic because of poor drainage on the peat soil. By the early 1960s, the investment of public money in the form of grant aid for improved agriculture had been written off; the area was by then under trees planted by the Forestry Commission using, of course, more public money. As they now approach maturity; many have blown over, others are twisted like corkscrews because of the infirm roothold and the rest have been clear felled. The value of the timber currently being harvested (2011) can barely be covering the cost of cutting and haulage, let alone the damage to the single-track roads which were never designed to take heavy traffic. The exercise was well meant and perhaps even well executed but, as an investment for the taxpayer, it was a disaster. In 2011, extensive areas of cut over forest are being replanted so setting off the whole sequence once again.

If ploughing for improvement was drastic in impact and only partly success-ful, a different tactic was widely adopted in the west and north Scotland and the Western Isles which was more successful. The criterion was low cost as land users tended to lack the means to carry out plough-based improvements on ground that was in any case often too rocky for heavy machinery. It had partic-ular relevance in crofting townships where much of the native vegetation was very short because it was over-burned, over-grazed and of naturally poor pro-ductivity. Local supplies of shell sand from the sea shore were applied at rates of up to 30 tonnes per ha (12 tons per ac) to the native vegetation followed by the scattering of a mixture of cheap 'seed cleanings' from grass seed merchants.

If the sheep are properly raked on a mixed-vegeta-tion sward, they will not damage heather.
(PHOTO: C. MACGREGOR)

These can contain useful forage species. It was trampled in to the peaty/sandy/shelly soil by closely folding sheep on it for a few days. A sward of varying quality soon emerged which was an improvement on the pasture it replaced. The technique was actually a sophisticated example of improving on nature because the result was not far removed from the 'machair', a natural dune land resulting from wind-blown sand borne inland and laid on top of blanket bog. The acidic moorland was converted naturally into a herb rich and relatively productive pasture which is a good agricultural subject. Floristically rich, it can support large numbers of breeding birds, particularly waders. It is one kind of surface treatment that can offer genuine, cost-effective improvements to local farmers.

End notes

1   Lance, 1983b.
2   Gimingham, 1972.
3   Anderson & Yalden, 1981.
4   Eadie, 1967; Eadie, 1969; McCreath & Murray, 1954.
5   Clarke, 1995; Clarke, 1996; Hart, 2004.
6   Milne & Grant, 1978; Lance, 1987.
7   Grant *et al*, 1963; Miles, 1988.
8   Mather, 1978; Hester & Sydes, 1992.
9   Eadie, 1978.
10 Compare also grouse hens and chick survival (Moss *et al*, 1975; Watson & Moss, 2008).
11 Scottish Government Agricultural Census Statistics.
12 Haldane, 1952.
13 Hart, 2004.
14 Hartley & Mitchell, 2005.
15 Grant *et al*, 1963.
16 Pennant, 1772; Thornton, 1804; Scrope, 1838.
17 Mackenzie, 1921.
18 Haldane, 1952.
19 Thornton, 1804.
20 Scottish Government Agricultural Census Statistics.
21 Holland *et al*, 2011.
22 Haldane, 1952.
23 Holland *et al*, 2011.
24 D-E. Darroch, pers comm.
25 Halnan & Garner, 1940.
26 P. Goddard, pers comm.
27 Maxwell *et al*, 1986.
28 Loe *et al* (2007) hold a different view.
29 DeGabriel *et al*, 2011.
30 Haldane, 1952.
31 Mackenzie, 1921.
32 Watson, 1989.
33 Grant *et al*, 1978.
34 Grant & Armstrong, 1993.
35 Palmer, 1997.
36 Welch, 1984; Welch, 1998; Welch *et al*, 1994.
37 In the last twenty years of the nineteenth century, the celebrated Aberdeenshire artist Joseph Farquharson of Finzean (1846–1935) painted many pictures of the local Feuchside Blackfaced sheep. His animals have little in common with the breed standard of the present day. They are depicted as small, thin-fleshed, fine-boned, dish-faced and goat-like. Similar stock of such 'unimproved' appearance can still be seen in the west of Ireland, suggesting that Farquharson's portrayal is an accurate record. It is reasonable to hypothesise that they had a lesser impact on the environment than today's big, bold beasts. Account of this should perhaps be taken when comparing stocking densities today with those of a century or more ago. It would be reasonable to suggest that two of today's animals have the same effect on vegetation as three or more in former times.
38 Budianski, 1992.
39 Mackie, 1924.
40 He noticed that a group of stags was loafing around on a low hill in May at NN697933 and that one individual was being attacked by a cock grouse. The stag ignored the bird and appeared to be eating something on the ground so he stalked and shot the stag. The oesophagus and rumen contained grouse egg-shell and yolk. This is not to say that deer eat eggs regularly but to suggest that individual animals are probably opportunistic egg-predators.
41 Hart, 2004.
42 Lance, 1983b.
43 Denman *et al*, 1967; Hardin, 1968.
44 Stevenson & Thompson, 1993.
45 Thorley, 1998.
46 Fuller & Gough, 1999; DeGabriel *et al*, 2011.
47 Loe *et al*, 2007.
48 Macdonald, 1883.
49 Colquhoun, 1840.
50 Stuart-Wortley, 1895.
51 K. Wilson, pers comm.
52 Condry, 1966.
53 Hodgson and Grant, 1981; Hodgson and Eadie, 1986.

Chapter 3
Grouse biology for the moorland manager

The significance of territory

An established territory is a prerequisite for a grouse to obtain a mate and breed. Without it, the bird is stateless. In its battle for survival, it is driven by the need to pass on DNA to the next generation; if it cannot achieve this, it is a failure in evolutionary terms. Until it has obtained space it can call a home range and can demonstrate unequivocally that it can hold on to it in the face of competition, no bird, male or female, is permitted the luxury of mating, nesting, incubating eggs and raising young.[1]

Factors influencing territory size

Territory sizes vary greatly between bird species. Those that nest in colonies, such as gannets, black-

A cock grouse ground-calling in late summer in a bid to establish his territory. (Photo: N. Picozzi)

headed gulls or rooks, tolerate neighbours in close, sometimes very close, proximity but, even at high densities, tolerance of one's neighbour is governed by the maxim 'thus far but no further'. Nor are all territories within the colony equally desirable. Birds in the middle generally have greater security against predators but have to devote more energy to maintaining all-round vigilance against intrusion from their neighbours. Those on the periphery are at greater risk of being eaten or having their eggs taken but do not have to devote the same energy to defending their space.

At the other end of the scale to colonial nesters, golden eagles establish exclusive use of wide areas because they need a large hunting range to be certain of obtaining enough prey to rear their young.

Grouse too follow territorial principles except that territory size varies greatly between regions,[2] habitats,[3] years[4] and individuals[5] with some territory owners showing high levels of aggression and others being relatively meek and submissive.[6] These variables jointly and severally determine territory size and thus breeding density.[7] Moderate to high levels of breeding stock are an essential requirement for the land manager if there are to be enough birds in August to provide a surplus for shooters to crop: management practices are geared to achieving this.

Breeding densities vary a hundred fold or more between different parts of the British Isles. On the best moors in the north of England at the peak of a population cycle, birds may be present over a wide area at a density of a pair to the hectare. At the other end of the grouse's range, the over-grazed and degraded blanket bogs of the west of Ireland or the sedge- and grass-dominated deer forests of northwest Scotland, may support only one breeding pair to the hundred hectares (250 ac) or more.[8] 'Grouse moors' range from, at one end of the spectrum, properties where birds share the moorland with red deer, are shot by two or three guns over pointing dogs and where a bag of five brace is rated a good day to top-notch properties in the Yorkshire and Durham dales, Northumberland and Derbyshire where five to seven hundred brace (sometimes more) are shot by six to nine guns in the course of four to six drives. The most productive moors can yield a season's average of up to a brace of birds per hectare year after year. No other wild game bird in the world is managed as intensely or shows such consistent results.

Though a manager's task is to manipulate habitat to support plenty of birds, his scope for influencing matters may be limited in practice. The presence of various species of grazing animals, their numbers and the time of year they are on the ground can be critical; a stretch of moorland overlying poor-quality acidic rock imposes limits on territory and brood size; another stretch, close to forestry, a nature reserve or unkeepered ground, may have continual problems with immigrant predators and so on – the potential varies greatly.[9]

Breeding success one year affects territory size the next. Because levels of aggressiveness vary, chicks from large broods tend to be less aggressive than those with only one or two siblings. It follows that birds raised in big broods tend to settle for less space when they compete for territory, leading to an increase in stock, a process that continues year on year until the moor becomes overcrowded and breeding success declines. The phenomenon is known as *density dependence*.[10] Once a decline begins, members of small broods are more aggressive and take larger territories. However, despite the increased space, there is no guarantee that they raise many young; populations may continue to fall for a number of years before bottoming out.

Territorial behaviour

Interaction between neighbouring cock grouse on a fine spring day in the uplands

Grouse over the stream.
(ILLUSTRATION: J. LOWES)

133

is one of the great sights in nature. Though the activities involve displaying, posturing, strutting, tail fanning and flicking, song flighting and generally showing off, they have a serious purpose. They are mechanisms for demonstrating fitness and the means of establishing ownership of territory. Although it reaches a climax in spring immediately prior to egg laying, the struggle between males actually begins the previous autumn: precocious young cocks start calling as early as late July. Even when still a part of the covey, they try out their voices and spend time displaying

An aggressive, territorial cock tail flicking; late March.
(PHOTO: N. McINTYRE)

in the early morning. They may leave the family group temporarily and make a tentative boundary reconnaissance during which they make song flights onto prominent lookout places such as knowes and large rocks. They are often challenged by other cocks although displays and challenges tend to be half hearted during August and early September.

With the onset of calm, frosty mornings in late September/October, territorial battles begin in earnest. Strife between cocks escalates and leads to the break-up of the covey as individuals attempt to carve out the niche that will become their home. Young cocks strive to establish a territory as close as possible to where they were raised (*philopatry*).[11] Familiarity with immediate surroundings acquired as a chick clearly has a survival advantage when birds come to be parents. An individual spars with brothers and father in its attempts to establish ownership. The male parent is often evicted and moves away to try to establish a territory elsewhere; if the stock on the moor is high, he is not given peace to settle, may have to travel a long way and end up trying to establish ownership on an inferior patch on the edge of the moor not strongly defended by other cocks. In general, if he fails to hold his original territory, he is unlikely to establish a new one elsewhere.[12] During the active period in autumn, when space is parcelled out, territory owners evict all other birds, be they mothers, sisters, aunts or unrelated.[13] The evicted birds, chivvied from pillar to post by successive territory-owning cocks, often flock up together and join with other groups suffering in the same way. These are the so-called packs of autumn which, in years of high production, may comprise many hundreds of birds. Membership of a large group provides a measure of security against predators but, to us humans, it seems strange that successive aggressive males are capable of moving on such a horde merely by using the weapon of threat.

The packs tend to attract the attention of predators because they are showy, mobile and often end up on bare ground with poor cover such as the middle of recent large fires or on grassy areas.[14] In contrast, though territory owners are displaying frequently, they are not moving far and are increasingly on familiar ground and thus at less risk. Of course, some get caught, particularly in early autumn, but the longer they succeed in holding a patch, the better they get to know each feature, stone and hiding place.[15] By late winter, they are familiar with every scrap of escape cover and how long it takes to get there, much as someone who, having lived in a house for years, can navigate unerringly from room to room without turning on the lights. Such a person is at a distinct advantage if attacked by an intruder.

As noted, the displays which the cocks put on every morning when it is not wet and windy have another function apart from establishing and maintaining ownership against the challenges of other cocks: attracting a mate.[16] An itinerant hen catches the eye of a cock and hangs around. If the cock notices her, he may start a courtship display; she may or may not be impressed. If he is too forward and pushy for her tastes, she will flee and try her luck elsewhere; if he is insufficiently fuelled up with testosterone, she will cut her losses and move on.[17] In times of population decline, many cocks fail to attract a hen even though there are plenty around. We tend to think of grouse as all being the same but the birds are strongly individual; they differ in posture, plumage and the way they relate to one another, just as humans do. As a pair bond becomes stronger, cock and hen tend to feed, grit and roost together. He can become quite possessive; if she gets up and flies to water for a drink on her own, she may trigger a sexual chase. He pursues her, tries to head her off and, like a collie dog, shepherd her back home. She may emit a distinctive call like a yelping puppy. When they return to the home range, he will almost certainly song flight on alighting. She sometimes makes the same call only an octave or so higher.

Breeding behaviour

As the time for egg laying approaches, the cock becomes ever more vociferous and the hen increasingly secretive, gravid and full of eggs. In the space of two weeks, a hen lays a full clutch of eggs averaging over 20 gm, an output that may represent up to forty per cent of her body weight – an enormous investment in the future. She becomes very choosy in what she eats, concentrating on sites with vascular plants and cotton sedge.[18] In 'early' years and at low altitude, she may begin to lay as early as the third week of March, normally laying an egg a day until her clutch is complete. Like the grey partridge, she covers her eggs until just before incubation begins and sits for twenty-four days; the first broods normally hatch about

Egg quality has considerable bearing on breeding success.

135

Bog pools and wet flushes are important for brood rearing.

10 to 15 May. In Scotland, it is reckoned that more chicks hatch on 28 May than any other day of the year though the peak hatch in the north of England is as much as two weeks earlier.[19] As soon as the chicks hatch, the cock's priorities change. From establishing and holding exclusive ownership of a patch which enables him to breed, his energies now go to maximising the chances of survival of the chicks his hen has hatched and on which both have staked so much. Display and song flighting virtually cease apart from an occasional call at dawn and dusk.

Territories desirable for breeding may not be good for raising young. Aggression between neighbouring cocks ceases as soon as the chicks hatch, leaving birds free to move to sites of their choice all over the moor. Because chicks grow fastest if they have access to plenty of insects, the priority foraging places support a rich supply, typically in bogs and wet flushes rather than on the dry heaths.[20] As chicks tend to grow slower if insects are scarce, broods move to moist places often a considerable distance from the nest site.[21] Hen grouse can move newly hatched broods many hundreds of metres to favourable places over the course of a few days without losing chicks provided the weather is good and both they and the young are of good quality.[22]

During high summer, the moor may appear to be empty of birds. All is quiet through the day. At dawn and dusk, cocks may occasionally pay a brief visit to their breeding territories and make one or two half hearted song flights before flying back to their families. If it rains and the vegetation is sopping wet, particularly in early morning, adults may take their broods onto bare ground to dry out, a time when the observant keeper begins to get an idea as to how well the birds have done. Gravel roads, newly burned heather, tightly cropped grasslands and sheep scars are all made use of at this time. By midsummer, early hatched chicks are six to eight weeks old and beginning to show up at dawn and dusk. Shepherds are on the hill gathering their sheep for the milk clipping, reports start to dribble in to the estate office of broods being encountered and keepers note the intensity of distraction displays by parent birds when broods are disturbed.[23] Throughout this time and in the run up to shooting, there is no territorial activity between adults; family parties live close together quite happily in those parts of the moor that provide suitable conditions for chick rearing.

Winter behaviour and flocking

The red grouse is an Arctic bird, the same species as the circumpolar willow ptarmigan (which is actually a grouse, not a ptarmigan). The reason it is classed as racially distinct is that, unlike its cousin, it does not turn white in winter. Our red grouse has plenty in common with the low-altitude populations of willow ptarmigan which live on the islands off west Norway as they do not turn white in winter either. The similarities are highlighted here because many think of the red grouse as in some way a delicate bird, poorly adapted to severe winter weather. Nothing could be further from the truth; it is well equipped to deal with the harshest winter conditions experienced on the hills of the British Isles. With feathered legs and feet, virtually the only patches of bare skin through which it can lose heat are on the soles of its feet and its combs. In normal autumn and snow-free winter weather, territorial birds display most days and keep other birds out of their patches. Their daily behaviour is the same on a frosty October day as it would be on a bright day in March. If snow falls but is of insufficient depth to produce total snow cover, most birds will stay on territory and hide among leggy stick heather where it protrudes above the snow surface. Only those without territory flock up and move about. The territory owner has no need to move as it has the necessary cover as well as access to heather shoots which supply a daily maintenance diet. If the snow cover is complete and no heather is visible, the birds all flock up because there is no identifiable territory to defend. They forget their mutual hostility for the duration and search for food as a group. In years of heavy snow such as 2009–10 and 2010–11, they may travel many miles in search of food and shelter.

If a deep snowfall does not drift, grouse burrow into the snow while it is still soft and live among or under the heather canopy, sometimes for many days.[24] They may move into woods or onto forest land where heather and blaeberry are available under the tree canopy. The advantage of woodland for winter shelter was recognised by one Mr Walter Spencer Stanhope in the early 1800s.[25] He planted trees around the edge of his Snailsden moor at Dunford Bridge for just that reason, supplying his birds with the same facilities as their Norwegian cousins. (His other claim to fame is that he probably originated driving grouse to butts or 'liggin holes' in the mid-nineteenth century.) If the surface of deep snow thaws slightly and then freezes hard, the icy crust can prevent birds from burrowing; they may then move off the moor. They can be encouraged to stay on their home range by breaking the crust mechanically with a suitable tool such as an agricultural harrow mounted upside down on its draw tree and pulled behind an Argocat or tractor. A contemporary head keeper in the Lammermuirs is quoted as having used a horse to pull an upturned harrow to enable grouse to gain access to heather;[26] the same technique was used in Swaledale in 2011. It also helps sheep to forage and they, by their passage to and fro, will further break the crust and so free up heather shoots for grouse to feed on. When they are very hungry, grouse follow close behind anyone or anything breaking the crust 'like gulls following the plough'.[27]

If a blizzard brings heavy snow which drifts, heather will be readily available on ridges from which snow has been blown clear. Only under extreme conditions do red grouse migrate long distances. They did so in the winters of 1962–63 and 2009–10 (nearly fifty years apart).This perfectly normal behaviour in response to extreme conditions became the subject of widespread remark on both occasions. After several feet of snow fell over the Christmas period of 1962–63, the Yorkshire/Durham dales lost all their grouse for many weeks. Packs were seen eating hawthorn berries on hedges in the Vale of York, twenty miles or more from the nearest heather. Schoolboys who shot grouse with air guns in the Skerne municipal park where the birds were eating cotoneaster berries appeared in Darlington Juvenile Court charged with shooting in a public place (rather than taking game out of season!). They were fined 5/- (£0.25); the 1963 shooting season was above average.

In 2009–10, there was a general exodus of grouse from hill land before the end of the year. A deal of correspondence ensued in the press, chiefly of an alarmist kind. The snow did not break until well into March. Most of the birds returned although there were reports of significant predation by raptors and mortality caused by birds flying into power lines and other obstacles.[28] Despite this, the 2010 shooting season was well above average in many places. Though even deeper snow fell in 2010–11, the same thing happened with the 2011 shooting season again being well above average, particularly in England. Clearly, the overall stock of birds did not suffer. Those alarmed by the heavy snow ignored the evidence from Scandinavia and North America where such conditions are the rule rather than the exception. Upland Norway is normally ice bound by mid-October and heavy snow falls usually arrive soon after. The willow ptarmigan flock up, abandon the montane plateaux and drop down into the valleys where there are extensive thickets of birch, willow, Scots pine and Norway spruce. They do not have the option of wintering on heather because the plant is not widespread but they do feed on blaeberry stems and tree buds. On a bright frosty day,

A pack of about 50 birds flocked up for security against raptor predation in snow.
(PHOTO: NEIL MCINTYRE)

it is a fine sight to see a score or more snow-white grouse perched in the crown of a birch tree against the background of an azure sky, looking like a crop of exotic fruit.

When numbers are high and the snow lies deep, red grouse move to woodlands and scrub to feed in places which are not normal, year round habitat. Heavy snow has the positive effect of causing them to feed on clean ground and prevents them ingesting worm larvae for the duration. Compared with a mild, open winter in which they stay on territory for six months, this behaviour may help to explain why grouse tend to breed well after a prolonged spell of hard frost and snow. Two snowy winters in succession (2009–10 and 2010–11) may well have contributed to the bumper bags obtained on some moors in 2011. In short, a snow-free winter allows grouse to continue their October pattern of behaviour through until spring but heavy snow falls and prolonged snow lie force them to behave as their ancestors did and their circumpolar relatives still do. A hard winter is no detriment to the population or its subsequent performance.

Summer movements

Once the hen has hatched her clutch, her ties to the home range to which she has been confined for months are at an end. She and her mate are at liberty to go where they will. In a normal summer with equitable temperatures, sunshine and showers, many family parties stay in or close to the territory established the previous autumn. Three things influence whether birds stay put or move. The first is the availability of surface moisture; moist vegetation in boggy places supports larger populations of food items and attracts overnight dew which both adults and chicks drink. Even in hot, dry summer weather, there is still a nightly fall of dew though it may not last for long after dawn, particularly if there is a breeze. As chicks can get heat stressed, broods trek as far as necessary to be close to reliable water sources such as bog pools, burnsides, hill lochs and peat hags which also provide shade. When carrying out brood counts in a hot July, it is common to find stretches of moorland apparently devoid of birds but suddenly, close to a loch or spring, to come upon several coveys all together in a small area. The second is when a population feels overcrowded. If there are many broods in a limited area, the parents may move their chicks immediately after hatching. Usually, the new place is relatively empty because it provides poor breeding habitat and was lightly defended over winter. It may be grass heath with only a little heather, an extensive stretch of short heather which has regenerated after a fire a few years before or boggy ground dominated by cotton sedge. Each can provide adequate food in summer but they are short of escape cover. The ground is 'clean' with plenty of room for immigrants.[29]

Come the autumn and the onset of covey break-up, the parents may leave their young in their new home range to fend for themselves, squabble, pair up and make the best of it while they return to the territory they held that spring where they try to re-establish their former ownership.

These brood movements in summer are frequent; shooters in August may be disappointed to find only a few birds on a piece of ground which held many pairs in spring. On the other hand, they may go to another beat where expectations are low and find plenty of grouse. The unpredictability of a shooting day can be upsetting to keeper and host alike, particularly if the day is being enjoyed by a team of paying guns with 'reasonable expectations' of so many brace in the bag. If adjacent moors are also stuffed with grouse, movement still takes place but the effect is less noticeable because birds moving off one hill face are replaced by immigrants from another. That such exchanges may be hard to detect does not mean they do not take place.

The third reason is food. Every sportsman knows that birds move to 'berries' in late July and August. As with water scarcity, certain areas are apparently empty of birds while a drive or part of a hill face elsewhere is full. Where hard rush abounds, such as in areas of over-browsed heather on the moor edge, birds can spend weeks in August foraging for the seed which is about the size of a small wheat grain. Again, heathery drives on the moor proper are found to contain disappointing numbers while plenty of birds are to be found on adjacent rushy fields and allotments. As they provide poor breeding territory, most birds normally move back to heather ground once the seed is finished.

Cowberry and blaeberry often grow in close association with each other.

In former years, when hill farmers grew oats for stock feeding on the edge of the moor, flocks of grouse, predominantly surplus birds evicted from the moor proper, would make a daily flight to the stooks and stubbles to feed on the grain. The cereal food was regarded as unhealthy and to cause disease.[30] All these movements are local and temporary. When the seasonal abundances are exhausted, the birds move back to their traditional home ranges – sometimes after the sportsmen, nursing their disappointment, have gone home!

Post-hatching weather: its significance

Most young shooters are reared on a good mixed diet. They are healthy: rickets, scurvy and beri-beri, endemic a century ago, are all but unknown. Today's youth are bigger and better developed physically than their grandparents. Diseases of overcrowding, dirty food and water such as tuberculosis, typhoid and cholera are rare. Children go to school where they receive a grounding in the three Rs and are raised with expectations that there will be fireworks on Guy Fawkes night, it will snow at Christmas, hens will lay coloured eggs in anticipation of

Easter Day and a poor grouse season will be explained away either because the birds got 'disease' or because the nests (or chicks – according to choice) got washed away or drowned by floods in May. What effect, therefore, does post-hatching weather have on grouse breeding success?

Grouse are not like partridges which, given half a chance, get tangled up, soaked and chilled to death in fields of young corn if it rains during Ascot week. Like ptarmigan, grouse are Arctic birds well equipped to deal with low temperatures when young. Of course there are records of weather extremes leading to adverse seasons. May snow certainly puts pressure on incubating hens and young chicks and a June 'cloudburst' can bring hill burns down in brown flood at short notice. However, summer rain is rarely of great account. First, grouse are normally excellent parents. If they are fit, fat and free of worms, they go to great lengths to look after their chicks and defend them by means of intense distraction displays if danger threatens; the quality of the display correlates with the number of chicks reared.[31] Second, the viability of the chicks reflects the quality of the parents; vigorous hens breed vigorous young and fit hens lay big eggs which hatch and rear well.[32] Third, even if there is a day of torrential rain on the hill when chicks are tiny, it is rare for a deluge occurring at first light when the chicks start to feed to continue without a break for the whole day. In Inverness-shire on 1 June, the sun has the potential to shine for over seventeen and a half hours; the probability of a break in the weather during daylight hours is high. Feeding patterns cope with weather windows of varying lengths. If the day is warm and dry, insects will be mobile and hard to catch so chicks have to forage hard to collect their requirements but do not need to be brooded for long periods. If the weather is cool and the rain intermittently heavy, insects are sluggish, slow moving and easily caught; anyone who doubts this should hide his face in the vegetation on a cool, damp day and just watch… A crop full can be gathered in a short time and the chicks returned to the warmth of the parental brood patch. Heather differs from grass when the rain falls in that it does not hold a heavy load of droplets that can transfer to a passing bird. When the rain stops, heather drips dry quickly; chicks scuttling about in the sward do not get soaked in the way that young pheasants and partridges do when foraging on a grass meadow or corn field. The test for assessing the severity of extreme, potentially damaging conditions is whether they are district wide or local and whether they affect several watersheds or just one hillside or valley. All too often, the proffered excuse for a poor season is a leaky one; a day of general rain and a severe gale on 23 May 2011 'ruined' good prospects on one Scottish moor so that only a token programme of walked-up shooting took place while a neighbour within line-of-sight a little over a mile away had several good days driving that yielded either side of a hundred brace each day. The poor weather must have affected both moors in the same way.

The significance of food quality

Although birds eat a wide variety of plants and fruits according to season and availability, the staple food of grouse is heather and their normal diet is made up of heather and blaeberry shoots. Though they have a considerable capacity to ingest and digest fibrous and woody material, they take only about five per cent of what is on offer.[33] Comparing what they consume with what is available shows that they are capable of selecting the best at all times.[34] This is obviously important for hens in the run up to egg laying but it is equally important at other times of year.

Heather burning improves habitat structure for grouse and makes room for more birds on a given area. The heather that regenerates after fire has a higher nitrogen and phosphorus content than nearby unburned material; birds seek out and perform better on the better food. As food quality reflects the underlying rock, nesting densities and breeding success rates tend to be higher on moors overlying limestone, basalt and soft metamorphic rocks which weather readily into fertile soils that supports more diverse vegetation.[35] Such moors also support higher numbers of hares and wading birds.[36]

It is a short step to evaluating the effect of artificial enrichment by added fertilisers to the heather sward.[37] Inexpensive nitrogenous fertilisers were plentiful in the 1960s and trials were soon in hand to test their effect on moorland.[38] Grouse breeding densities did attain high numbers locally but the improved palatability and nutritive status of vegetation attracted herbivores from far and wide. Sheep did raise bigger lambs but deer and hares also concentrated on the fertilised areas, damaging the heather swards and facilitating the encroachment of grasses and sedges. The experiment extended knowledge of the role of nutrient content in food plants but proved cost ineffective and had no practical relevance for the manager.

End notes

1   Jenkins *et al*, 1963; Jenkins *et al*, 1964a; Jenkins *et al*, 1970.
2   Watson *et al*, 1977; Watson & O'Hare, 1979.
3   Watson *et al*, 1984a; Watson & Moss, 1993.
4   Jenkins *et al*, 1963; Watson & Jenkins, 1968.
5   Miller *et al*, 1966; Miller & Watson, 1978; Miller, 1979.
6   Moss *et al*, 1995.
7   Moss, 1969; Moss *et al*, 1988.
8   Lance, 1978; Watson *et al*, 1995; Author, pers obs.
9   Picozzi, 1968; Parr, 1990.
10 Jenkins *et al*, 1967; Watson & Moss, 2008.
11 MacColl *et al*, 2000.
12 Jenkins *et al*, 1963.
13 Watson, 1967a.
14 Watson, 1967b.
15 Watson & Moss, 1990.
16 Moss *et al*, 1988.
17 Mougeot *et al*, 2005.
18 Lance, 1983a.
19 A. Watson, pers comm.
20 Park *et al*, 2001.
21 Myrberget *et al*, 1997.
22 Erikstad & Spidsø, 1982; Erikstad & Andersen, 1983; Moss *et al*, 1975; Moss & Watson, 1984.
23 Hudson & Newborn, 1990.
24 Chapman, 1924.
25 Stuart-Wortley, 1895.
26 Martin, 1990.
27 Stuart-Wortley, 1895.
28 Bevanger & Brøseth, 2004.
29 Watson & Moss, 2008.
30 Lovat, 1911.
31 Watson & Moss, 2008.
32 Moss *et al*, 1975; Moss *et al*, 1981; Moss & Watson, 1984.
33 Lance, 1983a.
34 Savory, 1978.
35 Miller, 1968; Picozzi, 1968; Watson & Miller, 1976.
36 Hewson, 1984; Pearce-Higgins & Grant, 2006.
37 Watson *et al*, 1984a.
38 Watson *et al*, 1977.

Chapter 4
Grouse and their management

Historical times

In the early 1890s, R. L. Rimington Wilson, owner of Broomhead moor near Sheffield, wrote to the sporting press to list four main factors affecting moor management: heather burning, driving versus shooting in other ways, the suppression of vermin and sheep. The last, he noted, was a farm topic to be dealt with from the estate office but the rest were under the influence of gamekeepers. All matters connected with them were his responsibility.

One must acknowledge that keepers had it relatively easy in those far off days. There were certainly plenty of them, perhaps ten times as many as today.[1] Because shooting and the employment it guaranteed were important to the countryside economy, keepers had high status. On most estates at that time, farming, in-hand or tenanted, took second place to shooting. Game was king: nothing in the late Victorian and Edwardian era got in the way of maximising yields. Today's sportsmen are astonished to read of the huge bags of grouse that were the rule rather than the exception in that epoch. Not only were they vast on a per day basis but estate records detail seasonal bags which stretch credibility. Many record books list grouse, blackgame and partridges in profusion in places and districts where there have been none for fifty years or more. Not everything has changed for the better over the past century.

Keepers who worked for a man with money would have staff at their beck and call for heather burning, draining, building butts and beating. In general they could do pretty much as they liked. (This is not to condone some of their practices which were indeed barbaric, the gin and pole traps being examples.) Strychnine was scattered about the countryside like salt despite being made illegal under the Protection of Animals Act, 1911 and the Protection of Animals (Scotland) Act, 1912. If a keeper wanted to gas a fox earth or a badger sett, he did so. The cuckoo was as much in his firing line as the nightjar, hobby falcon or golden eagle. There were no restrictions on the use of lead shot on wetlands. Neither he nor his employer would have understood the concept of biodiversity; there was no NE or SNH to make his business theirs, there were no SSSI or Special Protection Areas (SPA), no national parks, NNRs or Rights to Roam and no EU-based Single Farm Payments (SFP) with cross-compliance conditions. No police wildlife officer would have followed up spurious allegations of malpractice and no 'greenie' with beard and sandals checked his every move. There was no RSPB either, at least not in the form we see today. In short, it would perhaps have been surprising if game had not occurred at the high levels it did.

The Great War ended that hegemony. Britain came close to an industrial standstill in April 1917 as a shortage of pit props threatened to halt coal production and a general scarcity of timber made it almost impossible to prosecute trench warfare (crises that led to the formation of the FC after the war). The German U-boat fleet was sinking so many ships that were importing food that it was bringing the country to the verge of starvation. After the Armistice in 1919, farmers' views briefly got more of a hearing but, as ever, memories were short and imports of cheap grain quickly depressed agriculture once the post-war boom ended in 1921. It was to remain in the doldrums until the mid-1930s saw a gradual improvement in profitability led by the dairy sector. All was, of course, to change again with the outbreak of the Second World War in 1939.

There were some good grouse shooting seasons during the 1920s and 1930s and the mid-1930s generally saw bags comparable with those of the Edwardian era. The problems that modern keepers face began in earnest in the aftermath of the war. By 1945, the country was all but bankrupt and taxation was high. Landowners suffered punitive levels of death duty which led to the break-up of many estates, the FC was planting whatever land it could get hold of 'in the national interest', farming techniques were pushing back the boundaries of the wilderness areas and, saddest of all, many keepers who learned their trade in the 1930s had not returned. There was an acute shortage of skilled men in the late 1940s and 1950s.

The screw began to turn further with the Agriculture Act of 1947 and the Agriculture (Scotland) Act of 1948 which gave farm tenants a deal more power to ignore traditional crop rotations and farm their land without serious reference to landowners. In 1954 the Protection of Birds Act and the Pests Act regulated many keepering activities by protecting all birds of prey and proscribing various methods of predator control predators, notably the gin trap. Badgers became fully protected under the Badgers Act, 1973; the Wildlife and Countryside Act reached the statute book in 1981 with far-reaching consequences for the keeper. The Wildlife and Natural Environment (Scotland) Act became law in 2011, effectively consolidating many years of wildlife legislation in one statute.

The twenty-first century moorland keeper has to tread with care, not just with regard to wildlife statutes but also to his personal behaviour. Any infringement of the law, including motoring offences, be it proven or merely alleged, can result in the withdrawal of a shotgun or firearm certificate. Without them, a keeper cannot perform his duties. Some keepers have contracts of service stipulating that any infringement of the law will result in immediate dismissal. Carried to its logical conclusion, that means the loss of the roof over the heads of his wife and family as well. Small wonder that keepers feel victimised and unwanted. From being an important part of the economic and social fabric of the country-side, their status has declined greatly over the space of fifty years, particularly in the eyes of the urban majority.

Currently, the role of keepers is misunderstood, especially in Scotland, and often misrepresented, sometimes maliciously. They are portrayed as pompous 'men of the trees' toadying up to rich bosses or bloodthirsty killers of wild things generally. Many are regarded as heart lazy by the farming community because they are often at home during the day. The effective keeper's day is divided in two; he is out on his ground early in the morning and again late at night.

Many see the keeper as leading a comfortable life with considerable job satisfaction but in reality his is a precarious existence. He is permanently at risk of being smeared; the most blameless of conduct offers little protection from the 'planting' of illegal substances or corpses of protected birds on his ground followed by police being 'informed'. Though it is widely felt that this happens on a regular basis, it is hard to prove and the keeper has no easy defence against it. Vicarious liability is now a crime under the 2011 act; if a keeper is alleged to have broken the law, his employer can be held liable for the offence and may stand trial. It has not been tested in the courts at the time of writing. If a keeper is 'raided', he is assumed to be guilty until he proves his innocence, often at considerable emotional cost to him and his family. It is traumatic to have police, usually accompanied by 'experts' from the conservation bodies, descend without warning at night in platoon strengths of thirty to forty to turn his house, vehicles and outbuildings upside down in a quest for illegal traps or poisons. When the posse finds nothing and withdraws, apologies are rarely proffered. It takes a man of calibre to hold his nerve and not entrap himself with an unwise word or deed. It is small wonder that gamekeepers resent being told how to conduct their business by people from conservation and statutory bodies whose understanding of what keepers do seems at times as questionable as their record of conservation success.

Keepers have their own organisations which look after their professional interests. The National Gamekeepers Association and the Scottish Gamekeepers Association are active on their behalf, the British Association for Shooting and Conservation (BASC) supplies insurance cover, training and advice and the Game and Wildlife Conservation Trust (GWCT) keeps them informed on game science. Nevertheless, they find themselves increasingly isolated in the wider society. People seem reluctant to accept that the conservation of a broad spectrum of wild birds is in good hands if keepers are in charge. The public at large seems to prefer the marketing expertise of the corporate conservation sector and its bird or wildlife 'reserves' replete as they are with board walks, observation hides, and 'working groups'. It is comforting but, arguably, misleading to get one's wildlife information from a car park leaflet dispenser. The irony for the rank and file gamekeeper is that moorland reserves run by the conservation institutions rarely experience systematic management such as regenerative burning or pest and predator control. They tend to become havens for common predators and support disappointing levels of the declared wildlife interests which are often unproductive and rear few young.[2] In some reserves, the most common bird is the carrion crow because nobody has the motivation to control it.

The fruits of non-interventionism caught the attention of the press in July 2005 when a TV camera overlooking a hen harrier nest in the Clyde Muirshiel Regional Park (CMRP) filmed a fox cub killing four out of five hen harrier chicks in a nest. See p.146. The moor had been designated a Special Protection Area (SPA, the highest level of protection accorded to bird species under EU regulations) explicitly for hen harriers. Prior to 2002, when the core area of the park had been managed by enlightened keepers, as many as fourteen hen harrier pairs were recorded as breeding in the district. By 2010, eight years after keepering ceased, two pairs only were recorded as attempting to breed. Between 2005 and 2011, foxes were filmed predating harrier nests on three occasions: over forty per cent of the known hen harrier nests in the SPA were lost to fox predation.[3] A similar pattern emerged on the nearby Muirkirk and North Lowther SPA. At the time of designation in 2003, it was home to thirty nesting pairs, well over the one-per-cent criterion. In 2011, only six pairs were known to have attempted to nest, their breeding success unrecorded. SNH has made no attempt to correlate breeding success with whether the land was or was not keepered. It is open to doubt whether the current management regime has protected or further endangered the species and if the area continues to justify its status as an SPA.

The value of 'bird reserves'

A major problem with bird reserves is that managers take the easy option to assess and promote the worth of a particular site by recording the presence of species in spring when birds are 'showy' and easy to count. However, unless a reserve is solely for the benefit of passage migrants, the worth for birds which

Fox cub killing hen harrier chick, Clyde Muirshiel Regional Park – July 2005. (PHOTO: SNH, RSPB, CMRP)

use it is measured not by the number that try to breed there but by the number of young that those that do breed actually fledge. Reserve operators are coy about the figures and rarely, if ever, declare them. This is partly because obtaining them is not easy (though it can be done by competent field workers) but also because the data would highlight the fact that reserves are often sites of poor recruitment because of predation, particularly by crows and foxes. They act as sinks, drawing in surplus birds each year from elsewhere rather than producing a surplus to populate other sites. Ticking boxes – presence/absence of a species – is worthless in the long term if, as is common, young are not being reared. Bird reserves should be sites of excellence and maximum productivity for the target species that produce a surplus which disperses and colonises new areas. In practice, they frequently produce at a level below potential because those in charge are reluctant to undertake interventionist management. SNH and RSPB are constantly open to criticism on this issue.[4]

To create a resource that can earn money, the moorland manager must increase breeding densities and breeding success, a challenge that drives investment.[5] Increasing breeding densities means improving habitat while breeding success is influenced chiefly by predation and its control.

The sink effect

Red grouse are strongly territorial – territory size is a function of many factors, some of which interact. Because cocks squabble among themselves and parcel up space, the result is a relatively even spread of pairs over the area available to them. If an area is heavily populated in autumn, there will not be enough room for every aspirant to hold a patch close to its birthplace so some will be forced to emigrate to ground where the competitive pressure is less. This lower density area gets an influx of immigrant birds which tends to bring its population up to somewhere near that of the neighbouring ground. The effect is to smooth out breeding density over a wide area (see p.55). The term 'sink effect' describes an acute case of this process. An island of good production and high breeding density surrounded by poorer habitat and lower breeding density loses birds in autumn as part of equalisation. However good the management of the core area, there are not enough birds round about to hold in would-be emigrants; the breeding stock established on the good area the following year is nearly always disappointingly low. It is disheartening for a manager who does everything right in an attempt to improve his own stocks but is bled white by poor conditions on neighbouring ground. The only solution is to persuade neighbours to manage their ground better so that the whole area fills like a sponge and everyone benefits. (See below, *Grouse groups*).[6]

On a district scale, there is a 'majority factor'. If a group of twenty or thirty adjacent moors is predominantly well managed with, say, fifty per cent or more of them employing good keepers who control deer and predators, crop the population correctly and where sheep numbers are well farmed at a sustainable level, everyone has grouse in reasonable numbers. The converse is when under half the moors in a region are looked after properly; the larger half 'bleeds' the better run places round about to maintain stock. Because the displaced birds do not do well, nobody has anything to shoot.

Death by accident

Grouse which die from disease, displacement and predation are losses over which man has only limited control. Death by shooting can be planned and predicted but death by accident is often shrugged off as inevitable. This is not necessarily the case. Though losses from accidents are usually low and, if the population is large, have a minimal effect on breeding densities, it is a different story when populations are low, perhaps following a cyclic decline or on marginal moors where every bird is precious. Grouse are not subject to the same risks as grey partridges on arable land, where grass cutting, accidental disturbance and the destruction of nesting cover by grazing animals can account for nest losses in excess of forty per cent with predator losses on top.[7] But birds *are* lost as a result of farming activities, sometimes those of thousghtless shepherds and their rapacious collies but chiefly in connection with wire fences. Well over a hundred years ago, people were recognising the damage that the newfangled stock fences, telegraph and electric wires could do.[8] The situation today is rather different. Although stock and deer fences can and do kill birds, most telegraph wires are carried in multi-core cables which pose minimal hazard. Power cables for electricity distribution will always be with us and kill birds of many species but the most significant increase in dangerous obstacles stems from post-war enclosure of hill land using lightweight, high-tensile wire fences that can be hard for birds to see.[9] In former times, fences were made of plain or barbed wire as much as 3 mm in diameter with posts at 2 metre (6 ft) centres. Square-mesh 'pig netting' sold under the brand name Rylock came into general use after the Second World War and was obvious enough to be no more hazardous than plain wire fences. It was the development of the high-tensile fence using wire only 1.5 mm in diameter with posts spaced at ten metres or more and wooden or wire droppers to keep the wires spaced out that was to prove dangerous. What matters is not whether a fence is seven wires high or made up of four electrified ones but that, after a year or two, the wire starts to lose its glint. Increased afforestation in the Highlands saw the erection of many kilometres of deer fence two metres or more in height. Although usually of more robust construction than the lightweight stock fences, they were also that much higher and were to prove even more dangerous to birds.

The Banchory work showed that territorial birds survived well over winter and were less at risk from predation and accident compared with non-territorial and surplus floaters which drifted about all winter, often over ground with which they were not familiar.[10] Obstacles such as fences and overhead wires were more likely to account for these 'class two' birds than for territorial ones which became familiar with their home range, hazards and all. However, the relative invisibility of high-tensile fencing occasionally catches out even territorial (calling) birds.[11] They have been seen to fly into Rylock fences when flushed by a dog during spring pair counts.[12] It is plausible to suggest that a bird fleeing from a predator 'takes its eye off the ball', loses concentration and hits an obstacle with serious consequences. Collisions are also common in poor light when crepuscular cocks in low density populations concentrate their territorial activity. Losses are not confined to territorial adults in winter and spring; young broods, often raised accidentally by keepers or shepherds, can fly headlong into deer or stock fences. On driving moors, high-tensile fences lying athwart the contours (and so across the line of a drive) often catch birds unawares.

RIGHT:
Highlighting a deer fence with plastic strip.

BELOW:
Galvanised mild steel plates draw attention to the top wire of a stock fence Co. Durham.

Corks on a deer fence last for many years. (NN929301)

Proprietors used to insist on a clause in a wayleave agreement specifying that electric and telephone wires had to be corked or disced. When high-tensile fences became widespread fifty years ago, proprietors and keepers began marking them with corks, metal plates, bamboo canes or plastic netting. Careful study of collision incidents by many gamekeepers shows that it is fencing running up a nab end or traversing the top of a small knowe or hillock that causes the most damage. It seldom causes trouble in valley bottoms or when running alongside a watercourse. Costs can be kept in hand by corking or plating fences only where the risks are significant. The enthusiastic keeper on a budget can always resort to the ancient practice of 'bushing' a fence with bunches of heather. Aesthetically less pleasing but equally effective are strips of plastic fertiliser bags tied to the top wire.

Increased motor traffic has seen greater numbers of grouse ending up as road kill. Fast drivers can kill many birds in summer, particularly after rain when broods draw out of the wet vegetation to dry out on the bare road surface. Where busy tourist routes cross heather moorland with high bird populations, it is common to see dead grouse, adults and chicks alike according to season, lying on or beside the road. Though they never seriously reduce the numbers available for shooters, their loss can heighten the resentment some keepers feel when the tourist season is in full swing. (One well-respected keeper in Aberdeenshire told the author that he never minded road kills; when there were no grouse, there were none for drivers to kill; birds squashed by tourists usually heralded a good season.)

A shepherd's collie out of control can do a great deal of damage as man and dog are daily on the hill during the crucial time in June and early July. If young chicks are flushed in a strong breeze and before they have full powers of flight, they can be blown away like chaff. As they cannot easily turn into the wind to reduce air speed, they often crash land fast and are killed or seriously injured on impact. A bed of burned 'stick' heather is a forest of sharp spikes that can have a lethal effect on the soft body of a young bird. Hazards of this kind have always been present to some extent but have become more significant as more

149

people go to the moors in June when the chicks are small. Walkers can flush broods unwittingly with dire effect, particularly if they have dogs with them. Collision mortalities among adult grouse are probably of little significance when numbers are high in the autumn. The main damage occurs after Christmas when populations are lower, particularly in the context of a population crash or in marginal areas of naturally lower density such as the western half of Britain. When birds are scarce, they tend to lie doggo during daylight hours – there is little survival advantage in advertising one's presence when there is little competition for territory. At such times, interactions between neighbouring territorial birds tend to occur in the pre-dawn period and the gloaming. Man-made obstacles can easily be missed by birds taking part in a sexual chase or song flighting. It is particularly important to ensure the maximum survival of potential pairs when numbers are low and the population consists of a

Fences also snag non-game species. A short-eared owl is caught in barbed wire.

loose collection of individuals where every bird counts. When numbers are higher, the individual has to take the risk of displaying in broad daylight because, unless he joins the competing throng, he may not establish ownership of a territory and so ensure admission to the breeding population.

Ski pylons, cables, transmission lines and wind turbines

Ski clutter presents a menacing obstacle to birds in misty or blizzard conditions.

(NO142780)

Telegraph wires, ski pylons and steel electric towers can cause serious losses; large industrial paraphernalia on the open hill present as much of a risk for grouse and ptarmigan as fences. Their size makes them more visible than a fence but, as they are normally located on high ridges, they are often shrouded in mist and cloud. They cause many deaths.[13] In the Cairngorms in the Scottish Highlands, ptarmigan became locally extinct close to the ski development as they suffered death by increased predation as well as by accident.[14] Before the development, crows and foxes occurred at low levels but both species increased foraging efforts because of the availability of waste discarded by tourists and so they maintained artificially high populations. There was thus greater pressure on the breeding grouse and ptarmigan during the breeding season than before the development took place.

Transmission lines have been with us for many years and, without question, kill a percentage of grouse every year. Though no serious study on their impact has been carried out in Britain, the reality is that not much can be done about the losses. A study in Hemsedal, mid-Norway, showed that willow grouse and blackgame are frequent casualties. Conditions are of course different there: the grouse often move in blizzard and white out conditions and, because of the short day length from October to March, a larger proportion of 'daylight' hours offers only the poor light of dawn and dusk; collision risks increase as birds are forced to be crepuscular. Other studies showed that collisions accounted for 5.3 willow ptarmigan per km of power line per year and more than 2.4 times as many as were shot by hunters.[15] Extensive losses have been recorded elsewhere in Europe.[16]

Wind turbines are a novel hazard. As of 2010, few sites have been constructed on grouse moors but this is set to change. There are likely to be sites extending from Fort William to Elgin in an almost continuous band. Eye witness accounts of grouse flying among turbines suggests that the blades may not be a major threat. The money proprietors receive from power companies potentially provides substantial sums for investment in the moor. At one of the earliest sites, constructed in the 1990s near Halifax in West Yorkshire, packs of grouse were

Wind turbines are certain to kill birds of many species in the years to come. (NS720110)
(PHOTO: D. BRUCE)

seen flying freely among the towers but no more than ten metres above ground; the base of the area swept by the blades is considerably higher.[17] It is widely felt that golden eagles are at serious risk from blade strikes. If mortality from this source does reduce the number of breeding pairs in Scotland, blame must be laid four square where it belongs, not loosely attributed to the activities of gamekeepers. Whatever the pros and cons of the technology and its potential to affect populations of grouse and other moorland birds, the presence of wind turbines scattered over our wild landscapes is a visual intrusion greatly to be deprecated. Given the modest 'green' energy they yield, proposals should be assessed with greater rigour and opposed as a matter of principle by organisations such as Natural England and Scottish Natural Heritage. Instead, public bodies tend to nod through proposals without much in the way of objective scrutiny. Cynics wonder if the rigorous assessment of Environmental Impact Assessments is compromised by the need to protect budgets.

Turbines are often sited on ridges where the peat layer is deepest and aggressive excavation is necessary to accommodate them. Although protagonists make out that the life of the structures is a 'mere' twenty-five years, it is implausible that extensive industrial sites with expensive infrastructure including roads, sub-stations and grid-connecting pylons will be meekly abandoned at the end of the period.[18]

The impact of keepering

Heather moorland regularly produces bountiful grouse crops following a few years of improved keepering, particularly when an estate changes hands and the new owner motivates staff. The shooting bag can improve out of all recognition within two or three years due to the suppression of crows and foxes, grazing control, better standards of burning and, sometimes, draining and gritting. Most keepers and sportsmen put this down to reduced predator pressure but ecologists argue that grazing and burning are decisive. To investigate it properly would take at least six years of experimentation on the lines of the Game Conservancy split study on grey partridges on Salisbury Plain or recent work at Otterburn.[19] Given the weight of anecdotal evidence that moors improve after a management change and reports of grouse numbers collapsing on the cessation of serious keepering, it is hard to argue that such a study is a priority.

Evidence on a national scale was inadvertently provided by the outbreak of war; 1940 was a big grouse year in many places but there was little shooting because of the hostilities. Disease broke out in spring 1941, many birds died and grouse became scarce over large areas. After the war, keepering was resumed at first on the traditional driving moors of northeast Scotland and England and was paying dividends by the early 1950s. Moors in the western half of the country and in the Highlands generally were slower to recover as keepering was not resumed at the same rate; sheep farming became the dominant land use, extensive heather burning was performed in a manner unsympathetic to grouse and much moorland was afforested. Over most of the west from Argyll to Sutherland, keeper density remained low and grouse remained scarce, as they do to this day.

'Hard' shooting, syndicates and marksmanship today

It is easy to be an armchair critic of the shooting prowess of others. Keepers, flankers and beaters – even some owners, to their discredit – can be quick to pass judgement on guns that miss more grouse than perhaps they should. Only those who have shot grouse all their lives under every circumstance are competent to comment on what is widely recognised as perhaps the most difficult form of small-game shooting in Britain and, some would say, the world. That said, the importance of marksmanship must not be overlooked; grouse are harvested by a man behind a gun on the open hill, not by a slaughterman in a white coat. Human fallibility is an important influence on the grouse harvest. Grouse-shooting skills are sometimes poorly developed in those who can afford to come to the hills and bags are often low as a result. Under-cropping can become the rule rather than the exception particularly on moors where grouse are numerous such as the 'rich' ones of the North of England. Where a good home-bred team would shoot a hundred brace on a fine day in August for an expenditure of five hundred cartridges, an inexperienced group is likely to fire fewer cartridges and get a lower yield per discharge, ending up with, say, thirty brace for four hundred and fifty cartridges. A moor owner who regularly lets his moor to guns of this calibre during the high priced part of the season to safeguard his income (a common situation) will later face perhaps unscaleable obstacles to reducing his population down to realistic breeding levels. Though under-shooting has increased in recent years, it is wrong to put all the blame on incoming sportsman. Economic factors often dictate that a high proportion of the grouse on a moor are sold 'on the wing'. Besides, the owner's guests and syndicate members are not always better marksmen than first-time shooters from abroad.

Compare this with the years before the Great War. On 30 August 1888 on Blubberhouses Moor, near Ripon, Lord Walsingham shot 1070 birds in a day to his own gun with about 1510 cartridges. True, had there been a National Grouse Shooting Team, he would have been a certain choice but many of his contemporaries were his equal or only a short step behind him.[20] It was common in those days for a gun to go through a season shooting grouse at the rate of two cartridges per bird. Many people did nothing else but shoot from 12 August to early October. Of course they were good shots – and so they should have been. The effect of their accuracy would have been to cut the grouse surplus down to low levels by the time they left the moors to start on partridges in October. If keepers then started 'cocking' (if contemporary literature is to be believed, many did), the population would have been under heavy pressure and surplus birds scarce. Only in exceptional years did it get away from the shooters and raise the likelihood of a population crash. We should not despair because much the same is true today of individual moors shot and managed in the traditional way. Such places are not confined to the worm-free eastern half of the Britain. Where keepering is thorough and a core team of good guns shoots the moor year after year, bags can be remarkably consistent. Of course, a heavy outbreak of winter browning may produce a poor year but consistently hard shooting demonstrably tends to iron out population fluctuations caused by variations in breeding success, behaviour changes and worm build-up.

Similar results are achieved on 'dogging' moors where breeding densities tend to be lower.[21] In former years, most estates kept large kennels of pointing dogs, working as many as three or four brace in sequence during the course of a day. The guns would be fit and practised in this kind of shooting; the average covey would normally lose three or four of its number to two guns at the first point. The survivors would be found later in the day and put through the mill again. Over a four week season, every piece of ground would be gone over at least twice.

153

On a well shot moor, probably as much as eighty per cent of the August population would have ended up in the larder.

Where a group of friends forms a syndicate to manage and shoot over a piece of ground, the criterion for membership is usually as much the ability to pay the subscription as to shoot straight. Syndicate shooting began in the last quarter of the nineteenth century when groups formed 'Shooting Companies'.

> There were four of us and for about £200 a year each which included all expenses, we passed a happy ten weeks in the Highlands.[22] The bag averaged seven hundred brace of grouse while blackgame, partridges, hares and snipe made up two thousand head [so], according to present prices, our little company had invested their money wisely and received good dividends in the shape of health, happiness and sport.[23]

Counting

Before any livestock manager harvests an annual crop, he should know how big that crop is so that exploitation is confined to the annual surplus and does not dig into capital. The grouse manager is no different from the shepherd and his flock and has to equip himself with hard information both on how many stock birds he has and then, in a second operation, how many young that stock has produced. Only if he does this can he be sure that he is managing the population correctly and maximising his yield in the medium and long term.

Spring counting

This is normally carried out in the second half of March. Sample areas are selected which are representative of each vegetation type on the moor. The size of the area is a matter of personal choice; some people count a whole grouse drive and others map out a kilometre square. The area has to be delineated using either natural features such as streams, roads, fences and walls or be pegged with prominent markers. Starting on the downwind side, the operator hunts out the whole area by quartering back and forward with dogs, preferably pointers or setters.

A setter on point; a sight to make the heart race.

154

Birds are flushed off the area wherever possible to avoid double counting. If any fly forward onto ground which has not been covered, the place where they alight is noted. By the end of the operation, the position of each bird and whether cock or hen, will be recorded on a map. Back home, the count can be summarised as pairs, trios and single cocks and hens. These data provide the basic information for that sample area. When combined with other counted areas, the figures can be used by extrapolation to arrive at a stock figure for the whole moor.

July counting

Brood counting is designed to assess breeding success which obviously has a direct bearing on the crop of birds which can be harvested. It is not necessary to count each plot with the accuracy of the spring count. Many people use transect walks and start at one end of a drive and walk through it – sometimes with two or three people in line. Data from a transect are perfectly adequate provided a) the same line is followed exactly each year (i.e. counting effort is standardised) and b) an adequate sample of the hens counted in spring are recorded. Most operators are happy to base this count on fifty per cent of the hens present in March. This set of figures provides the manager with a data set consisting of three columns – cocks, hens and young. Additional notes will record the state of growth of the juveniles which may have a bearing on the timing of the shooting programme. The critical young/hen figure (i.e. breeding success) can be deduced from these data.

Interpretation

The whole point of counting is to avoid surprises, not only those which yield disappointment on a shooting day when birds are found to be scarce, but also to enable the manager to plan his late-season shooting programme with confidence if there are more birds on the ground than expected. It also enables him to match his programme with what is on the ground, avoid over-shooting and leave an adequate breeding stock for the subsequent year. It ensures that in the good years, the stock is shot down to a 'safe' level.

The grouse's life expectancy is short and few birds breed twice. The rate of turnover in the population is high and this takes place whether the birds are shot or not. This means that in an average year the upper limit of what can be cropped safely is around sixty per cent of the July figure. A 40–50 per cent take accommodates winter loss and provides a safety net for the nervous manager who anticipates heavy winter predation. If the breeding population is close to full capacity in spring and young per hen is high in July, then a heavy shooting programme is in order and attempts can be made to shoot over seventy per cent of the July standing crop. When production is poor, half the July figure can still be taken because the birds themselves see to it that subsequent breeding numbers will fall whether shooting takes place or not.

Like so many aspects of wildlife management, the basic data provide a guide for the intelligent manager prepared to back his interpretation of what he sees on his ground. It may not always be perfect but it is a substantial step forward compared with the 'wait and see' approach which was normal until the middle of the twentieth century. Provided the work is carried out in exactly the same way on the same area and with the same search effort each year, a sound basis is provided for comparison between years.

Cropping the population

The point of managing a population is to maximise yield on a sustained basis. If grouse populations were left entirely to nature, the crop would be much smaller and the economics of moor management undermined by the threat of reduced economic activity in places where alternatives are few. For managed game populations such as reindeer and European moose, the end product is the meat.[24] The trophy and the sport are, on balance, of secondary priority to those who rely on hunting to fill their freezers. With grouse shooting, the situation is reversed; only a minority shoots grouse to eat. For most sportsmen, it is a sport first, last and all the time with no 'trophy' except perhaps a photograph or two of the day's shooting. The meat, though a nutritious natural food, does not dictate how the sportsman eats in winter. It was not always thus. Barns in the north of England to this day hold nets used, sometimes by poachers, sometimes legitimately, to catch grouse which flew off the moor onto adjacent fields and 'allotments'.[25] The birds had a cash value to their captors – some were sold live to other moors 'to change the blood' or increase the August stock and hence the bag of shot game, others had their necks unceremoniously wrung before being packed into hampers and sent to Leadenhall in London or provincial markets. For a hard-pressed fell farmer, they were a cash product just like fat lambs, suckled calves or the turkeys and geese his wife fattened for Christmas. The adjacent moor owners would have fulminated at their activities but there was little they could do about them. Long term, the practice probably did the moor more good than harm. Then as now, the average moor owner typically shot fewer grouse than he should for the good of his stock. (See p.160, *Cocking*.)

Though harvesting grouse starts on 12 August, what happens next varies according to the policies of the moor's owner. Many start driving immediately, others have 'a little walk around to see what's there' and some walk the moor systematically in the first week and start driving in the last week of August. Different policies suit different types of moor. In the productive north of England, driving starts at once. Nobody thinks of walking up because numbers tend to be high and in any case the birds do not sit well. In parts of Scotland, there is a long-standing tradition of walking up before driving, perhaps because the birds hatch slightly later. The third approach caters for uncertainty in places where the keeper may spend little time on the hill or are close to being abandoned. A walk in line on the first Saturday of the season with 'a row of boys' is no way either to assess what is on the ground or to crop the birds on a moor. It has long been a sound maxim that half the season's predicted bag should be in the larder by 31 August, i.e. within the first sixteen shooting days. Once September arrives, the weather roughens up and birds tend to form bigger groups. Driving a hundred groups of ten birds obviously yields a bigger bag than driving twenty groups of fifty.

It is important to understand how shooting bears differently on different sections of the population. Driving on a hot day in August sees the birds of the year getting tired and 'tucking in' particularly if the heather is long. It is the adults who make long and repeat flights and get shot disproportionately higher as a result. In contrast, walking up with a team of four to six guns in line bears more heavily on young birds. (In former years, there were some, the author's father among them, sufficiently sharp-eyed regularly to select and shoot the old birds in a covey, a skill one seldom sees nowadays.) Walking up usually results in single cocks, either unmated or widowers, escaping as they tend to sit in prominent places, are alert to danger and get up wild or out of shot. Shooting over dogs (pointers and setters) is different again. With steady dogs, though the guns arrive at the point keyed up and, perhaps, excited, they are also balanced and ready to shoot. They are not taken by surprise as can happen with walking up, marksmanship tends to be better and a higher proportion of shots find their mark. As the day goes on and coveys break up, the dogs find individual birds which would be walked over by an in-line team. A high percentage of the birds present are shot, an effect which tends to bear evenly on young and old alike. (Old cocks will often 'hold' to a dog and allow the shooters to come in range when they would flee the rather more obvious group of shooters walking in line.)

Bag analysis

Knowing what has been shot is a management aid – there are few moors where no heed is taken of the make-up of the bag; it is not difficult to separate old and young in the game larder at the end of the day. In any case, a proper host will try to ensure that guests go home with young birds. Early-season driving bears differently on different age classes and the bag is not usually a random sample of the population.

A worked example may prove useful here. Imagine a 200 ha (500 ac) section of driven moor that holds fifty pairs and ten single cocks (ninety birds in total) in the April counts. Then assume that a breeding success of six young per hen is recorded in the July counts giving a total population on the ground on 12 August of sixty old cocks, fifty old hens and three hundred young. The moor is then shot three times and a bag analysis undertaken at the end of each day.

Table 8.

Bag analysis – a worked example.

	No. of old cocks shot	No. of old hens shot	No. of juveniles shot	**Total bag**	Young per old bird shot	Young per hen shot
First time over, Week 1	25	15	80	**120**	2.00	5.33
Second time over, Week 2	10	10	64	**84**	3.20	6.40
Third time over, October	13	13	80	**106**	3.08	6.15

The ground is driven in the first week and yields sixty brace, twenty-seven per cent of the birds present. The bag is fifteen hens, twenty-five cocks and eighty young birds. More adults than young birds are shot because, as noted, it is the adults who make long flights; more adult males than females are shot because some old cocks are living on their own and fly singly. Eighty young birds to fifteen old hens is a ratio of 5.33 : 1, slightly lower than the July count.

The next shoot a week later harvests the same proportion of what is there. This time, the bag is eighty-four birds, made up of ten old cocks, ten old hens and sixty-four young. The ratio of young to hens has widened to 6.4 : 1, more than the July count because hens have been shot out of proportion to their number because they are still behaving differently from their young.

Theoretically (making no allowance for winter loss and assuming a stable population), a hundred birds are required for stock making a hundred and six still available for shooting. If those birds are cropped on a day in October, the coveys will have broken up and the bag will be close to being a random selection reflecting what is on the ground, namely thirteen old cocks, thirteen old hens and eighty young. The young-to-hen ratio of 6.15 : 1 almost exactly matches the July count, indicating a properly cropped population.

The breeding population of a hundred birds remaining on the moor after the shooting programme comprises of six old cocks (12 per cent), six old hens (12 per cent) and seventy-six birds of the year (76 per cent).

The physical analysis

In the example above, the first day's bag is made up of two young per one old bird. Concerned by this ratio, cautious managers sometimes curtail shooting instead of holding fast and believing the July count figure of six young per hen. Accurate data *must* be collected and the really important information derived from bag analysis in terms of management is that of the number of *young shot per hen* as this is the only valid way to measure breeding success.

To do this, split the bag into three piles, old cocks, old hens and young birds. Ignore the old cocks; what matters is the ratio of young birds to old hens. As the season progresses, the proportion of old birds in the population declines just as the young birds are becoming strong enough to match the flying performance of their parents. The shooting programme starts to bear evenly on the different age classes and bag analysis yields figures that more closely match the July count.

The two outermost primary feathers are sharp and spotted on young birds (LEFT), rounded on old ones (RIGHT).

Cock (left) and hen: the throat feathers are foxy-red on the cock and speckled on the hen.

Division of the bag into young and old is often poorly done. Some still try to age birds by crushing the skull or seeing if the lower mandible will support the weight of the bird (which indicates an adult). In some years, this test is unreliable even on 12 August because the mandible of an early-hatched bird may well have ossified and will support its weight as a result. Another test which does not work (though it is in common use) is the 'short third primary' – an old bird in delayed moult (usually a hen) can look like a young one.

Grouse have ten primary feathers. Young birds moult their first set in sequence at approximately weekly intervals, replacing them with adult ones rounded at the end which look quite different from the pointy ones they replace *but they do not moult and replace the outer two feathers.* By the time the birds are about eleven weeks old, the process is almost complete – the wing shows eight inner feathers with rounded tips and two outer ones which are sharply pointed, often with a dusting of ginger or cream-coloured spots or pale tips on the last two millimetres. These characteristics identify a young bird reliably and do so until the moult takes place the following July, by which time it is, of course,

ABOVE: *The wing of an old hen on left, with two juvenile primaries from the previous year which are still to be shed. Young cock on right.*

LEFT: *No way to age grouse.*

159

an 'old' bird aged about fourteen months. Like the young, adults replace their primaries from the inside out. The process is usually nearly completed by the start of shooting though two or three of the outermost feathers may still be growing. In that case, they are shorter than the others and the outermost quills are soft and bloody. When the moult is complete, the full one-to-ten sequence of feathers have the same glossy texture while the tips of the outer two are blunt and rounded.[26] The test scientists use to be doubly certain of distinguishing an adult bird from a juvenile (less than a year old) is to establish the presence or absence of the Bursa of Fabricius, a short blind passage a millimetre wide and about 1.25 cm deep on the dorsal side of the anus. It is present only in juvenile birds up to about six months old.

It has taken forty years for spring and July counts to become widely used as management tools even though they enable predictions of singular importance when shooting is being let. Surprisingly, it is still almost universal for bag figures to be qualified with the ratio of young birds to old and uncommon to find owners or keepers recording the bag in the terms of young birds to old *hens*. For shoot planning and administrative purposes, *the young-to-old ratio is a worthless statistic* with no part to play in moorland management.

'Cocking'

Cocking is rarely practised nowadays though it was done as recently as fifty years ago on a number of moors in the north of England. Based on the theory that old cocks are less desirable as breeding stock than birds of the year, keepers, once formal shooting was over, stalked and shot as many 'old' cocks as they could in October and November. In some cases, they continued out of season into the new year. The practice, whose efficacy was never subject to scientific study, relied on the 'observation' that old cocks were more aggressive than young ones. As a good many keepers cannot reliably tell an old bird from a young one in the hand in mid-autumn, they are no more likely to know the difference in the field; the theory is flawed from the outset. Besides, aggression in cocks is a function of the year. When a population is declining, young cocks tend to be more aggressive than old ones but the converse applies when numbers are increasing at the start of a cyclic upswing. Selectively shooting unmated cocks which are holding territory from November onwards might well take competitive pressure off adjacent mated pairs but there is considerable room for error in deciding which birds are unmated. In autumn, hens do not always stay in close proximity to their mates in the way they do as winter progresses.

A former beat keeper from 1960s Yorkshire described the operation.[27] In the autumn of a big grouse year, he used to account for several hundred birds, stalking and shooting calling cocks and making a cache of the slain in a hessian-lined pit dug into a peat bank on the edge of the moor. Because keepers had no transport other than push bikes, it fell to one of the estate workers to collect the birds with a vehicle for market twice a week. As an operation, it was closely akin to 'becking', a widespread mid-nineteenth century poaching activity prevalent in mining valleys in south Northumberland, Co. Durham and North Yorkshire.[28]

Where cocking was thoroughly carried out after a heavy shooting programme and continued to the end of the season (10 December) or beyond, it may well have reduced the number of territory owners. But, if surplus birds were scarce, the shot birds would not all have been replaced and the breeding population the following spring would have been below capacity. The result would have been the same as if the shooting programme the previous autumn had cropped the population to a level slightly below the holding capacity of the moor. An experiment by the Banchory team on its moor in Kincardineshire proved beyond doubt that a population did not need to crash every few years and could be prevented from 'boom and bust' cycling by shooting the right proportion of the stock.[29] At Bolton Abbey in Yorkshire, keepers regularly shot five hundred cocks a year over the twenty-five years from about 1880 at a time when the annual bag averaged over three thousand seven hundred brace.[30] 'Cocking' was typically contributing an additional take of almost seven per cent of the conventionally shot bag.

The significance of bird quality

Our hypothetical bag analysis began with a spring stock of fifty pairs including only six old cocks and six old hens; the majority would have been birds of the year. As worm burdens are partly a function of age, this kind of stock structure sets up a sound basis for future breeding. Every authority who ever wrote about grouse management stresses the importance of young birds though the emphasis was often based on the notion that old birds chase off young ones or even that they are infertile! In truth, aggression in young cocks varies from one year to the next; birds tend to be more aggressive in autumns following poor breeding but are more tolerant of peers and neighbours after productive seasons.[31] When breeding is poor, shooting should be hard, early and over by the end of August. That tends to remove not only old, wormy birds but also young, aggressive cocks from small coveys that are claiming territory early. The effect is to leave space for surviving members of bigger, later-hatched coveys that are typically more tolerant of neighbours. This goes against the gut feelings of owners and keepers. If breeding is poor and there are some late birds on the ground, some postpone the start of shooting until September. But, by this time, the bag is a random sample of the birds on the ground, not one that allows manipulation of the population to ensure that quality birds comprise next year's stock. Performance is jeopardised before the new season has even started.

The same principle applies to hill sheep. No flockmaster holds on to old, broken-mouthed ewes with weeded udders and chronic nematodirus infections. They go 'down the road' in autumn to ensure a young, active flock in tip-top health. Grouse are no different and the moor owner is always better off having a less-than-full moor populated by quality birds than one stuffed with querulous, aged, wormy rubbish.

Rearing and releasing

Managers of low-ground shoots understand the necessity of rearing pheasants and partridges to ensure consistently large bags; for many, their working year revolves around a release programme. Why have grouse escaped this degrading fate? After all, the financial inducements to maintain good bags are even greater with the bird-in-bag figure for driven shooting at up to £100 per bird (2011), more than double the price of the best pheasant shooting. To understand why, it is necessary to examine the biology of the bird. Grouse are adapted to live on poor-quality food containing indigestible tannins and lignin (see Chapter 3, *Grouse biology*). Compared with the nutritious, easily digested seeds and grains of the vascular plants which support low-ground game, a fibrous, woody diet of heather requires a digestive system that can extract enough nutrients to maintain the bird's body weight and condition while it copes with the strains of establishing territory, courtship and parenthood. The gut of a grouse is distinguished by the size of the two caeca or blind guts which discharge into the posterior end of the ileum. They operate in a similar way to the rumen in cattle and sheep or the large intestine in horses. For grouse, this fermenting vat comprises about forty-five per cent of the weight of the alimentary tract compared with twenty-six per cent in pheasants and fourteen in hand-reared red-legged partridges.[32] Its full and complete development is essential if the bird is to get the most from a poor diet. Its importance is illustrated when the caecal function is impaired by strongyle worms. The damage they do to the delicate lining of this part of the food canal can reduce bodyweight quickly because of the loss of blood and body fluids and lead to impaired breeding and even death (from 'disease') in extreme cases.

In the wild, grouse chicks feed largely on insects for the first three weeks of life but start to eat small quantities of vegetable matter almost from birth.[33] Moss capsules, tender, lettuce-like blaeberry leaves and the tips of growing heather shoots are all readily available by the time the chicks hatch. By the end of June, young birds are on a diet of almost one hundred per cent vegetable matter. They are selective and, of course, free to move over substantial areas of moorland to obtain a wide choice of food. They not only eat the best food available but a great deal of it; the caeca develop rapidly and take on their function of extracting nutrients from vegetation which becomes increasingly fibrous and indigestible as summer progresses. By the time the birds are fully grown in August, they have a developed and functional gut which enables them to survive on heavily lignified food.

Contrast this with the hatching of grouse eggs in captivity. As numbers are probably small, a bantam or small hen is used as foster mother. The young chicks are supplied with the best of processed food, perhaps in the form of a turkey or partridge starter crumb containing twenty-six to twenty-nine per cent protein. After ten to fourteen days, they are offered cut bunches or turves of heather and

Polytrichum *moss capsules are eaten by grouse chicks from an early age.*

blaeberry which they eat avidly though their main food is still grain based. They grow very fast on a diet which is more nutritious and lower in fibre than that of their cousins in the wild. By the end of July, they are almost full grown, sleek in plumage and fat in body, but the appearance conceals the unfitness of the inner grouse to cope with a natural diet. If the birds are put on heather, they eat large quantities; a brood of eight or ten hand-reared birds strips all the green shoots in a 3 m by 2 m pen (10 ft x 6 ft) in a matter of hours. We know that only a small proportion of the heather that grows naturally is suitable for grouse maintenance and growth; birds confined to a pen are forced to eat material they would spurn in the wild.[34] The deficit in nutrient intake from heather is made up by an *ad lib* supply of growers' pellets. The moment of truth approaches. The keeper decides that the birds are big enough to be released so he takes pen, poults and foster mother to a convenient piece of heather ground. After a couple of days to acclimatise, the poults are released a few at a time. Because the foster mother is normally retained in confinement, they 'hold' fairly well, have an unlimited choice of natural food and rapidly forsake the pellets on which they were reared. But, within a few days, their condition starts to decline and, within a month, most are dead, if not from malnutrition then from predation or accident. Because of their treatment in early youth, they are poorly equipped to survive as adults.[35]

The keeper vows that he will do things differently next time – the chicks will be reared much 'harder'. He spends endless hours cutting and bunching heather for his pets and the birds eat everything put before them. After the first ten days, he cuts back the supply of crumbs or pellets. Within a few days, he notices that they are not growing as fast as they did the previous year. A few days later, he handles a bird and finds it surprisingly thin. Within a few days they start to die but *post mortem* findings show no signs of clinical disease – just emaciation.[36]

All this begs the question of what happens when the birds are released into the wild. Will they compete successfully with wild birds for territory and establish their own? Is the piece of moor where birds are to be released home to any wild birds? If there are natives about, the incomers will be given a bad time and few will establish a territory. If there are no natives about, the area is probably unsuitable for birds and the plan should be abandoned. Anecdotal evidence provides clues. Grouse are an aviculturist's dream; they are placid and docile in the rearing pen and become tame and confiding. In the autumn, the cocks put on a show of displaying at any visitor who comes near their 'territory'. Any birds that survived the summer release show the same attributes at liberty, with the cocks often terrorising the keeper's dogs and attacking intruders. Because they tend to come towards, rather than retreat from, people, they are useless as quarry for the shooter. So far as is known, they are close to useless as stock birds as well. There are exceptions but there is generally little evidence to support the notion that a worthwhile proportion establish territories and breed.[37]

Can this dilemma be resolved? The answer is a qualified 'maybe'. A method has been described that appears to steer a middle path.[38] Rearing was done in an extensive manner with the foster bantam confined but her brood at liberty to forage widely on the moor from day seven or so. The researcher did not use marked birds but the system described, though labour-intensive, seems plausible with a good chance of working. It seems unlikely that many hundreds of birds could be reared by this method but it might have value where there had been a collapse in the wild population to such low levels that a moor is virtually devoid of birds. If eggs could be obtained and, say, two hundred poults released, it is reasonable to argue that the absence of competition from resident birds might allow some of the released cocks to establish breeding territories and produce an input of young birds in the autumn of the year following the crash. It might raise breeding densities to a level where significant shooting was possible earlier than if the owner had relied solely on nature. Were an enthusiastic owner to try the method, he would hopefully mark the released birds, cost the experiment and publish the results. It is worth noting that this technique is being used to reintroduce grey partridges on low ground where they are presently extinct.[39] In our present state of knowledge, rearing has no part to play in the management of grouse moors. That may change if releasing for stock could be achieved but the innate behaviour of the bird forever precludes releasing them for driven shooting on the scale now common with low-ground game. For this, all true sportsmen should be devoutly thankful.

Transplantation of adult birds

In the early years of the twentieth century, well before population dynamics were understood, many proprietors and lessees of poor-quality shootings went to great trouble and considerable expense to catch up birds from high-density moors, particularly from the north of England, and release them where birds were scarce. It was done in the belief, later shown to be mistaken, that birds from a prolific moor were intrinsically superior and would improve less productive popula-tions.[40] None of the releases appears to have been documented properly; supportive evidence tended to rely on anecdotal claims of improved bags. However, the activity was widespread and would not have been done on the scale it was unless there was some evidence that it improved matters.

There have been other occasions where owners have caught birds in one place and let them go on another which, while appearing suitable for grouse, holds only a few or even none. Some cases involved birds from the north of England (which are rufous red) being put in places where the natives are much paler (Scottish islands). Others involved the dark, almost black form from northeast Scotland. Even if the birds are of the same colour phase, 'changing blood' is unlikely to be much use because there is *always* a reason why the receiving moor is short of birds and incomers are *always* at greater risk from predators than incumbents.

Some birds do survive when moved in this way.[41] A cock caught in February on one moor, ringed and released on another sixty kilometres away within six hours of capture, was shot in August three seasons (forty-two months) later. It is reasonable to suppose that he paired and bred and so made a contribution to the moor's improvement though hard evidence is lacking.

A larger programme took place on the Island of Coll.[42] A century ago, Coll was described as

> a wild shooting extending to over 14,000 acres…with some grass and heather and arable land. The bag should include about eighty brace of grouse, two hundred brace of partridges, six hundred snipe, 150 woodcock…[43]

The author visited the island in the late 1960s at the request of an owner because grouse were deemed to be extinct. The habitat was found to be good enough to support some birds but early morning listening showed no sign of any. Shortly afterwards, he sold up and a new owner later took over a large area of the island. He released two adults and six young (two coveys each of four birds) in autumn 2000 followed by forty-five adults (age and sex unknown) in December 2001. No systematic counts were made and no predator control carried out but occasional coveys consisting of up to a dozen young were seen every year up to 2007, indicating that breeding probably took place in each of the five years 2003–07. The last recorded bird was seen by an RSPB warden in the talons of a peregrine in 2010; it is assumed that grouse are again extinct. The suite of predators on Coll includes greater black-backed gulls, ravens, hooded crows, peregrines, hen harriers and great skuas (bonxie), all of which breed on the moorland. There are no ground predators (foxes or small mustelids). It seems that the sheer weight of avian predators bearing on this island population ensured that, sooner or later, grouse would be unable to sustain a presence. On the neighbouring island of Colonsay, grouse and blackgame were plentiful at the end of the nineteenth century but are now extinct.[44] It is understood that the present owner is considering a reintroduction but the evidence from Coll suggests it is unlikely to work. If the project does go ahead, it is devoutly to be hoped that released birds are marked, that spring and July counts carried out and the results written up.

Plumage and size variations in grouse have been well known to sportsmen since the nineteenth century.[45] Grampian birds are darker than those from the Southern Uplands or North Highlands; Western Isles and Irish grouse are yellower than the reddish type which predominates in the north of England. These quasi-races have each evolved a plumage shade which gives the bearer the best chance of survival against the background of the natural vegetation. An Irish bird, hard to see against the yellow sedges of the bogs of west Mayo, would stand out a mile on a black hill in Angus and *vice versa*. Transplanted birds are likely to have less chance of survival than home-bred natives for reasons of visibility alone. If a grouse population increased and shooting improved after release, it is unlikely to have been due to the genetic influence of the introduced birds.

The notion of mixing the sub-races of grouse is controversial. Birds in the west of Ireland grow more slowly than those in highland Scotland. This may be due to a longer growing season but cannot be related to day length because the chicks in Scotland, being further north, experience more daylight hours in the first twelve weeks of life. The Irish birds do not grow faster if fed unlimited high-protein food so their slower growth rate must be genetically determined.[46] Observant sportsmen say that the grouse of the central Scottish highlands are smaller and have a more racy appearance than birds from north England. There may be some truth in this. Samples of forty birds of similar age (as measured by primary wing feather development) from the Highlands and Northumberland showed a small difference. The Scottish birds had primaries 1.91 per cent longer than the Northumberland birds; English birds of the same age were between 3.5 and 5 per cent heavier.[47] Even if this difference is dismissed as anecdotal and, given the small sample size, insignificant, it does at least suggest that it is wrong to translocate bird populations.

If 'changes of blood' led to improvements in the past, the explanation may be that someone enthusiastic enough to take birds halfway across Britain was probably enthusiastic enough to manage them properly. Foxes and crows would have been strictly controlled, burning standards improved and gritting carried out. Altering an unfavourable grazing regime, changing the burning and exerting control over crows and gulls would have been enough to improve grouse numbers in Mull, Lewis or Kintyre without importing strangers from Yorkshire even if the latter got the credit for enhanced bags.[48] The idea that inbreeding results in a loss of vigour was once widely held but work on partridges and grouse demonstrated that brother/sister matings do not occur in the wild.[49] The closest observed pairings are uncle/niece and nephew/aunt and these are seen but rarely. The two sexes behave differently at covey break-up and prior to pairing. Cocks are philopatric, taking territories for the next season as close as possible to where they were raised while hens disperse and float about over a wide area before selecting a mate often far from their juvenile home range.[50]

Transplanting birds has generally done nothing to improve the performance of the receiving population though the interest it engendered may well have been reflected in improved management regimes that proved valuable for the population as a whole.

The predator trap

Situations arise where the moor manager seems to do everything right but the grouse population does not respond as predicted. Burning is properly carried out, grazing is maintained at the right level, foxes and crows are strictly controlled and there are no ticks. To all appearances, there should be grouse at a density which permits some serious shooting. Instead, numbers stay stubbornly low. This is particularly common on 'island' moors where the area under good management is surrounded by poorly managed or unmanaged ground.

Under normal conditions, a reasonable number of pairs establish territory at a density of perhaps a pair to 8 hectares (20 ac) in October with few, if any, surplus birds floating about in small packs or coveys during the winter. If over-winter losses are low, the breeding stock is adequate though there may be significant emigration losses if there are few birds on adjacent moorland. Winter losses from predation and accident mean that the spring sees fewer breeding pairs on the core area – the previous season's surplus was too small to make good over-winter losses. The birds breed well, often spectacularly so if crow and fox control is thorough. By August, there are a few large coveys and almost no single cocks or pairs without young – proof positive that predator control over the summer has been effective. There are sixty to a hundred birds per sq km on the moor in August, enough to permit a modest shooting programme and leave enough stock for the next season. But, had it not been for emigration losses engendered by the 'island' moor, the results would have been substantially better.[51]

The peregrine is an efficient grouse predator.
(PHOTO:
N. MCINTYRE)

The over-winter decline in territorial birds is caused mainly by raptor, particularly peregrine, predation.[52] Because there is no surplus, losses cannot be made good from non-territorial flock birds. Breaking out of this impasse appears impossible unless a much more extensive area is put under improved management. Even when a break-out is achieved and numbers rise to a level which permits serious shooting, small populations are always vulnerable particularly if they fall below a certain threshold.

Grit

Because birds have no teeth, they rely on a specialised organ at the beginning of the alimentary tract to prepare food for digestion. The gizzard of the grouse is about the size of a walnut and consists of a thick muscular capsule with a hard, leathery, abrasion-resistant lining inside. By alternately contracting and relaxing, the gizzard muscles apply a squeezing movement to the contents which, with the aid of ingested grit particles, grind up food into pieces fragmented enough to allow the digestive juices to work on them. The macerated pulp passes out of the gizzard into the duodenum, ileum and caeca where the main digestive activity takes place. Because grouse food is bulky and high in cellulose and lignin, it is essential that the food is well broken down physically before it leaves the gizzard. Grouse must have access to grit of a suitable size and hardness to pulverise their food if they are to get the best from what they eat.

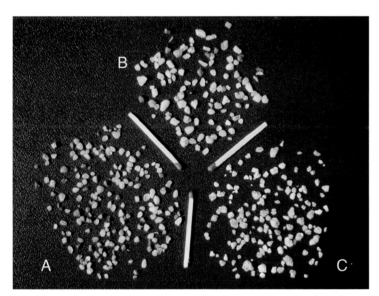

A
Grit from a shot grouse

B
Grit from a shot blackcock

C
Stannon No. 1 Cornish grit

The washed-out gizzards of grouse shot in autumn yield grit of a mean size of about seventy pieces per gram. A healthy bird in mid-October, when grit loads tend to be at their highest, may have one thousand two hundred pieces in the gizzard though up to a thousand is more normal. The largest pieces are the size of a match head but most are smaller. Birds choose the hardest material available; if they can get it, knobbly pieces of quartz are the first choice. Grit is rejected when it becomes worn and smooth but the rate of turnover is affected by the availability of fresh material. In the depths of winter, when the main sources of grit may be locked in ice or under snow, birds can hold on to what they have for weeks and come to no harm but they increase turnover sharply when it is plentiful. Willow ptarmigan in Scandinavia stock up on grit in autumn and eat very little after October even if it is plentiful simply because normal northern winter weather conditions mean it is unavailable. Red grouse show the same general pattern though they take it if the opportunity arises. It is a useful exercise to slice open a gizzard in late summer and wash the contents in a jam jar under a stream of water. The mineral particles can be dried on a piece of blotting paper. The normal base load per gizzard in August should be five to six hundred pieces.

Grit of good quality occurs naturally on most moors. In some places, every sheep path is littered with suitable material and every roadside quarry or rock outcrop provides a supply. If it is rumoured that the natural supplies on a moor are poor, a glance at a handful of fine gravel taken from an eddy in a burn reveals the grit status: it is often surprisingly good. At the time they are gathering most grit and a good supply is important, birds are not confined to a territory and will fly a mile or more (> 1.5 km) to collect it from a suitable place. Even in spring, when birds are aggressive, it is common to see birds fly into a neighbour's territory if there is a favourable grit source such as a shingle bank in a burn or a hill road. The birds, aware they are 'trespassing', slip in quietly, do not song flight on landing and pick up what they need quickly without drawing attention to themselves. They are rarely chased off by 'the man in charge'. Only on moors where peat is thick and stretches unbroken for many miles, such as in the west of Ireland, could grit could be classed as scarce. Even in these places, territory sizes and breeding success are no different where grit is plentiful than where there is hardly any. There are no documented cases of grouse doing badly because of a shortage of grit or well because they had plenty. The mystique that surrounds grit is a product of the observation that grouse regularly eat grit in quantity and the belief that providing it is a simple, harmless, relatively inexpensive and perhaps positive thing to do. In an owner's mind, if he has gritted his moor well and grouse do badly the following year, one variable at least can be discounted when examining the reasons for a disappointing season.

People are often unimaginative in how they put out grit; it is common to see it spread on paths, sprinkled along roadsides and placed on rocks outcropping on the moor. These are where it is most likely to occur naturally; the places that need attention are flat, boggy areas with continuous vegetation cover and where, even on dry ground, the soil is not exposed. Fire sites are excellent places to grit as the material will not be over-topped by vegetation for a year or two. If burning is done properly, gritting every newly burned site at a convenient time of the year ensures good distribution – unskilled workers can be told to put grit on all the black patches. Inverting a turf and placing grit on top will extend the time it is available. With Argocats, quad bikes and Land Rovers now virtually universal, there is little of the backbreaking toil that was the lot of the keepers fifty years ago. They were often expected to carry a game bag full of grit on their back every day they went around their beats.

Some resourceful people plough a long, shallow furrow with a wheeled tractor through unbroken vegetation and put grit on the bare ground so produced. On moors with an average natural supply, a tonne to 2000 ha (1 ton to 5000 ac), set out in four hundred piles of 2 kg (5 lbs) each is about right. There is little point in putting out forty heaps of half a hundredweight (25 kg) each but this is often how it is done. Sometimes it is strewn on hill roads, themselves a main source of grit! (Keepers have been known to put grit out on butts which are favourite lookout places for birds but this should be recognised for what it is – a crude attempt to impress the guns. It risks damaging the action of an expensive weapon should grit particles get into the action.)

The best material is Stannon No. 1 from the Cornish China Clay Company. It is the right size (<2 mm) and shape and readily eaten by grouse. It comes in 20 kg (50 lb) bags which makes distribution easy. Flaky, sharp quartz grit with the superficial appearance of broken glass or any grit containing crushed shells should be avoided at all costs. Oyster and cockle shells may assist domestic poultry to lay strong-shelled eggs but it is an actively harmful material on the hill. Grouse live on a diet in which the balance between phosphorus and calcium changes little in the winter; a sudden increase of calcium in the hen bird's blood, such as occurs if she starts to eat cockle shell grit, can have harmful biochemical effects. It is a matter of wonder how unsuitable is some of the grit marketed to unsuspecting owners and keepers. A common fault is that the particle size is too large; some products have as few as twenty pieces per gm. No self-respecting bird can swallow material like this. Examination of samples from different suppliers showed that, with some, the proportion of over-sized particles was more than fifty per cent by weight. These pebbles will never weather down to a more suitable size before they are buried in vegetation. The true cost of such material is twice that of more suitable size. It is beneficial if the material has at least twenty per cent by weight of pieces between 0.5 and 1 mm diameter as these are eaten readily by chicks. Before buying grit, owners should examine a number of gizzard samples and compare them with what is on offer from the merchant.

Dusting

Birds love to dust bathe; the action serves to help maintain feathers in top-class condition. In former times, shelters were erected on partridge manors to keep the soil dry underneath. They were usually well used; some owners even sprinkled insecticidal derris dust in the depressions to remove ecto-parasites. The practice was rarely followed on grouse moors (though it was mentioned in the twentieth century literature) but it might well increase the birds' comfort, if not their health, were it to be used on today's micro-managed moors.[53]

A natural grouse dusting site, Buskerud, Norway.

169

Hill loch shores as well as bog pools are much favoured by broods in hot summer weather. (PHOTO: D. PHILLIPS)

Drinking water

Water is essential for sheep and grouse. Sheep drink every day, copiously in hot weather, and will travel far to do so. Grouse also take water daily but in a variety of forms. In hard winter weather, they ingest snow and rime incidental to eating heather. They eat snow directly if they have to but that carries an energy penalty of eighty calories per gram of water ingested as the snow has to be melted by body heat. Summer droughts tend to result in the precipitation of heavy dews. Droplets which condense on vegetation during the night are sipped in the early morning and supply all the birds' requirements however hot the weather becomes during the day. In early summer, broods may congregate in damp places but it is unclear whether this is because the water is essential for their well being or incidental to their quest for insects. Victorian authorities laid great store on water and encouraged owners and keepers to make available a plentiful supply of clean spring water.

> Every trickle of water should be puddled up and a succession of small drinking places made… In fact, the spade has as much to do in obtaining good results as the gun has. What with looking after the springs, altering and repairing the butts, the spade is not out of the keeper's and watcher's hands from April to August.[54]

Studies on Kerloch moor in Kincardineshire showed that, even on the driest ground in summer, surface water was never more than a quarter of a mile (400 metres) from any point; even very young grouse chicks can cover such a distance in a summer day.[55] As all sportsmen know, grouse move in hot, dry weather in late summer to places with a constant supply of surface water. They inhabit wet places whether for drinking, bathing, insect abundance or just because they are cool. We can safely say that they like a plentiful supply of water even if it is not critical to their well being.

As do all birds, grouse bathe to keep their plumage in good order. They do so in streams, rills, pools and lochs. Individuals and pairs will both move off their territories if water is not available just as they forage for grit on land held by a neighbour. If they are seen at close range by the local territory owner, they may be challenged and evicted but they usually take care not to draw attention to themselves by crowing on alighting or raising their combs, both of which could trigger an aggressive response.

Dew pans

In the heyday of grouse management in the nineteenth and early twentieth centuries, many owners placed galvanised-iron dew pans on dry slopes and ridges. These work on the principle of condensing moisture and conserving it against subsequent rapid evaporation. They are still to be seen in northeast Scotland and eastern England, some of them still working many decades after they were placed on the moor. It is likely that their considerable popularity arose because owners and keepers associated improved bags with their use. We know that hens take their young to bog flushes soon after hatching as they provide surface water and are rich in insects. We also know that the flushes can be dangerous for chicks if ticks are present on the moor because they often harbour high concentrations.

Galvanised-iron dew pans were widely used in the nineteenth century. They are now museum pieces.

If hens and chicks use a dew pan for their water requirements and have ready access to adequate quantities of insects, this might prevent them putting themselves at risk by moving in vegetation where there are tick concentrations. An observant keeper reporting good broods in the vicinity of a dew pan would have been reason enough for an enthusiastic owner to deploy many on his moor. Good grouse bags subsequently would be attributed to the dew pans rather than, say, better burning or vermin control.

Artificial water holes

At Castle Grant, Moray (NJ055375), bog flushes were home to concentrations of ticks in the 1980s and 1990s. Grouse broods which moved to them immediately after hatching lost many chicks in a short time. Golden plover were affected in the same way, typically losing all their chicks within a week of hatching in all the three years 1992 to 1994.[56] The head keeper thought that constructing simple pools of standing water on the dry heathery slopes where grouse were nesting and away from the flushes might yield dividends. A chain of pools was dug with a JCB. Within weeks, the piles of spoil were being used as lookout posts by individual birds and chick droppings were seen at all sites in the first summer.

Like golden plover, dotterel need short vegetation for nesting. (PHOTO: N. MCINTYRE)

The pools were clearly well used and appreciated by grouse. Golden plover used the surrounding areas if the heather was newly burned or short.

Simple field studies were later conducted by leaving panes of glass smeared with lubricating grease out for twenty-four hour periods to see if insect abundance was greater beside the pool compared with thirty metres away. Adult craneflies (tipulids) are a popular, large and valuable food for chicks. Though they are mobile, the panes beside the pools caught more than double the number than did the controls.[57] It is reasonable to conclude that constructing pools on

A cheaply excavated bog pool. (NJ055380)

dry slopes has a quadruple effect; they hold broods close to their hatching place by supplying water, they discourage them from moving into areas beset with ticks, they enhance the supply of a preferred insect species and are appreciated by golden plover.

Contour drains and bird dykes

Hand dug in the 1920s, this bird dyke is still working well nearly a century later. North Yorkshire.

Water has another, indirect, effect on grouse. A century ago, keepers and owners paid as much, perhaps more, attention to the creature comforts of grouse than to their own. A copious supply of clean water was regarded as a passport to good health for bird and man alike. Many natural springs were opened up by hand and kept clean by attentive keepers. In some areas, notably the drier northeast of Scotland and the Scottish borders, carefully aligned contour drains were hand dug to transport water along otherwise arid slopes. They often provided water for man and beast alike. Examples can also seen in Yorkshire and as far south as Dartmoor. There is no doubt that birds would have used these places to drink but their real importance was perhaps not fully understood at the time.

Spring water bubbling out of a hillside is likely to have percolated through rock since it fell as rain or snow and dissolved salts, particularly those of calcium, on the way. This raises its pH, making it more valuable than rainwater as a supply of plant nutrients. If it is channelled along a hillside, sometimes for hundreds of metres, it fertilises extensive areas of vegetation downhill of the channel because of seepage through the drain bank. As the bank starts to weather and leak, one can see a series of green streaks downhill from it; examination of the plants which often make up the 'green' usually shows a spectrum of dwarf vascular calcifuges such as saxifrage, lady's mantle, wild white clover, violets and pansies. These are all indicators of fertile soil and provide high-quality food for the hen grouse in spring. Heather growing alongside the green ground is also likely to be higher in nutrients. | 173

Keepers and keepering

"As with partridges and other game, I have always observed that where there are first rate…keepers, there is always a pretty good stock of game."[58] wrote Sir Joseph Nickerson, a famous shot and shoot manager in his day and he linked the state of the keeper's hand at the morning greeting with the prospects for the day; the hornier the hand, the more game was likely to be shown to the guns.[59]

What does the keeper do and why? The question is rarely asked but the answer is critical to understanding the profession's role. Everyone accepts that habitat management is the number one priority but creating a shootable surplus also requires the removal of predatory species which compete with man's interests. In 1886, a noted writer, observer of nature and practitioner of game-keeping wrote of predation that:

> Notwithstanding all that certain theorists may have written about the 'balance of nature' adjusting itself, the destruction of vermin is an essential condition to good sport. This truth is so thoroughly borne out by experience that we can only express surprise at its being called in question *by any that have had the means of informing themselves upon the subject.* [Emphasis in original].[60]

Any contemporary game manager worth his salt who enters the frequently sterile debate on predator control with members of the conservation movement would say 'Hear, Hear' to that. Hill keepers manage habitat, burn heather and may control weed grasses. They frequently cut invasive scrub where it is unwanted, destroy pests such as crows and foxes and control deer. In so doing, they improve circumstances for a wide variety of non-quarry species.[61] Ring ouzels are neither shot nor are they a human food item but they are more numerous on keepered than on unkeepered moorland; greenshanks do better in the absence of crows; lapwings and golden plover are also many times more numerous on keepered moorland.[62] These are spin-offs for members of the public interested in conservation from land managed primarily for game shooting.[63] Keepers' actions create conditions favourable for producing a surplus of target species.

The bad old days of keepering: a gibbet in Kincardineshire, 1968.

It is a truism that it is impossible to kill something twice; if the conditions encourage a pair of birds of whatever species to take a territory and the hen bird sits on a clutch of eggs, hatches them and raises young, the birds are more likely to be there later in the year for the tourist to admire or the sportsman to shoot (depending on species) than if a crow had eaten the eggs in April or a fox or badger had killed the sitting hen on her nest.

In short, keepers are there to create that artificially high number of game animals and birds known as a shootable surplus so it can be harvested by people, many of whom pay a lot of money so to do. His activity drives diversity in the countryside and is certainly responsible for the maintenance of extensive areas of heather moorland. Though his prime product is the game crop, he also creates a rich and diverse collection of non-quarry species to be enjoyed by specialists and public alike.

The hill keeper's worth is ultimately assessed by accountants. A man who devotes all his time to producing grouse has to provide a season's bag of three hundred brace per annum (all let) to cover his costs. If he is regularly producing four hundred brace single-handed, his job is secure. A moor yielding a thousand brace on a ten-year average should yield sufficient revenue to cover the employment of at least two and probably three men full-time. If a keeper is expected to attend to other duties, such as deer stalking (open hill or woodland), a low-ground shoot or servicing fishing parties, employers have to take the value of these enterprises into account when sizing up the added value which accrues from that part of the job.

Public access

Low-ground game preservers have sought traditionally to keep shooting grounds quiet and free from disturbance. Perceived wisdom has always been that the shooter is not keen to have the world and his wife trailing through his pheasant coverts in autumn because this can scatter birds and reduce bags on shooting days. At nesting time, walkers and their dogs can do harm unwittingly when birds are sitting or have small chicks and good records exist from sixty years ago which show the harmful effect of this on grey partridges.[64]

On the other hand, red grouse and ptarmigan are much more tolerant of human disturbance and interference and take little notice of walkers – particularly if they see them regularly. Grouse living alongside the Pennine Way or near ski centres in highland Scotland ignore the crocodiles of people which pass constantly. Breeding density generally tends to be no lower and breeding success no different in places where there are many people compared with other locations where few people go[65]. Truly 'Familiarity breeds Contempt'.

When the Countryside and Rights of Way Act (CROW) was passed in 2000, many English moorland proprietors were apprehensive and anticipated that there would be an increase in conflict between walkers and shooters but problems have been few. There is always a potential conflict with unruly dogs disturbing and scattering young broods in bad weather which may lead to losses. Fortunately, fewer walkers go to the moors when the weather is bad, so the anticipated problem has proved to be less than anticipated.

What is certainly a matter for concern in some locations is food waste and general refuse which some walkers and tourists leave behind. This can support larger-than-natural populations of crows, gulls and foxes which then prey on the eggs and chicks of game and moorland waders[66].

ABOVE AND
RIGHT:
*The grouse
shooter is the
person who
puts the price
on heather.*

Grouse groups

Moorland owners have been familiar with the concept of Deer Management Groups (DMGs) for decades and few now doubt their worth. Deer are highly mobile; one man's treasure can be his neighbour's toxin. However, deer, particularly hinds, tend to occupy massifs and heft to a home range. Though stags can travel far during the rut in autumn, they too have discrete winter and summer ranges. The way an estate owner manages his deer impacts on others round about. DMGs provide a forum for the development of common policy, improved management and the dissemination of knowledge. Some may regard them as talking shops and bone yards for the subjective assessment of antler quality but there is little doubt that groups have value in that they bring together neighbours on a regular basis, many of whom live isolated lives in a working environment that can be downright lonely.

The same principles hold for grouse even though they are essentially sedentary with regions tending to hold discrete populations. The consequences of one moor owner's actions may not be as dramatic as they can be in deer forest country but sloppy keepering, casual burning practices and under- or over-shooting can impact on neighbouring ground in a negative way.

In 2002, the Heather Trust examined the possible advantages of forming a grouse group in west Perthshire. The idea was certainly not new, having been put forward nearly a hundred years earlier.[67] The scheme involved collaboration between five moors in four ownerships south and east of the A822 with the intention of killing red deer down to the lowest possible level to assist in controlling sheep ticks and of working out a programme of prophylactic treatment for the hill sheep flocks. People on the ground were enthusiastic, owners thought it worth a try and costs looked set to be modest. It collapsed within a year because of a failure to put in place the adequate gathering and exchange of information. Had it been properly coordinated, it could have developed into a group which, for example, shared setter dogs for grouse counting and collected sheep data in a uniform way. It could have agreed targets for the numbers of birds to be shot each year, thus helping to iron out fluctuations. Cost savings from bulk purchases of sheep medicaments, cartridges, vehicles, grit, Larsen traps and so on would have been significant. Ultimately, the group could have carried out joint marketing of any surplus shooting. Groups of proprietors are better placed to defend themselves against hostile pressures from those who seek to change the *status quo*. Perhaps in a few years time, some enthusiast will start an Upper Findhorn or Lammermuirs group and take this torch forward. The small west Perthshire group died in infancy solely because of a lack of leadership from the coordinator. It is therefore encouraging to hear that neighbours in a number of districts throughout Scotland have started to band together to form groups.[68] The cause of grouse shooting will be well served if more people start taking such a proactive stance.

The surge in capital and rental values

During the 1950s and into the mid-1960s, grouse letting values were modest. Plenty of places could yield two thousand brace in a fortnight.[69] Teams of Americans came to shoot in August, augmenting many wealthy people from Europe and the UK who would take a lodge for a shooting holiday. On a lesser scale, a September fortnight in 1964 in a partially staffed lodge in Banffshire (NJ335286) consisting of four days driving and four days walking up and yielding eight hundred and fifty brace to eight or nine guns was to be had for about £1000 including staff, beaters and the use of estate vehicles.[70] In the twenty-first century, if such bird numbers even existed there, one would have to pay two hundred times as much for the same product. In those days, grouse were everywhere in numbers to suit all pockets. You could get any number of driving days of forty to sixty brace on hill farms in the southern uplands for modest money. Nobody put much value on the grouse; they were just 'there' in all levels of numbers to suit all pockets and that was that.

Much the same pattern prevailed in many deer forests. If there was a scattered grouse stock, enthusiastic proprietors began keepering again. Grouse numbers began to rise and more owners of isolated moors in the west and north began to revive their grouse interest so that by the early 1970s, some regular shooting was taking place on widely scattered estates.

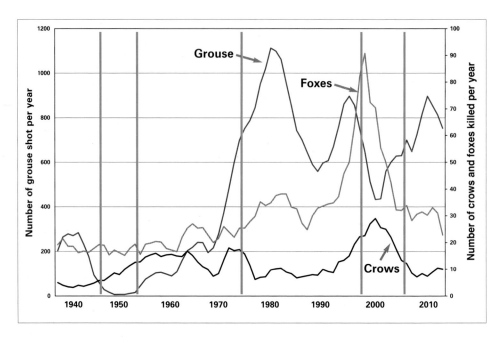

Changes in numbers of foxes and crows killed and grouse shot over seventy-four years and the incumbency of six head keepers (Central Scottish Highlands). Management strategy changed to stress grouse rather than deer alone from about 1970. Grouse bags often reflect the interests of the estate manager and the head keeper. (A change of head keeper is shown by a vertical green line.) The first three incumbents had little interest in grouse, unlike the latter four. (GRAPH: D. BRUCE)

A typical example was Kinlochewe – an extensive deer forest of some 14,000 ha (35,000 ac) lying athwart the spine of Scotland between Loch Rosque and Loch Maree. Much of the estate is steep and rocky and clearly just deer forest and some of it is green and grassy but there are about 4000 ha (10,000 ac) of good grouse ground on the Bruachaig beat (NH090610) covered in short, wind clipped heather mixed with sedges and blaeberry and dissected by small streams on the banks of which heather grows more vigorously. Slopes are gentle and the ground is ideal for shooting over pointers and setters. Most of it is above 300 m (1000 ft) which puts it above the tick limit. At that time, many red deer, particularly hinds, wintered on this ground at a peak density of about one animal to 4 ha (10 ac). A few sheep grazed the land in summer. The desirable mixture of tall and short vegetation was maintained naturally with no burning needed on the open moorland. Grouse were present at a dogging density of a pair to 10 to 16 ha (25 to 40 ac) in spring. There were resident peregrines and eagles (both species unmolested), occasional pine martens and an annual influx of grey crows each spring. The take of foxes was normally fourteen to eighteen a year with one or two dens in traditional sites dealt with as a matter of routine in April.

Between 1965 and 1988, about 2,500 brace of grouse were shot over dogs, an average of over 100 brace a year. The biggest bag was in 1984 when 286 brace were got, the smallest was 1970 with 16. The biggest one-day bag was 35.5 brace (1981). In 1989, staff changes saw an end to grouse keepering. That year, spring-immigrant crows nested unmolested and the resident foxes were barely disturbed. By August, a predictable pattern emerged. The grouse ground was occupied by many single cocks, some amalgamating into small 'bachelor' coveys of up to five individuals. Pairs without young were commonplace and there was a scattering of broods of eight to nine well grown young following the excellent summer of 1989. The bag was well down on expectations (about sixty brace) with numerous signs of crow predation to be seen. Sucked eggshells and fox marks were encountered frequently.

In 1990, the spring and summer were again favourable and breeding should have been good but no crow killing was done, three fox dens got away untouched, cubs were run to ground by collies at the summer sheep gatherings and flocks of twenty to twenty-five crows were commonly seen quartering the moor in June and July. Ten days dogging produced a best day of seven brace and a season's total of less than thirty-five. For all practical purposes the grouse interest on this splendid moor had vanished in the space of eighteen months along with its 1990 capital value of £200,000. Things were no better in 1991 and the owner sold up in 1992 partly because there were no grouse to shoot any more. In mid-September 1992, a two hour walk with setters on what had been the best grouse ground on the beat yielded sight of three grouse, all single cocks. Although this story is sad, it is sadder still in terms of grouse fortunes in the Highlands. Kinlochewe was the most northwesterly estate in the UK to produce consistent and worthwhile bags, a reference point for many who aspired then, now and in the future to have grouse on what is widely (but erroneously) classed as marginal land for the bird. The loss of Kinlochewe as a serious dogging moor, even if it is only temporary, is a tragedy.

Now listen to a story with a happy ending. Burnfoot Farm (NN675905) lies on the rich basalt ridge making up the Gargunnock Hills which run southwest from Stirling. Of its 800 ha (2000 ac), about 500 ha (1200 ac) are heather in mosaic with grass and sedge. During the 1970s, it yielded consistent bags of between

Evidence of corvid predation.

150 and 270 brace per annum under a system of DIY keepering grafted onto a commercial sheep farm. Towards the end of the 1970s, advancing years saw the proprietor hand the farm over to a relative who farmed it for a decade. As he had little interest in the shooting, few grouse were shot and keepering inputs declined to low levels. He sold the farm in 1989; the new owner leased further areas and increased the size of the 'island'. In the spring, a pair count indicated a grouse breeding density of about a pair to 32 ha (80 ac). He then employed a full-time keeper who systematically killed foxes and crows. Three seasons later (1992), the moor yielded bags of over fifty brace a day and a season's bag of nearly two hundred. This yield has since been increased to a season's average of about eight hundred brace. Even though it has no advantage of next-door grouse neighbours, substantial keepering inputs ensure that grouse thrive in good numbers. By the mid-1990s, bags of over a hundred brace a day were got most years. One can be confident that the result reflected good management rather than a cyclic or weather response because a comparable moor (also an island) of 600 ha (1500 ac) on the other side of a mile-wide valley which grouse rarely cross collapsed in 1992 from a regular three hundred brace per annum to nothing due to an outbreak of disease followed by a cessation of grouse management.

Increased bags tend to follow good keepers when they change jobs. That this can happen even in the first year of a keeper's tenure suggests that the effect of tightly controlling foxes and crows is immediate. It is followed by a second, more gradual lift as the effects of better burning take hold between three and eight years later. Though most keepers and managers know of places where this has happened, the results are rarely, if ever, documented.

Moorland acres nationwide are not a particularly valuable land type. Where blanket bog sells for £250 to £500 per ha (£100 to £200 per ac), best quality arable land may change hands at £20,000 per ha (£8000 per ac), forty to eighty times as much.[71] Land values reflect potential output with traditional wisdom holding that moorland can support only a few sheep of low productivity and so it is valued accordingly. Grouse shooting on the area does add value but prices rarely reflect potential income (hope value) from grouse shooting. Even the farm value at £500 per ha is influenced by the SFP, a social subsidy designed to keep people living and working in upland areas many of which are remote and short of 'services'. Land value can be influenced by government schemes such as the SRDP. For example, plantation forest policy, native woodland creation, the conservation of specific associations of flora or a policy aimed at improving the status of a rare bird affect land use decisions and thus land values. Under a 'single purpose' management regime, the land may end up as degraded, eroding blanket bog because it is asked to carry too many sheep and cattle; under another, it may become dominated by scrub woodland because grazing is light or non-existent. But with competent management, it is possible to have sheep farming *and* shooting on the same area with both enterprises turning in regular profits.

When an estate or large hill farm with an area of moorland is put on the market, the heather component is valued on the basis of how many sheep it carries on a sustained or historical basis; a grouse or deer interest adds a significant premium. Both are based on a multiple of annual rental known as 'years purchase' (YP), normally twenty years but sometimes longer. For a dry heather hill in southeast Scotland or eastern England capable of supporting breeding ewes at 0.8 to the ha at an annual rental value of £12 to £20, the capital value of the land for sheep farming at 20 YP is £240 to £400 per ha (2011). If the area produces a ten-year average bag of three hundred birds per 400 ha and the capital value of a brace of driven grouse is £4000, the additional capital value is £1500 per ha. In parts of Scotland where red deer are harvested, the capital value of £20,000 to £30,000 per stag can add £100 to £150 per ha.

It is a simple enough exercise to put meaningful cash values on heather moor but one rarely carried out by conservation bodies who propose land use changes without making any effort to understand the position in which land owners or farmers typically find themselves. The following examples illustrate the 20 YP capital value of four contrasting properties. (Grouse bags are a ten year average.)

Example 1
A 1000 ha stock farm in upland Britain; unenclosed 'hill' either 'white' or 'grey' in character; supports 1500 hill ewes; rental value £18,000 to £30,000; no shooting interest; over-browsed in winter; no predator control; little or no conservation interest.

1500 sheep:	£360,000 to £600,000
Value per hectare:	£360 to £450

Example 2
A 1000 ha stock farm in eastern Scotland or north England; unenclosed upland either 'grey' or 'black'; supports 1000 hill ewes; rental value £12,000 to £20,000; located within a shooting complex; good predator control; local over-browsing; 250 brace grouse driven; significant conservation interest, especially nesting waders.

1000 sheep:	£240,000 to £400,000
250 brace grouse @ £4000/brace:	£1,000,000
Value per hectare:	£1240 to £1400

Example 3, 1970 conditions
A 4000 ha moor in highland Scotland; unenclosed upland, all 'black'; 1000 hill ewes; one gamekeeper; 500 brace grouse part walked up, part driven; ten stags per year; some conservation interest.

1000 sheep:	£240,000 to £400,000
500 brace grouse @ £1000/brace:	£500,000
10 stags @ £5000/stag:	£50,000
Value per hectare:	£197.50

Example 3, 2010 conditions
Unenclosed upland mostly 'black' but some 'grey' below 300 m (1000 ft); many immigrant deer in winter; 500 hill ewes; one gamekeeper; 200 brace grouse walked up; 20 stags per year; few or no nesting waders; many sheep ticks.

500 sheep:	£120,000 to £200,000
200 brace grouse @ £2000/brace:	£400,000
20 stags @ £25,000/stag:	£500,000
Value per hectare:	£255 to £275

Example 4
2000 ha moor in Yorkshire or Co. Durham dales; unenclosed upland mostly 'black'; 1000 ewes (tenanted); no deer; two gamekeepers and a trainee; ten-year average bag; 4500 brace grouse driven; high conservation interest; many nesting waders.

1000 sheep:	£300,000
4,500 brace grouse @ £4,000/brace:	£18,000,000
Value per hectare:	£9150

The unit value in Example 4 is over thirty times that of a similar piece of ground without a grouse or conservation interest such as Example 1. Obviously, these figures do not take account of the additional and variable costs attributed to the shooting, such as keepers' wages; in Example 2, additional costs for four days' shooting would be in the region of £2500; in Example 4, £20,000 or more. Nor do the figures reflect direct benefits to the local economy such as hotel occupancy, 'spend' in local shops and casual employment on shooting days for flankers, beaters, vehicle drivers, dog handlers and so on.

The figures are no less startling on a rental basis; current letting values of £150 to £170 per brace before VAT are readily obtained by estates selling even modest days of fifty to seventy brace. The revenue from a pair of typical days' driving goes a long way to paying a keeper's wage. The same is true for walked up grouse; a modest day of twenty brace costs a team of guns £1500 or more.

When estates can key into such sizeable income streams, it is surprising that it should be a matter of remark when an incoming tenant or owner with money to invest turns game numbers round in a few years. But moors where this happens are the exception rather than the rule; the majority of owners manage their moors reactively rather than proactively – many whose families have held estates for generations are sleep walking into oblivion. When the money runs out and they lose house and home, they should attract scant sympathy. In most cases, they have only themselves to blame.

On the other hand, an incomer who knows what he is doing and has the resources to invest in moor management can reap large rewards. Though it undoubtedly makes it easier, it is important to recognise that it is not essential to have a long pocket to aspire to owning a grouse moor. Every year there are opportunities to buy substantial blocks of upland for little more than its farm valuation and use proven techniques over a modest time span to get rid of grass dominance, grow productive heather and create a moor on which grouse thrive and can be driven. If such a land block lies close to or against existing moorland, well and good but it is not essential.

The decision to invest boils down to size. A thousand acres of white land overlying limestone in a northeast of England dale with established moors nearby will quickly yield several hundred brace a year under top management whereas five times that area is needed if the site is isolated from other, keepered ground. An even larger block is needed if the underlying rock is poor and neighbouring land uses are hostile.

'Put and take' partridges

Moor owners the length and breadth of the country have succumbed to siren voices promising cash flow from driven red-legged partridges on a moor edge. If a 'bolt on' project of this type is properly managed and adequately staffed, it can be good for estate finances. Like hand-rearing pheasants, results are predictable, cash flows can be ensured, the sport can be exciting and the financial uncertainty of grouse performance over the years is diminished; there is always something to shoot. But, as no man can be in two places at once, work on partridges can push management of the moor into second place.

There are various grades of 'put and take' shooting. In some cases, the 'moor edge' is entirely 'white' and the released birds go nowhere near the heather; in others, the birds are released, held and shot on what is actually grouse ground – or was until circumstances changed and the grouse declined. Using the latter for partridges is a mistake. Superimposing reared birds of unknown origin and health status on a grouse population, however small, is foolhardy because it is easy to introduce pathogens to which native grouse and blackgame have no resistance.

If the grouse have declined or disappeared from an area where they were plentiful in former days, it is in the owner's interests to address the problem at source. Of course it is acceptable to use partridges to generate money which can be invested in the moor but it should not be an end in itself. Those who make it one should do so in the certain knowledge they run the risk of losing the grouse, probably for good.

End notes

1 Phillips, 2011.
2 Merricks, 2010; Olstead, 2011; Walters, 2011a.
3 Pers comm, protected source.
4 Olstead, 2011.
5 Duckworth-Chad & Thomas, 2011.
6 See Chapter 1, *Burning patterns in former days.*
7 Potts, 1986.
8 Macdonald, 1883; Speedy, 1886.
9 Miquet, 1990; Bevanger & Brøseth, 2004.
10 Jenkins *et al*, 1963.
11 J. Grave, pers. comm.
12 Author, pers obs.
13 Watson, 1979.
14 Watson & Moss, 2008.
15 Bevanger & Brøseth, 2004.
16 Miquet, 1990; Novoa *et al*, 1990.
17 Author, pers obs.
18 A SNH board member boasted at a conference on the issue some years ago that 'wind-power is here for a hundred years at least'. (D. Bruce, pers comm.)
19 Fletcher *et al*, 2010.
20 Gladstone, 1922.
21 Cazenove, 1936.
22 Worth approximately £20,000 in 2011.
23 Grimble, 1894.
24 Reimers, 1989; Andersen & Sæther, 1996.
25 Stuart-Wortley, 1895.
26 Parr, 1975.
27 J. Grave, pers comm.
28 Stuart-Wortley, 1895.
29 Moss *et al*, 1996.
30 Lovat, 1911.
31 Watson & Moss, 2008.
32 Author, unpub.
33 Park *et al*, 2001.
34 Moss, 1972a; Savory, 1978.
35 Moss, 1972b.
36 Savory, 1978.
37 Waddington, 1958.
38 D.S. Wintle, pers comm.
39 Phillips, 2011.
40 Hamilton-Maxwell, 1832: Lance, 1975.
41 C. Stuart Menteth, pers comm.
42 R. Wainwright, pers comm.
43 Watson Lyall, 1912.
44 Gathorne-Hardy, 1900.
45 Hall, 1885.
46 Watson & Moss, 2008.
47 Author, pers obs.
48 Watson *et al*, 1984b.
49 ICI Game Services, 1953.
50 Warren & Baines, 2007; Moss and Watson, 2008.
51 www.basc.org.uk/en/departments.
52 If peregrines are identified as the main problem [they usually are], buffer feeding with captive pigeons is a reliable solution (Phillips, 1990b). See Chapter 5, *Prey and predators.*
53 Parker, 1929.
54 Stuart-Wortley, 1895.
55 A.Watson, pers comm.
56 Author, pers obs.
57 Author, pers obs.
58 Stuart-Wortley, 1895.
59 Nickerson, 1998.
60 Speedy, 1886.
61 Fletcher, 2006.
62 Pearce-Higgins & Yalden, 2003; Fletcher *et al*, 2010.
63 Tharme *et al*, 2001.
64 ICI Game Services, 1952; ICI Game Services, 1953.
65 Picozzi, 1971.
66 Watson, 1996; Storch & Leidenberger, 2003.
67 Lovat, 1911.
68 K. Wilson, pers comm.
69 Waddington, 1958.
70 A. Studd, pers comm.
71 Bell Ingram, 2011.

Chapter 5
Prey and predators

The great predation debate

In 1952, the ICI Game Services[1] noted in its annual review, *Your Year's Work* that:

> We know of no instance in this country of a completely 'natural' shoot which has provided a reasonable surplus of game to be shot over a period of years. The fundamental basis of game conservation and protection is the production of an annual surplus which can be harvested, and this can only be done by improving the chances of survival and reducing the mortality factors continually influencing the game population. One of the most important of these is vermin.

A grouse eaten by a raptor. Note the notches in the breast bone.

Though the topic was low-ground game, the remarks are apposite to any strategy designed to increase game or wildlife: to raise the numbers of a target species, *every* factor that may limit its productivity must be examined *in context*. If it is to be effective, the strategy inevitably involves intervention, the control of predators. Because prey and predator relationships are complex, they can be hard to grasp and our understanding can be clouded by the simplistic attitude of previous generations which saw species as either wholly harmful or entirely beneficial that should be killed or preserved accordingly.

In 1857, the *Gazetteer of Scotland* published a list of predators killed in the name of game preservation on the estates of Aberchalder, Glengarry and Glen Quoich over the period 1837 to 1840.[2] As the figures (see Appendix, *Glengarry, 1837–62*) were endorsed by the owner in 1931 and so seem accurate, they can tell us much about the Glengarry estate from the mid-nineteenth to the early twentieth century.[3] They might seem to suggest a powerful predatory force lined up against the game and wildlife population, are certainly cited often enough as 'proof' that the gamekeepers and watchers of the time were assiduous, even relentless, killers and are used to back the implicit assumption that all Scotland's keepered ground is managed the same way today. They weren't and it isn't. They understandably offend twenty-first century sensibilities but the serious commentator must place them in perspective.

The numbers killed are a four-year aggregate and so need to be divided by four to obtain an annual total. Extending over 60,000 ha (150,000 ac), the area from whence they came was heavily populated and biologically rich. With every available corner cultivated, it must have been teeming with prey, particularly small mammals, to support such large numbers of mammalian and avian predators. Smaller predatory mammals such as stoats and weasels, chiefly dependent on microtins, were in turn the food of polecats, pine martens and wild cats. Of the avian predators, merlins, kestrels and buzzards would have benefited from the plethora of small mammals; large numbers of kites, crows and ravens would have lived off dead farm animals, particularly cattle, and household refuse on the dung middens located by every dwelling. In the first half of the eighteenth century, the economy was based on cattle – there were few sheep and almost no deer and the hill was rarely if ever burned – but sheep numbers were increasing by the third quarter of the century. Everyone in the community had an interest in killing predators; the crofter saw some species as enemies of his primitive crop-based agriculture, others as predators of sheep. The new wealth of the Industrial Revolution, fuelling a demand for field sports, provided employment for people for whom the alternatives were destitution or emigration. The glen was well wooded, particularly at its eastern end, and a habitat for goshawks, buzzards and long-eared owls. Sparrowhawks were present but, not being seen as serious game predators, were killed only incidentally. The district was home to many crofters which explains the house cats, crows and kites and the scarcity of foxes. Ravens would have been numerous. Though cattle numbers declined, sheep numbers increased (though annual mortality was perhaps as high as ten per cent) and with increasing quantities of deer grallochs in the season (as their number also rose), there would have been enough carrion to support large numbers of scavengers.[4]

Set properly in context, the numbers of predators killed become less remarkable. As there were over 40,000 ha of grouse moor and deer forest, the take of eagles, harriers and peregrines is hardly surprising. The high number of kestrels suggests that there were numerous small mammals for them to feed on and that the shrub zone and herb layer on the open hill must have been floristically rich. Stoats and weasels were scarce initially and may not have been pursued with much vigour though, had they been numerous, they would have been caught in traps set for polecats and pine martens. As the number of larger predators fell, so stoats and weasels increased. Even with the high crofting presence, it is astonishing that a mean of only three foxes per annum was taken. Assuming (reasonably) that they were adults, the figure represents a pair per 40,000 ha. Breeding densities of a pair to 1300 ha (3200 ac) were recorded in Glen Derry in the 1950s and in Glas Maol/Cairnwell in the early 1960s – thirty times as many.[5] Truly, fox catching must have been efficient. It is also noteworthy that the take of badgers was nearly six times that of foxes, a figure which could easily be replicated today in many areas in the north Highlands. The grouse bag is surprisingly modest considering the size of the area. If the 851 birds came from 16,000 ha (40,000 ac) of moorland, that was a mere twenty-one birds per 400 ha (1000 ac). Even today, many similar Highland areas could produce an equivalent yield if hunted with good setter dogs as was the norm at that time. By 1912, grouse were down to six hundred birds per annum while the number of stags shot had risen to forty-five.[6]

Ninety years after the predator figures were published, the 1931 edition of Ellice's *The Place-names in Glengarry and Glenquoich* listed the birds then nesting in or visiting Glengarry. The list (see Appendix, *Glengarry bird numbers, 1931 & 2011*) was compiled by Murdoch Matheson, a keeper/stalker who, born in 1870, lived in the glen all his life. This list was checked and updated in 2011 by Derek Moffat, an amateur ornithologist and frequent visitor to Glengarry, and Lea McNally, a resident for over forty years. Matheson recorded ninety-three species as 'regular' and thirty-one as 'occasional' breeders. Eighty years later, Moffat and

McNally recorded seventy-five 'regulars' (down nineteen per cent) and sixteen 'occasionals' (down forty-eight per cent) as breeding and that 'visitors' were down from thirty to twenty (thirty-three per cent).[7] Again in 1931, the estate owner recounted that:

> ... when it was determined to sell the estates [in 1843], a great raid on all vermin was commenced... till 1863... The tenants who followed, without exception, preferred the deer stalking to the grouse shooting, hence the heather was burnt out in an endeavour to improve the grazing for... the stags. The vermin was allowed to increase as it was found that the grouse...frequently put the deer on their guard... These measures were successful in eliminating the grouse. The best grouse years were in 1846 and 1854 in which years 1845 and 1983 grouse were shot... It is doubtful if fifty brace could be killed over the whole area at present. If, however, the causes of this wholesale killing of grouse were ever removed, no doubt the district would soon show as good records as in the old days. In the winter and especially in snowstorms, hand feeding has to be relied on to support the stags where heather in former days might have sufficed.

Hand feeding supports enhanced numbers over winter but often at the expense of vegetation quality.

The ecology student seeking historical and anecdotal evidence of moorland loss over the last one hundred and fifty years need look no further than the decline of Glengarry from a relatively prosperous crofting community (at least by Highland standards) through the upheaval caused by the mid-nineteenth century clearances to what is now a degraded and impoverished deer forest. As Matheson's list shows a biological richness which beggars belief today, something has gone badly wrong. The regular burning of large fires coupled with intensive grazing by sheep and deer has led to widespread damage to the moor; compared with former times, the glen is home to few people drawing their living directly from the land. There has been an increase in plantation forest on sheep walks, inundation of fertile strath land following hydroelectric development during and after the Second World War and a steep increase in the number of red deer as the sole significant herbivore. Vegetation on the hill is scanty, the herb layer is thin and patchy and the shrub layer is relatively unproductive so a reduction in the numbers of small mammals is inevitable. As part of the hydroelectric development, an island on Loch Loyne was submerged when the water level was raised. It had been an important nesting site with nearly fifty nests of twenty-five species recorded on it in 1931.

Given that keepering in Glengarry is no longer performed on anything like the scale of former times, it verges on absurd to blame gamekeepers and shooters for the current paucity of wildlife. It is equally absurd to lay universal blame for reductions in wildlife interest on gamekeepers and sportsmen on other estates, particularly those further east in Scotland and the north of England where degradation has been less dramatic. The truth is that the management of extensive tracts of the western half of moorland Britain over the past one hundred and fifty years has proved unsustainable and has led to catastrophic declines in wildlife.

Though game records from other estates show that grouse were much more numerous than today, it is an oversimplification to suggest that they achieved the reported levels only because many predators were killed. The tight keepering documented at Glengarry certainly kept nest and chick predation low but other factors must have been at work; the bags of those days *could not be produced* from the stocks the land supports today. We can discount atmospheric pollution and poaching as reasons for the collapse; we are left with only factors working in concert – habitat quality and soil degradation.

Further evidence of the richness of former times comes from contemporary writers, one of whom described the prey base of peregrines as including 'sea gulls, guillemots, plovers, curlews, ducks, land rails, partridges, pigeons, rooks and grey crows' as well as 'hares, rabbits, rats, moles and other small quadrupeds'.[8] Even if this is not strictly accurate or applicable to Glengarry (which is a long way from the sea gulls and guillemots of the west coast), there were clearly many more species available for peregrines to select as prey than reside in the area today.

As now, many in the nineteenth century held strong, even antagonistic, opinions about keepers and keepering, particularly with regard to killing birds of prey, and debated the issue hotly in the press of the time. Many surmised that the birds were good for grouse as they acted as natural sanitary inspectors by selectively killing weak and diseased birds.[9] The hypothesis was also advanced that the general health of a grouse population might improve if foxes were left un-molested. There is probably a grain of truth in that claim; work done by the Game Conservancy (GCT) suggests that foxes are more likely to find incubating hens hosting large worm burdens than those which are not. Even so, it would be a foolhardy manager who banked on foxes being *that* selective in their choice of grouse prey.[10]

Getting a feel for what habitats were like a century and a half ago is not easy because there are few good nineteenth century photographs of moorland but one does come across old photographs with a moorland background from time to time. Where the site can be identified, it can be instructive to look at the same place at the end of the twentieth century. The differences are usually marked. Few sheep can be seen in the photographs whereas the only time they are not seen on most moorland today is when they are gathered for a handling. This is true of much of the British Isles, the west of Ireland as much as Perthshire, Sutherland or Wales. The causes of habitat deterioration vary but the common thread is reduction in the vigour and cover of the heather and its replacement with purple moor grass, white bent and deer sedge. It is a rash commentator who argues either that avian predators are the chief cause of the decline in red grouse nationwide or that the claimed decline in predator numbers can be attributed to illegal killing by keepers. High raptors numbers might be a local contributor to bag reduction but they are not the sole or even the major cause of grouse scarcity nationally.[11] Only a fifth of the uplands in England and Scotland are heathered and keepered; the remaining four-fifths are essentially a desert containing neither a thriving grouse population nor raptors in significant numbers.

Intraguild predation

Before leaving old anecdotes, it is worth reminding ourselves that predators eat each other, something that was probably important in mid-nineteenth century Glengarry.

> The practical moor manager is not concerned with the laws of natural selection and survival of the fittest, but rather with the adaptation of these laws to his own special requirements... Birds and beasts of prey are not wholly good or wholly bad, in the destruction of mice, rats and voles they also play a useful part and the extermination of the greater vermin entails the (additional) duty of keeping in check the lesser pests, which might otherwise become too numerous owing to the destruction of their natural enemies.[12]

Although the term had yet to be coined, Lovat was describing what is nowadays called *intraguild predation* – the interaction of one predatory species with another.[13] In natural environments such as those of Iceland and the high Arctic, the many mammalian and avian predators interact in such a way that one species keeps another in check with smaller predators falling prey to bigger ones. Nearer home, foxes eat hen harriers as part of a suite of ground-nesting birds. (On well-burned moors supporting many hares, the fox, like the eagle, takes many as they are easy to catch, particularly the leverets. Encouraging hares should be a key component of moorland management despite the tick risk under certain circumstances.) Though eagles are apex predators because they eat hen harriers, fox cubs, buzzards and crows, they tend to hunt hares in preference to grouse or ptarmigan because they are easier to catch and provide a bigger meal. Wolves and lynxes eat foxes, foxes eat pine martens, pine martens eat stoats and weasels, peregrines eat merlins, sparrowhawks and kestrels and the bold, fearless goshawk takes anything it can master. Where there is a broad spectrum of predators, the smaller ones – which pose the greater danger for grouse and ptarmigan – are often at relatively low densities because they are themselves preyed on and the larger ones, having a wide choice of prey, bear less heavily on game. In short, predators can be so busy eating and being eaten by each other that a high proportion of game birds manage to rear their young successfully because they make up only a part of the prey base. If this were not the case, how did sporting visitors like Pennant and Thornton at the end of the eighteenth century make such immense bags over land where keepering was unknown? When apex predators are scarce or absent, the lesser ones often increase dramatically and may come into serious conflict with mankind. When goshawks were reintroduced to Kielder Forest in Northumberland with the intention of enhancing biodiversity, they killed pigeons, crows, kestrels, tawny owls, short-eared owls, long-eared owls, sparrowhawks, weasels and stoats as well as struggling populations of red squirrels and black grouse despite Natural England and Scottish Natural Heritage according them the highest conservation priority there.[14]

ABOVE:
The red squirrel.
(PHOTO:
N. McINTYRE)

LEFT:
The pine marten.
(PHOTO:
N. McINTYRE)

In a truly natural situation with a full suite of predators typical of grouse range in Scandinavia and North America, it is easy to see how predators live off each other. 'Quarry' species are only a part, often a small one, of the total crop. In situations like that, cropping of prey species by man-the-hunter occurs at low levels over extensive areas, making him merely a part of the suite of predators. The shooter accepts that the birds for which he accounts are, in percentage terms, insignificant against the numbers taken by others despite the fact that in good years, he may kill many birds. He cannot undertake a management regime designed to increase numbers further in the way that is normal in Great Britain. The impact on game populations of this type of shooting is minimal; the hunter is happy to leave matters to nature and accept her bounty. The bag is heavy in some seasons, light in others. The true sportsman accepts disappointment with a shrug of the shoulder.

In Britain, however, the moorland is not natural; for many predators, game birds, particularly grouse, are the most numerous prey available. Prey has been boosted to artificially high levels by habitat management practices designed to increase the numbers of 'prey' birds for sport shooting.

A final point in the keeper/predator debate is the fact that there are only around two hundred full-time grouse keepers on the uplands in Scotland and three hundred and fifty in England. The total area of keepered hill ground within the LFAs is about a million ha in Scotland and 350,000 in England: keepered ground makes up approximately twenty per cent of national moorland.[15] Of the fifteen species of eagle, falcon and hawk native to Britain, only the kestrel is in decline – five species are stable and nine are increasing.[16] As *all* the species perceived as a problem for game management are increasing, it is illogical that no mechanism can be devised whereby keepered moorland can be identified and a management regime devised and implemented which allows the killing of certain species under licence and using approved methods. The measure would at once safeguard the economic base of fragile communities and, because unkeepered ground would continue to be 'left to nature', render the vast majority of upland a 'nature reserve'. In our overcrowded island with its extensive, heavily modified 'natural' habitats, the do-nothing option would quickly be shown up for what it is – a disaster for wildlife. The only way forward to encourage populations of target species, be they lapwings, curlews or blackgame is to adopt a policy of intervention.

The canton concept

Uplands are highly variable; a species, be it predator or game which is common in one locality may be a matter for remark in another. Because Britain is a small country in which management prescriptions often need to be varied to accommodate local circumstances, regulations enshrined in statute as to what can or cannot be killed in the name of game management and applied nationally can be the source of sharp conflicts between shooters and the conservation bloc. Each group must learn to respect better the position and problems of the other. To that end, moorlands could be divided up into cantons and administered under separate rules (bylaws) which answer the shooters' needs and satisfy the objectives of conservation. A solution is urgent because managed moorland and the species it supports are declining fast, particularly in Scotland and largely because of predator pressure. When well over ninety per cent of grouse moors have disappeared within the last sixty years, a rethink is urgent. Those with a strong hostility to field sports may rejoice at this loss but the hard fact is that the areas no longer keepered now support a fraction of the flagship species beloved of bird watchers compared with when they were under competent game management. Social prejudice is a poor engine with which to drive conservation strategy.

One approach could be to designate a number of general areas and, within each, agree on which species should receive positive management with a view to enhancing numbers and which should be controlled in addition to those, such as foxes and crows, which can be controlled by legal means at present. To address those areas where the problems are most acute, an initial grouping could be:

North Highlands – all land north of the Great Glen
    Priorities: golden eagles, merlins, peregrines, pine martens, red kites,
    ground-nesting birds generally; Permit to kill: badgers, buzzards.

Mid-Highlands – Monadhliaths, Cairngorms
Priorities: black grouse, capercaillie, eagles, hares, moorland waders, red grouse; Permit to kill: badgers, buzzards, goshawks, hen harriers, pine martens.

West Perthshire
Priorities: black grouse, eagles, hares, moorland waders, red grouse, red squirrels; Permit to kill: badgers, buzzards, hen harriers, pine martens, ravens.

West Scotland and Hebrides
Priorities: choughs, golden eagles, hares, moorland waders, ravens, sea eagles; Permit to kill: buzzards, hedgehogs on islands.

Lammermuirs, Moorfoots, Lowthers
Priorities: black grouse, red grouse; Permit to kill: badgers, buzzards, hen harriers

People hostile to game management will throw their hands in the air at this proposal but, as long as game shooting is a legitimate activity, its considerable economic and social importance needs to be protected and developed. The present impasse benefits nobody.

Foxes

Their food and spacing behaviour

There are approximately 250,000 adult foxes in Britain.[17] In a 'natural' situation in highland Scotland, they occur at a breeding density of about a pair to 1200 ha (3000 ac) but this is low by lowland standards – Macdonald found a breeding pair to the hundred ha (250 ac) in Oxfordshire.[18] Because density is a function of the year-round availability of food, numbers are lower in the hills where times

The fox is a key predator. (PHOTO: D. KJAER)

of plenty tend to be intermittent. There is usually a supply of deer gralloch from August until Christmas but carrion becomes progressively scarcer thereafter until March/April when deer mortality from natural causes increases. It peaks in May. Thereafter, summer-nesting birds and young mammals are the main food source. The quantity of mid-winter food has a major effect on both the size of the breeding population and how many ova the vixen releases, how many embryos implant and, ultimately, how many cubs a vixen bears and rears.

The fox has to be an opportunist. Perhaps a good year for rabbits or hares follows a hard winter when carrion is abundant or a plentiful rodent year supplies a substantial source of food. Taking one year with another, there is usually enough food to enable roughly the same number of breeding foxes to survive – keepers find the same number of occupied dens each year almost irrespective of how many adults they kill during the rest of the year. When foxes are at low densities and there are many more dens than pairs to occupy them, the same dens tend to be used, suggesting a preferential ranking. This may be important if or when fox numbers start to increase, for example because of neighbouring forestry. Keepers can become complacent; having dealt with the usual dens in April, they may overlook further litters only to have a rude shock in late May or June when cubs are seen outside some hole in an unlikely place, gambolling about among the remains of grouse, mallard and curlews and pieces of lamb and deer calf carcases. Sooner or later, of course, such a den is overlooked all summer and a litter gets away completely. The odd litter within a district can soon supply many extra foxes for the keepers to catch by snare or 'lamp' during the following autumn. The 'odd' litter is all it takes because, when foxes are depressed to low densities by thorough keepering, vixens tend to bear and rear many young. It is not unknown for a hill fox to rear double figures of cubs to maturity in one litter; the cry goes up that a district is 'crawling' with foxes. Indeed it is compared with historical times.

These circumstances generally hold good for land in northeast Scotland over 600 m (2000 ft) without a serious farming interest. The situation changes dramatically on sites overlying richer rocks as their soils support vegetation that encourages high rodent and lagomorph numbers, allied perhaps to more intensive hill agriculture, increased carrion and with tree plantations nearby. Fox populations may ultimately be limited by territorial behaviour between pairs but not, apparently, until densities approach those typical of urban districts. There are fox populations on hill land in Britain that are out of control, particularly in parts of the southern uplands and Wales, to the extent that a dead sheep will be completely scavenged within forty-eight hours and ground-nesting birds rear so few young that breeding numbers dwindle year on year because of a lack of recruitment. Eventually, local extinction occurs. This is not the simplistic scenario described above of the high hills with predator and prey in a rough balance but one where man's influence indirectly boosts the fox population because he creates much of the food supply (sheep carrion and rodents living in young plantations) closely interspersed with plenty of cover in thicket woodland (see Chapter 8, *Fragmentation*). If these land uses of careless farming and tree growing are allied to a reduction in keepering interest, hill land within a whole district soon becomes the victim of a downward spiral. More food means more foxes; more foxes lead to fewer grouse; fewer grouse makes keepering uneconomic; fewer keepers means more foxes. Before long there is no incentive to conserve heather; over-stocking with sheep, over-burning by shepherds and increased tree plantings are the inevitable outcome.

Where sheep have been cleared, land fenced and native woodland created, large populations of microtins usually follow and fox numbers increase accordingly. In large areas in the western half of Britain, keepers regularly kill an adult to 80 ha (200 ac) or less over extensive areas between February and May when foxes are confined to a breeding territory. Such a take suggests breeding densities comparable with the richest of lowland and urban habitats.

Foxes devour almost anything. They eat dead fish on the seashore, raid dustbins behind suburban houses, enter cattle courts and eat recently dropped placentae and uneaten cattle concentrates, scavenge around slaughterhouses in the middle of cities and take whatever they can get whenever they can get it. They can eat a kilo of food at one sitting.[19] A feed of this magnitude twice a week is ample to sustain them though they feed more often if the opportunity arises. Their flexible feeding habits make them capable of living almost anywhere; not so long ago, there was a flourishing population of four hundred-odd individuals in London's docklands (before they became a STOL airport and skyscraper development). There can scarcely be a hectare on mainland Britain which is not visited by a fox at some time of the year.

There have been many misleading statements on foxes and their food. Tables showing the percentage of different foods making up the fox diet over the year can be suspect because they are based on an analysis of scats. What appears in a scat is the indigestible part of what the fox has eaten; an animal eating low-residue food like fish entrails or blubber off a rotting seal on the foreshore is not going to excrete much of a clue as to its last meal. Because foxes are opportunists, they may take a lot of a plentiful and easily caught food, such as pheasant poults near a release pen for perhaps a month or more in July and August. Even if an individual preys on poults as a main source of food for two months or so, a scat analysis may show that there is five per cent or less 'game' in its annual diet because of the general difficulties of finding and collecting scats in summer. The analysis of scats tends to bias the data but the figures are often used to argue incorrectly that foxes do not harm game interests.

A similar argument is used about fox predation on lambs. Studies on the effect of foxes on lamb survival in Lochaber and Sutherland (both marginal districts for lamb production on the open hill), showed that many lambs allegedly killed by foxes were either born dead or were so weak that they died soon after birth from causes such as mis-mothering, chilling and starvation.[20] The conclusion was that foxes were not to blame for lamb losses, that better husbandry and ewe nutrition would reduce lamb mortality and that increased efforts to kill foxes were unnecessary. The conclusions might well have been valid in those study areas but farmers running hill flocks in districts such as the Southern Uplands, where sheep are generally well nourished and husbandry is good, lose lambs when foxes are plentiful and do not lose them when foxes are scarce. There are data to support these observations. In North Ayrshire/Renfrewshire, thirty-seven farmers who between them kept 19,200 breeding ewes (1992) on 22,270 ha (55,000 ac) of hill and upland marginal ground experienced losses attributable to fox predation averaging about three hundred lambs a year (eight per farm, worth about £250 at the time) in the years up to 1985 when few foxes were killed. In 1985, they formed a fox club. Over two hundred adult foxes were accounted for annually between 1985 and 1992 and lamb losses fell to less than twenty per year (<0.5 lambs per farm). During the period, farm ownerships, structures and management did not change materially; the only variable of consequence was that good fox catchers were employed where there had been none before and reduced breeding fox numbers to low levels. The effect on other wildlife was equally significant; mountain hares and blackgame increased and grouse numbers responded to the point where the employment of two full-time game-keepers was justified.

It has been argued that fox predation tends to remove the weaker or less fit members of a population.[21] Studies in Swaledale showed that foxes found incubating grouse hens carrying large burdens of strongyle worms more easily than hens carrying few worms. It was postulated that predation removed birds likely to be poor breeders and a hazard to the population overall and that their loss would have little deleterious effect on the future of the population or on bags the following year. Again, the findings may have had some merit but it would be an imprudent manager who allowed foxes to run riot in 'disease' years in the hope that they would trim out affected birds and ignore 'good' ones. In truth, foxes take what they can find; though they may take more 'bad' birds than 'good', the loss of any 'good' birds in years when the population is experiencing problems with parasites is likely to have a negative effect on its rate of recovery.[22] The Banchory team found over a ten-year study that an average of thirty-eight per cent of radio-tagged hen grouse living on an unkeepered moor were killed by foxes. Some argue that the figure is misleading because loading a bird with a transmitter, however small and light, may disadvantage it and predispose it to being caught.[23] The team certainly found that radios on adult grouse changed their behaviour; cocks became less aggressive and hens moved around less when birds were dispersing in autumn. However, assuming that the figures do more-or-less accurately reflect what happens in nature, nearly four out of ten sitting hens can be lost to foxes on moors where little or no fox control is practised.

The interaction between foxes and grouse

The relationship between foxes and grouse (and moorland wildlife generally) is complex but high numbers of foxes bear heavily on the breeding populations of many moorland species of wide conservation interest, some of them endangered and purportedly enjoying SPA protection.

Consider a theoretical example of a moorland area with a pair of breeding foxes per 800 ha (2000 ac), a density typical of large areas of Scotland and north England. In good years on good ground, grouse are present at a pair per 2 to 3 ha (5 to 8 ac) or better – there are thus between two hundred and fifty and four hundred pairs as potential prey within the breeding territory of the pair. Imagine that they and their cubs have been overlooked and escaped the attentions of the keeper and are preying on this food supply from early April to the end of June. If the foxes take one hen, her eggs and chicks every day for that ninety-day period, they are removing somewhere between a third and a quarter of the standing crop of breeding birds. If it was a 'good' year with survivors averaging five fledged young per hen, there would be between 1120 and 2170 birds on the 800 ha. If half of these are shot – a possibility given good shooting manage-ment and marksmanship – the bag would be 280 to 540 birds per 400 ha (1000 ac), a moderate to high yield and a good result. The effect of the foxes on the bag would have been small.[24]

Descend the scale a bit to an area of similar size where grazing control is poor and burning unsympathetic to grouse. A pair to 8 ha (20 ac) is a reasonable spring stock and presents our resident foxes with a hundred pairs of birds. Though it is unlikely that the foxes would succeed in finding ninety hens out of this reduced population, anecdotal evidence from many sources suggests that they hunt grouse selectively where other prey is scarce. It is not fanciful to predict that sixty hens are killed by a persistent pair of foxes which have become grouse specialists. That leaves an August stock of three hundred birds (forty pairs, two hundred young at five per breeding pair and twenty single cocks), a poor prospect for the sportsman. In this plausible case, foxes have made the difference between what would have been a modest driving moor and one with barely enough birds to stand a day or so walking in line or shooting over dogs.

Continuing the descent down the scale, we come to the typical 'dogging' ground of north and west Scotland. A spring density of a pair of grouse to 20 ha (50 ac) can be reckoned pretty good as it results in our 800 ha moor carrying forty pairs in spring. If thirty hens are taken by foxes, only ten coveys are left by August. In a good year that saw each hen rear five young (a high figure for ground of this nature), the August standing crop is at most one hundred and twenty birds – twenty pairs, fifty young and thirty single cocks. Eighty of these are required for stock leaving forty birds for the sportsman to shoot.

The calculations take no account of losses to raptors, stoats, crows, nest flooding, disease, etc. and assume an above-average breeding performance but they serve to illustrate the impact that foxes can have, particularly on low grouse-breeding densities. It is apparent that the impact of foxes on grouse (in fact of any predator on prey) is a function of grouse density – thriving populations on quality habitat are better placed to resist predation. The higher the population, the less effect predators have on it and thus on the shooting results the following season. Where predators are prevalent and grouse numbers are allowed to fall below reasonably high levels, the situation can quickly become parlous; one bad production year can push a respectable moor into the vulnerable category. If the following year's breeding is not good enough to prompt a population recovery, numbers plummet. As shooting is not possible, a hard-pressed management may decide it can no longer justify keepering (see Chapter 4, *Keepers and keepering*). Making a keeper redundant may solve a cash flow problem but it sweeps longer-term problems under the carpet – no keepering means more foxes; more foxes mean a higher rate of loss of breeding birds. A respectable middle-grade grouse moor can be written off in the space of two to three years with serious consequences for neighbouring moorlands which suffer a domino effect.

Fox control methods

When keepers were plentiful in the nineteenth century, foxes were scarce and not ranked as serious predators. It was the sheep farmers as much as the grouse keepers who kept up the pressure on them. Many landowners were actually quite tolerant, feeling that their depredations were exaggerated but when numbers collapsed in both Scandinavia and parts of Britain, following outbreaks of sarcoptic mange, the response in small game numbers was marked.[25] On many moorland areas outside the Highlands, foxes were almost unknown. The author grew up on an estate in Ayrshire; one of his earliest recollections is being taken aged two to see a dead fox laid out in the yard at the back of the house – it was such a rarity that outside workers and indoor staff alike came to have a look. In the twenty-first century, over four hundred adult foxes are killed annually within twenty miles of that estate. In former times, the keepers often had the assistance of the local 'tod hunter', a man who hunted foxes with a bobbery pack of collies, hounds as well as terriers which dealt with those which had gone to ground. (Their romantic image made them a common subject for Victorian artists.)

Other control methods were widely used; strychnine was sprinkled on carrion like salt, killing many crows, ravens and eagles; the gin trap was used to catch adult foxes and cubs not only at dens but also at water holes or 'drowning sets' – a method widely employed within living memory.[26] When the gin trap was made illegal in England in 1958 and Scotland in 1971, many keepers felt they had lost a limb. The wire snare took its place although it was initially regarded as of limited use on the open hill where there was a risk of catching sheep or deer. At about that time, people started using powerful spotlights and centre-fire rifles from vehicles; foxes could be called within range at any time of year. The snare and the lamp, coupled with the traditional round of the dens in spring, served well as long as foxes were not too numerous but the 1970s population explosion, which was associated with the expansion of large-scale afforestation, was to expose existing techniques as wanting. Unreliable marksmanship made foxes lamp shy and, though 'lamping' still accounted for many animals, the percentage of the population killed started to decline. In the mid-1980s it was common to find keepers saying that they'd 'been out with the lamp, seen seven foxes and got four', news aired with a note of self-congratulation. Had they stopped to think about it, it would have been obvious that, if seven were seen, others were not – four out of seven may sound better than four out of ten but the latter figure probably reflected reality better. People who relied too heavily on lamping soon found they were over-run in the spring and that lamb and grouse losses were unacceptably high.

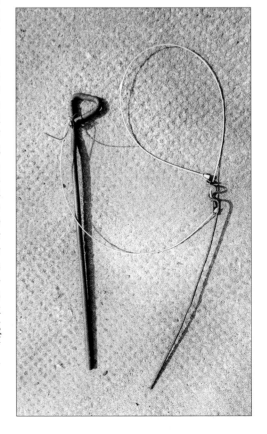

It is essential that the man on the rifle is an expert shot, not someone who just thinks he is. Everyone misses from time to time but the missed fox rarely stands for the light a second time. To get it, keepers may have to spend many cold, boring nights at a den in the spring, waiting, often in vain, for the lamp-shy animal to come within range. It took Robert (Rab) Kerr, a Dumfriesshire keeper, to show the way forward. During the 1970s, he worked at Glenlee, New Galloway where there were many foxes and extensive tree plantations around his beat. He it was who developed the technique of catching foxes in snares at middens, graves or stink pits (the terminology varies according to district) which consist of dumps of

A simple fox snare with stop, a steel anchor peg and tealer made from scrap fence wire.

carrion, deer gralloch, salmon kelts, dead foxes and so on. Today, the law requires that the rubbish is buried but incorporating a vent pipe allows the all-important smell to escape. During his nine years at Glenlee, Rab caught over eight hundred adult foxes and built up an estimable wild pheasant shoot before moving to a grouse beat at Drumlanrig in 1984. Since then, the method has become widespread throughout hill ground in Britain. It is so efficient that many keepers find the numbers of occupied dens in spring are comparable with those of fifty years ago. In many places, lamping has become a secondary technique although it is still useful for checking whether foxes are moving on the ground.

The midden technique operates on the same principle as the old drowning set which worked because foxes like to investigate dumps of carrion. The difference is that middens are used primarily at the bottom of the hill, ideally in a bed of rushes or patch of thicket woodland, whereas the drowning sets were usually located in bog pools on the hill tops. Middens can be used on ground grazed by sheep provided the snares are fitted with an efficient stop to prevent them closing on a sheep's foot. If hill farmers are concerned about the risk of catching sheep or lambs, the snares are equally efficient if the site is fenced with large, square-mesh netting of the Rylock type. They work best if baits are varied during the winter; nothing beats adding a dead salmon kelt or a bag of fish heads to the carrion pile early in the new year to rejuvenate a site that has been successful in the autumn but has gone quiet. A dead fox of either sex is a most useful addition as the year goes on – the smell of another fox appears to be almost irresistible to a travelling animal in search of others but a 'brimming' or clicketing vixen is worth her weight in gold as she is an irresistible draw to any travelling dog fox which comes within smelling range.

For the technique to be legal, three points must be observed. First, it is an offence under the Dogs Act, 1906 and successive Animal By-products Orders up to 1069/2009/EC, to leave the carcase of a domestic animal unburied. A dead sheep cannot be left on the surface; turves and moss must be thrown on top of it so it is out of reach of scavenging animals. (The regulations do not apply to wild animal remains so deer gralloch, sea fish offal, rabbits and game entrails do not need to be buried.) Second, under the Wildlife and Countryside Act, 1981 (largely superseded in Scotland by the Wildlife and Natural Environment Act, 2011), it is illegal to use a locking snare and all snares, not only those set in middens, must be inspected at least once every twenty-four hours. The keeper may find himself in a dilemma because, if he walks all over a midden every day, he catches little. However, with a little ingenuity he should be able to site catching places in locations where they can be overlooked from a distance, perhaps with the aid of field glasses. Third, it became law in 2011 that a snare has to carry an identification tag so the person setting it can be held responsible for its correct use.

Although many methods of taking predators used by earlier generations of keepers are now proscribed – and rightly so – fox catching is one instance where a modern development has proved at least as effective as methods of yesteryear that are now illegal. A comprehensive network of middens takes out almost all resident foxes before the end of March and makes doing the dens in April much less arduous for the keeper. That said, there are some situations where middens do not work well. They are at their best where foxes are hungry and numerous in relation to their prey base. Where they are few and, for example, rabbits, white hares or released pheasants are plentiful, results can be disappointing though cases such as these can be dealt with easily enough by conscientious work at the dens in springtime.

Finally, locking snares are illegal. Unfortunately, whoever drafted the legislation failed to heed practical advice – a free-running snare, the only legal design, is not always efficient at holding a fox. If an animal is tentative and, having put its head in the noose, finds itself held, it will often freeze. The natural spring in the wire can cause the snare to open, the snare falls off and the animal walks free. With a locking snare that cannot happen.

The most efficient snare is a locker with a stop placed in such a way that it cannot close up completely – whether it is triggered by a fox, a sheep's leg or a protected animal. A 'locker' with a stop is a holding snare only and cannot kill its captive, however hard it may struggle. The free runner, on the other hand, cannot be fitted with an *efficient* stop because, if it has one and the animal stops pulling against the fastening, the snare is likely to open. It may injure the legs of domestic and wild alike and is likely quickly to kill any animal it catches – in the event of a protected mammal accidentally getting into such a snare, it is in danger of being killed. The snaring of foxes is under constant scrutiny with evidence being systematically collected by those hostile to the practice to try to demonstrate that protected animals are being killed; pressure continues to mount to outlaw snaring altogether. Anyone who doubts this should refer to the Holyrood debates preceding the passing of the 2011 Act. A substantial body of opinion irrevocably opposed to snaring will not rest until it is made illegal.[27] Should this happen, the entire basis of upland management as presently conducted would be jeopardised; the present number of keepers, shepherds and fox catchers would never be able to control foxes to a level where upland game and other wildlife would survive in quantity without the wire snare. Not only would quarry species decline but moorland waders and passerines would be reduced to levels lower than those of today with a knock-on effect of reductions in the raptor prey base. Lamb losses in many hill districts would also be likely to rise, jeopardising the viability of farm communities.

To safeguard snaring, the simplest of amendments to the legislation is required, legalising locking snares provided that they are fitted with an efficient stop to prevent them closing to less than a stated circumference (25 cm/10 in). Not only would keepers and farmers be keen to accept the change but a major animal welfare criticism of snaring would be removed at a stroke.

In one respect, fox control in the twentieth century has come full circle. In former times, the tod hunter and his assortment of hounds was a valued member of the rural community. Specialist dogs are still widely used today for fox control. Some keepers, taking a leaf out of an ancient book, take loose-running foxhounds and lurchers with them when going their rounds, particularly in summer. People can be found doing this from Inverness-shire to Yorkshire. A head keeper from Co. Durham, now retired, tells how the two hounds he had with him at all times 'were worth another beat keeper' because they were so quick to pick up a stale drag, follow it and put the fox to ground.[28] Hounds which have served their time as members of a pack of foxhounds are cheap to keep and safe with sheep.

There are ten traditional foxhound packs in Scotland. Four foot-packs in the Highlands – the Argyll fox control association, Atholl & Breadalbane, Lochaber & Sunart and the Three Straths – hunt extensive forestry blocks and rough ground on foot to posted guns and rifles with considerable success. The activity is regarded as outwith the scope of the Protection of Wild Mammals (Scotland) Act, 2002 insofar as it relates to the hunting of wild animals with 'dogs'.

Fox clubs

It has been recognised for many years that isolated moors are always more at risk from high predator pressure than those surrounded by keepered ground. (See Chapter 4, *The sink effect*). Because of changes in land use and the resulting fragmentation there are now many more island moors than there were meaning that a keeper often finds himself in charge of an area which has no keepered ground round about. He is always going to have a more difficult time keeping on top of the common predators than a colleague located in the middle of a keepered block. One method of addressing this problem in Scotland is to form a fox club approved by the Scottish Government Rural Payments and Inspections Directorate. It has to be spearheaded by farmers whose sheep graze the hill land and who have much to gain if foxes are scarce. There are thirty-one fox control associations in Scotland (2011).[29] Many farmers become keen on catching foxes once they are satisfied that the midden technique works and have learned the necessary skills. For a hard-pressed keeper looking after an extensive acreage single-handed, the help is invaluable. Farmers can set up catching places and service them while on their daily livestock rounds. Even if the keeper has to devote a day or two helping them to set up middens in autumn, it is time well spent. A club with a proper constitution and budget can attract discretionary grants for costs incurred over part of the winter for items such as vehicle costs, snares, dog subsistence and rifle ammunition. Where farmers get practical support for a modest annual subscription usually based on the number of sheep they keep, they become enthusiastic supporters of do-it-yourself keepering. It is crucial not to lose sight of the fact that the club exists for the systematic destruction of foxes, something that can be used to cajole support from a reluctant farmer who occupies a farm in a crucial location. Middens should be plotted on a map and sited so that no catching place is further than a mile or so from the next one.

Badgers – the unacknowledged menace

Badger cubs emerging from a sett. (PHOTO: D. KJAER)

For decades, badgers have been regarded by the general public as a species which deserves sympathy and protection. In major part, the sentiment stems from the brutal exploitation of wild-caught and often disabled animals by criminals who, with bestial cruelty, test the ferocity of their dogs against a courageous animal which is always the loser in such encounters. Another reason is that 'Badger' is seen by generations of children as the bumbling, loveable, harmless, bear-like and kindly creature of an enduringly popular nursery book (with an implausible entourage of rat, mole and toad, not to mention two young hedgehogs as servants!)[30] and as a benign character on children's television. The third reason is a matter of concern: the animals' undoubted role in maintaining and spreading bovine tuberculosis and the passionate debate that local control engenders. TB not only has serious implications for the economic welfare of farmers in districts where cattle herds react positively to mandatory tests but is also a human health hazard. Even though badgers have statutory protection, to oppose killing them because human welfare requires it seems almost to outstrip rational thought. Nevertheless, objectors lobby politicians who, especially those with an urban background or representing an urban community, tend to give in to those who oppose culls on grounds of political expediency. Vigilant people also regularly monitor setts to detect any interference with the animals; woe betide the farmer who puts a lamb-killing fox to ground in a badger sett and goes after it. In no time, he is apprehended by police and brought before a magistrate.

As a direct result of their protection, badger numbers have soared and reached saturation point in many locations in England. Over the past fifty years, badgers have achieved hallowed status and their numbers have responded accordingly. The population is no longer cropped for the durable hair which was sought after for the manufacture of shaving brushes. In the middle of the twentieth century, there was scarcely an affluent person in the land who did not use such a brush daily. Now nothing but the ubiquitous motor car dishes out sudden death to them. They have taken over huge tracts of Britain as the chief predator and scavenger. Traditional fox earths which are the first port of call in spring for many highland keepers are found routinely to be occupied by badgers. In many districts, their numbers are so high that it is common to find them foraging in broad daylight. Badgers are omnivorous and are as likely to be recorded inverting cowpats in their quest for earthworms on a Somerset dairy farm as they are to be scavenging the carcase of a dead seal on the shore on a highland sea loch or searching every last piece of potential nesting cover for a partridge nest on the Wiltshire downs. Although rare on moorland in northeast England, they are nevertheless recorded as predators of grouse nests[31]. If they are doing that, they are also inevitably predating other species. Although normally nocturnal, the marks that they make when foraging for invertebrates, airing their bedding or digging their latrines are everywhere and give away their presence. They are implicated in the scarcity of hedgehogs in southern England and declines in populations of solitary bees. In northwest Sutherland, animals living in setts on moorland within easy reach of the coast are responsible for the extermination of local breeding populations of shelducks in the nearby sand dunes over the past twenty years and snuffle up wader chicks efficiently when foraging on the sea shore.[32] On a Speyside estate, a contract fox controller employed to help protect endangered capercaillie encountered protected badgers on occasions too numerous to record when lamping. Sooner or later, those in power are going to have to address the issue of ground predators in the round and managers will have to manage the effect of the spectrum of predators on chosen species in steep decline. The badger has to take its share of the blame for this. In the informed opinion of many, badgers are heavily implicated in the widespread decline of many ground-nesting birds and are threatening the very survival of some scarce and endangered species in both the uplands and lowlands of Britain.

Gulls

Black-backed and herring

The three species (greater and lesser black-backed, herring) are powerful scavengers and predators and some of the largest birds around; the greater black-backed gull stands 70 cm (2 ft 6 in) tall, the herring gull about 10 cm (4 in) shorter. People in farming districts often refer to them as 'meat' gulls because of their fondness for fish offal, carrion and garbage and for a well-developed propensity to attack any prey they reckon they can overpower. In coastal districts, they frequent lambing parks in the certainty of finding placentae or the hope of finding a sickly lamb or 'couped' ewe which will rapidly be blinded and pecked to pieces. All three species forage at land-fill sites and increasingly conflict with man on account of a growing habit of colonial nesting on roofs and ledges in cities. In summer, they forage on hill land, exploiting up-draughts and gliding slowly along hillsides in almost effortless flight, scanning to right to left as they go. The young of all ground-nesting birds are at risk as well as juvenile hares. The balance of power between grouse and gull is finely balanced; if a black-back launches an attack, it is likely to be successful but a fit cock grouse will stand up to a herring gull. It is not uncommon to see a grouse take the initiative and attack a questing gull which will usually back off. If the grouse has young and sets off in pursuit, the gull normally beats a hasty (and noisy) retreat.[33]

As predators, they are hard to control because the only legal method of killing the adults is to lie up and shoot them on their flight lines, which are often well-developed and used daily. English keepers do better here than Scottish ones because there are so many more of them. Three or four men concealed in a line and spaced fifty metres apart are much more likely to succeed than a lone keeper; it is hard for a man single-handedly responsible for 4000 ha (10,000 ac) to set an effective ambush. In Scotland, gull depredations on moorland often go unchecked.

Common and black-headed

In summer, both species are primarily birds of the fields and moors and nest colonially. Occasionally their nests are mixed up together.[34] The birds are often found nesting on bogs and lochs surrounded by grouse ground. A few larger colonies may consist of several hundred pairs but many are much smaller. Over the past forty years, many gulleries have been abandoned; it is noticeable that those located on lochs without islands are the most likely to disappear. Some bogs holding a hundred or more pairs have lost the colony completely in the course of two or three years, invariably to fox predation. Even colonies on sites where the birds have been confined to islands have gone over the past seventy years. One can only conclude that their progressive disappearance is related to a decline in the biomass of invertebrates produced by both the hill land and the water over which they forage. Land use changes may be involved as well; where cultivation is reduced, food availability in spring tends to fall. Spring ploughing may have been important in that it provided a large supply of easily harvested food in the run up to egg laying in early May. The change in numbers may also be influenced by the general increase in otter populations as they are perfectly capable of predating eggs and fledgling birds in an island colony.

The general increase in otters may be affecting the breeding success of birds using littoral zones such as divers and Slavonian grebes. (PHOTO: D. KJAER)

As grouse predators, these gulls are unimportant. This is not to say that a sickly chick being reared by parents in poor condition will not be at risk; there are eye witness accounts of gulls taking such chicks but they cannot be blamed for grouse declines in the vicinity of gulleries. Grouse have often been observed successfully incubating their clutches while surrounded by nesting black-headed gulls.[35] Provided there is no antipathy between grouse and gull, it is advantageous for the grouse because the mob rule which prevails in a gull colony keeps crows away.

Hooded and carrion crows

Most moor keepers dislike crows even more than foxes. Whether of the carrion or hooded race, crows are conspicuous in their behaviour and leave obvious signs of the damage they cause; the skulking fox may go unnoticed but no countryman about at dawn or dusk in spring can miss the raucous call of the crow advertising a territorial presence from tree tops or other eminences. Though crows can be embarrassing for the keeper, comparing their predatory effects on grouse with that of foxes is scarcely to compare like with like. Crows may take many eggs and even chicks in spring but they do not kill adult birds – hen grouse usually re-nest soon after losing a first clutch. In years when the hen is in good order, a second brood may be almost indistinguishable in size from a first one by mid-August. A study on an island in north Norway involved the killing of all the breeding crows on one half of the island throughout the spring but leaving them undisturbed on the other where they took many willow ptarmigan eggs. The numbers of young willow ptarmigan fledged per hen did not differ significantly between two sites. Where the crows were allowed to remain, the ptarmigan re-nested and, although the chicks obviously hatched later, there were as many of them in September as on the area where the crows were all killed in spring. No effect on breeding density in the subsequent year could be attributed to the predation either.[36] Workers at Banchory showed that crows spend many hours sitting motionless on a vantage point watching the behaviour of potential prey. Immature non-breeders which stayed in flocks in spring did not go 'egging' and some individuals which held territory did not spend time seeking eggs and nests. This confirms that nest finding is a learned skill and that older crows are likely to be more damaging to nests than first time breeders. It follows that moors thoroughly purged of crows every year are less likely to suffer nest losses in the event of the odd breeding pair being overlooked by the keeper in April.

Crows also hunt for eggs and squatting chicks by flying slowly at a low height above the moor. In an elegant trial designed to test how successful birds were at nest finding, dummy grouse nests were made up using artificially blotched bantam eggs. To enable sites to be re-visited, the position of each nest was identified with a bamboo cane aligned on a compass bearing a standard distance away.[37] Within a few days, all the marked nests were lost to crows; they had cracked the code. It is not fanciful to assume that the regular presence of a pair of grouse in a certain small area of the moor in spring sends a signal to a local pair of crows of the likelihood of a nest nearby. The nest-finding skills of individuals were shown in the same trial by observing marked crows. Some birds appeared to be specialist nest-finders.

A multi-catcher crow trap, Argyll 1968. The trap would be illegal today because the sheep carcase is open to the air.

Some will ask 'why kill the crows at all if re-nesting results in little change in ultimate brood size?' The answer is that managers want to produce as many full-grown young grouse as possible for the start of shooting on 12 August. Late birds may strengthen the shooting in September but owners would rather have them taking an active part in the programme in August as well as later on. This on its own is reason enough to kill crows, particularly as certain individuals in any population kill grouse chicks. In years when grouse are not doing well because of poor hen nutrition or 'disease', the debilitated hen that loses her first clutch often does not have enough energy to produce a second one. It may be that one clutch is all she can manage and that, having failed with that, she gives up any intention of breeding in an attempt to ensure her own survival. If plant growth in June is poor and the weather cool and wet, the viability of and prospects for a second clutch might always be poor but the manager would rather let the hen lose eggs or young through her own failing rather than have any potential for recruitment, however limited, pre-empted by a crow. The Banchory team showed that the effect of crows was more significant in years when birds were doing well than when they were doing badly as the eggs they took in poor years contained chicks that were less likely to survive to adulthood anyway. However, as the manager needs all the grouse he can get in bad years, achieving an average of one extra fledged chick per brood is worthwhile if it means that the total population is larger in the autumn. In any case, the conscientious keeper is proud of his moorland birds be they quarry species or not and rejoices to see good broods of curlews, golden plovers, lapwings, dunlins, dotterels, common sandpipers, snipe and so on fledging on his ground. This is much more likely to happen if the hill pastures, moor edges and hill tops are kept free of crows in spring and early summer.

The status of crows

After rooks and wood pigeons, crows are probably our third most common avian species over large parts of the country. To drive around unkeepered countryside in early April is upsetting for most keepers or sportsmen. In pastoral areas, densities of a breeding pair to 10 ha (25 ac) are commonplace and the birds are conspicuous if the woods and hedgerow trees are mainly deciduous, as is typical of lowland Britain. Because there are so many, it is reasonable to pose the question 'what are they living on?' The answer lies yet again in the way that man influences the environment; lowland agricultural practices tend to maintain crows at high densities. Extensive areas of Britain are today more fertile than they have ever been, something which helps all corvids. Intensive grassland management involves heavy doses of nitrogenous fertiliser which increase the availability of soil invertebrates. Slurry spread on grass fields supplies undigested grain from cattle and pig yards and it also contributes to increasing invertebrate numbers. Permanent grass pastures may not make up a high proportion of British grasslands but they support abundant populations of flies, spiders and beetles. Food spilled at sheep troughs, feed blocks and cobs provides food during the winter and road kills can be plentiful locally. In March and April, many districts provide copious amounts of carrion in lambing fields. From early June onwards, modern early-maturing varieties of winter-sown cereals with panicles entering the 'milk' growth stage provide nourishment and moisture at what can be a critically dry time of the year. This aids the survival of newly fledged young birds. Farmers increasing their self-sufficiency grow these crops on land running right up to the moor edge. Late-summer outbreaks of myxomatosis deliver many disabled rabbits. All these factors alongside reduced keepering effort on many estates have contributed to this increase of crows. Many crows never see a bird's egg in their lives but some territorial birds can be hard on hedgerow-nesting species and can impact severely on ground-nesting birds. Crows are a serious hazard to rare species; observers and scientific workers alike have noted the effects on waders when a nearby woodland or forest contains breeding crows.[38]

Elsewhere, problems caused by crows stem directly from man-the-tourist with places like the Cairngorm plateau and Ben Lawers (popular with picnickers in summer) attracting and supporting flocks of immature birds and so indirectly increasing predator pressure on species nesting locally such as ptarmigan and dotterel. From the manager's point of view, the importance of flock crows is that they provide a large surplus each year which disperses into areas where crows have been systematically controlled; it is doubtful if there is ever such a shortage of over-wintering birds that the spring breeding population is limited. Crows are little affected by hard winters which tend to increase the mortality of other species and so make more food available. The manager has to accept that all he is likely to achieve with his spring offensive is a reduction in breeding numbers that lasts long enough to let his grouse hens hatch their eggs and get the chicks started. Fortunately, this is all he needs to do. Once chicks are hatched, the parents have less to fear from crows although parents in poor condition do not defend their young with much enthusiasm; a bold crow will take grouse chicks up to three weeks old.[39]

Timing the control effort

Crow breeding is synchronised with the breeding cycles of potential prey; most crows start laying towards the end of March and are usually sitting by mid-April. Their young, hatching towards the end of the month, can be fed on the plentiful supplies of carrion from lambing fields as well as on the eggs of upland birds. If the pair bonds are disrupted by the killing of at least one parent and as many crow nests as possible are destroyed in early April, re-nesting does not get underway until early to mid May. By the time these crow eggs are hatched and the chicks are demanding large amounts of food, most grouse and waders have hatched and the worst of the danger is over. Desynchronising the crows' breeding cycle is a sophisticated way of helping prey species get through the vulnerable incubation period. The effect is not confined to grouse and ptarmigan but holds good for all the hill-nesting waders such as dunlin, golden and green plover, dotterel, redshank, greenshank, curlew, oystercatcher and common snipe. As a solution, it may not appeal to a hardline keeper to whom the presence of any crow on the moor or moor edge in summer is anathema but it is a useful tactic for the overworked man responsible for a big acreage.

Crow control methods

Legal methods of crow control are confined to shooting, nest destruction and live trapping in multi-catch cage traps and the more portable Larsen cage (the latter developed by a Danish keeper of that name). Until recently, poisons and narcotics were widely used even though they were (and of course still are) illegal. Most keepers devote maximum effort to crows in spring when territorial birds

Hooded and carrion crows are both easily taken in Larsen traps, and one species will decoy the other.

are preparing to nest and raise their young. Methods vary according to preference but most conscientious keepers run a network of large cage traps in the winter and early spring in an attempt to reduce the overall population to manageable proportions and then clean up the territorial crows using Larsen traps in March and April before the grouse start to nest.

Multi-catcher cages and Larsen traps

Large cage traps are easy to make at home and catch well if certain principles are followed. Pre-baiting is good practice; the roof with its funnel entrance should be left off for several days and food placed both inside and outside the trap. A high T-shaped perch should be mounted on one corner; perches inside the cage enable captives to display themselves well and so be more visible. Cages should have a ground-entry funnel as well as a top one as many crows walk round and round the perimeter before trying to enter. Some keepers fear they may catch game but, as it is obligatory to check the cage daily, the risk is more imagined than real. Cages should be made in sections so that they can be dismantled and moved. Ringing the changes in terms of site and bait increases catches. Most crows caught in big cages are flock birds of low social status that, if left alone, would probably not get breeding territories that year and so have only a small potential to damage nesting birds. Large cages can catch territorial birds but, as the effort of moving them from one territory to the next is formidable, it may not be possible to put one close enough to the centre of a bird's nesting territory to be sure of catching it quickly.

As one crow looks very much like another, decoy birds should carry a ring on one leg to make them easily recognisable. Crows caught in a multi-cage should be despatched after dark with the minimum of fuss and disturbance as their distress calls can alert others some distance away. Crows are intelligent and soon come to associate cages with danger unless this is done. Because they eat a wide variety of food, baits are best selected according to what they are eating at the time. During winter and early spring, carrion is good but baiting a cage with a domestic animal carcase is illegal. By the end of March, hens' eggs are useful; in the vicinity of places such as country parks, picnic sites and so on used by many people, slices of white bread with pieces of sandwich paper attached work well. Flock crows using sheep and cattle food will draw to pellets or cobs.

A word of caution about large cage traps. For reasons which are obscure, birds of prey ranging from eagles to kestrels sometimes find their way into them even when they contain no decoy bird. Once the crow-catching season is over in May, it is important to make sure that the door is tied back securely and the funnels are closed so that nothing can be caught inadvertently.

Towards the latter end of the winter catching period, attention should be paid to collecting a stock of vigorous birds and keeping them in a large aviary. They will be needed as decoys for Larsen traps. As decoy birds are valuable, they should be looked after properly on practical as well as humane grounds. If treated well, they soon become semi-domesticated and, if fed suitable fresh food to appetite, will work for the keeper for the whole season. If one is lucky enough to have an aggressive bird which is a good caller, the target bird immediately tries to evict what it sees as an intruder and so gets caught quickly – sometimes in as little as five minutes.

Widespread use of the Larsen trap has greatly assisted keepers as it is possible to transport one fully assembled in the back of a vehicle. It is good for taking territorial birds and a worthy successor to the old-fashioned but illegal system of open traps placed beside a caged decoy bird close to an occupied nest. The law is explicit on the care of captives: they must not be tethered or injured, must be in a cage large enough for them to stand upright and fully extend their wings and provided with food, water and shelter. Although not a legal requirement, they appreciate a perch to stand on – contented crows do their job best.

The double-catching Larsen trap is light and easily carried. It needs no bait as the decoy bird is the attractant. With Judas in the central compartment, the trap should be placed close to perch trees or lookout points where calling crows are seen regularly. If nothing happens within twenty-four hours, the trap should be moved. There are prime catching places in every crow territory which work every year but it takes trial and error to find them. The circular Larsen, although more cumbersome to move about than the double catcher, has three side entrances and is effective in catching the occasional bird which is reluctant to enter through the top funnel. In former times, once the hens were sitting, keepers would stalk and shoot them off their nests or hide close by and wait for the sitting hen to return but this is no longer necessary with Larsen traps. If the trap is placed within sight of the sitting bird, one or other of the pair will be caught quickly.

Once the crows start to nest, the observant keeper has to spend time diligently searching stream sides where isolated trees occur as well as watching large woods within hunting range of the moor. Though time consuming, it has to be done properly if serious damage resulting from an overlooked pair is to be avoided. A large-scale map is useful for plotting the locations of all nesting pairs; as crows tend to space themselves evenly over an area, a gap in the map might suggest the presence of a pair that has been overlooked.

The law as it affects the use of crow traps

With effect from 1 April 2008, would-be users of crow traps must apply to the police for a general licence under the Wildlife and Countryside Act, 1981 and attach a tag to traps to identify the user and provide a police contact number. The police record the name and address of the user and allocate a licence number which is written on the tag. There is no standard form of tag but suitable examples can be obtained from the National Gamekeepers' Association, the Scottish Gamekeepers' Association or BASC. Once issued the licence is valid until it is no longer required.

Roost shooting

Crows roost communally in winter. Large numbers can sometimes be shot as they flight in for the night particularly in wild weather but that tactic suffers from the same criticism as devoting time to catching flock birds in cage traps in winter. Removing large numbers during the short winter days simply improves the survival chances of the remainder and, as many crows are migratory, local dividends tend to be small.

Poisons and narcotics

For many years, employers turned a blind eye to grouse keepers using poisons and narcotics despite their having been illegal since 1954. Their use gave, and by association still gives, game shooting a bad image. This is a pity as there is little doubt that the narcotic alpha-chloralose applied on bread specifically for killing crows and gulls is both humane and safe where other species, particularly protected ones, are concerned. When pressed, both RSPB and ITE are bound to agree with this as both have used it for controlling egg predators in the past and the technique is still used for dealing with pests such as feral pigeons in urban locations.[40] The controversy has always stemmed from keepers and farmers putting out meat baits designed to kill foxes or crows. Meat baits used in this way are indiscriminate and responsible for killing eagles and buzzards every year and putting the scavenging red kite, the subject of regional reintroduction programmes, at risk. Forty years ago, DAFS staff held informal discussions (with the author among the parties) and suggested that the law should be altered to provide a spring 'window' – say 15 March to 15 April – during which egg baits in numbers reflecting the size of the land holding could be bought from licensed suppliers such as DAFS regional offices or county veterinary laboratories. The eggs would have contained an accurately measured dose of alpha-chloralose sufficient to kill a crow. The purchase of a hundred eggs per 400 ha (1000 ac) of moorland would have enabled crow numbers to be depressed to low levels for the crucial period when hill birds (not only grouse) are nesting. Then and now, opposition never came from landowners or farmers but from those opposed to any use of 'poison' in the countryside even though the technique could be licensed just as readily as snares and crow cages. Professional gamekeepers do not have cause to use eggs in this way. They can control their crows by trapping but areas under the 'management' of conservation bodies would cease to be the corvid reservoirs they are today if those in charge, who could surely be trusted not to abuse the technique, were to conduct a systematic crow control campaign in spring. SNH encourages the private sector to kill crows in the name of conservation but does not appear to confront the issue with the same enthusiasm in public or semi-public places where the management is under its influence. Such a campaign would not only make a positive contribution to biodiversity over extensive areas but would also win them many friends in the private sector. If seasonal use of the material were to be permitted, the illegal use of poisons and narcotics would decline to the point of extinction and game managers would be unwilling to risk severe penalties for possession of alpha-chloralose or organo-phosphatic chemicals for which they had no legitimate use.

Ravens, jackdaws and rooks

The author is unaware of any published data on egg and chick losses to jackdaws and rooks in upland bird populations though there is plenty in the literature on damage to pheasants and partridges in arable areas. Rooks often flock to the moor edge in June and come under suspicion as a result. Usually, they are foraging for leatherjackets (crane fly larvae) but two nineteenth century observers reckoned that rooks and jackdaws both took grouse eggs, particularly in a dry year.[41]

Ravens are a different case because they are protected and because their numbers and breeding success are related closely to the amount of available carrion within their range. Ravens have declined in many parts of Britain not because of persecution but because sheep have been cleared and sheep walks afforested. In other parts, they are plentiful to the point of being a farm pest. Their situation is anomalous in that they enjoy full protection under the WANE Act, 2011 but suffer adversely if conscientious shepherds observe the 1906 Dogs Act and the Animal By-products Orders. Where the standard of sheep husbandry and sanitation on the hill is high, ravens are scarce. If they are to be encouraged, they need access to a plentiful, year-round supply of dead animals if they are not to conflict with shooting interests. To most hill people, 'the bird of ill omen' is at least greeted with respect if it is not actively welcomed. The wide range of their calls and the astonishing aerial acrobatics they seem to do for the sheer fun of it always make people pause, look up and marvel. They may take an odd grouse nest or kill the occasional lamb or deer calf but, by and large, they have small economic effect. It is hard to find many people with strongly entrenched feelings against ravens though it has to be admitted that, on a driving day, grouse are restless and difficult to control if over-flown by them; the silhouette of the bird is quite raptor-like, particularly if it is flying in a strong wind but that is weak evidence for making out any case against them.

Raptors

Sportsmen often see themselves in competition with hawks, falcons and owls, feeling that predators taking their quarry are taking birds that are rightly theirs and would be theirs to shoot had the predators not got there first. However, raptors must be put in context with everything else that affects the sportsman's bag. Legal issues aside, the effect that raptors have on grouse populations is often less than commonly thought. It is a mistake to lump all raptors together because different species have different effects according to different sets of circumstances. Individual hawks or falcons often become specialists and hence prey-specific; they may have little economic effect on an estate if they specialise in hunting lapwings or young mountain hares but it can be a different story if grouse or grey partridges are their preferred prey species.

Buzzards

The buzzard is Britain's commonest raptor with a breeding population reckoned to be somewhere between 31,000 and 44,000 pairs.[42] After juveniles have fledged at the end of summer, there are perhaps 250,000 individuals on the wing nation-wide. Breeding in every county in Britain, they are a common sight with significant impact on the national fauna. Strong enough to tackle live prey up to 500 gm, their main foods are said to be rodents, reptiles, amphibians, larger insects (species undefined) and earthworms. Carrion is also important; outbreaks of myxomatosis in late summer make for a plentiful supply of moribund rabbits which, although weighing well over 500 gm, provide luxury food for adults and young alike.

Because of their habit of using an observation perch and launching gliding attacks from it, many birds and mammals are at risk from their predation. There is no doubt that they have contributed to the decline in waders on farmland along the moor edge. They are effective predators of juvenile oystercatchers and lapwings up to the time of fledging.[43] In fact, almost any creature with a body mass under 500 gm is at risk. Buzzards are in constant conflict with game managers. Like all predators, they are opportunistic. Pheasant poults are within the 500 gm band for about six weeks after they are put into a release pen at six to seven weeks and most red-legged partridges are within that weight band all their lives. Wild grey partridges are always at risk and grouse poults are vulnerable until the end of July.

Supporters of buzzards make great play on the fact that they live on earth-worms throughout the winter and that their impact on game is therefore small. However, over large areas of their range, worms are all there is for them because the 'larger insects', reptiles and amphibians are hibernating and the 'small rodents' have been all but eaten out. It is a common sight in arable areas to see desperate birds following the plough like seagulls (NS765011)[44].

Common frogs are often eaten by buzzards.

During a spell of hard frost in January 2008, seven dead buzzards were found on a three ha field of winter wheat (NS749013). The cause of death appeared to be starvation as the breast muscles had atrophied and the sterna were prominent.

A mechanism administered by SNH permits the control of buzzards under licence. Though this was confirmed in a debate in the Holyrood parliament in June 2010 and despite problem cases being submitted and requests for licences being made subsequently, none have been granted to date. The shooting fraternity is of the opinion that the licensing mechanism is a smokescreen and that the authority has no intention of granting licences in case doing so starts to change the public perception of raptors. If one licence was granted, others would surely follow. The likes of RSPB could be expected to argue that the floodgates would open and that buzzards would be exterminated – perhaps even that sparrowhawks and hen harriers would follow. Though buzzards are susceptible to intraguild predation from the goshawks and eagles that exert a measure of control in some areas, they are in plague numbers over swathes of the country, particularly in Scotland. No case can be made for the continued protection of buzzards over extensive swathes of the UK (See *The canton concept* p.192).

Golden and sea eagles

A day spent shooting grouse over dogs in the north and west Highlands is affected quite frequently by the presence of golden eagles. Eagles move grouse *en masse* from one area to another perhaps a mile (1.6 km) away particularly if the grouse are at a moderate or high density. As this brings an element of uncertainty to a day's sport, both keeper and host can become upset when a beat where many birds were known to have bred proves almost empty when the shooters arrive. The converse also happens; an area can be filled up on the morning of a shooting day by birds moved from nearby with an unexpectedly heavy bag being the result. Over a season, one effect cancels out the other. Eagles on a driving moor have a more dramatic effect, often shifting grouse in droves with one brood lifting another until the whole hillside is cleared. The panic response is greater even than that created by a line of beaters and can upset plans when previous drives have concentrated birds in an area with the objective of achieving a decent bag in the second half of the day. Due to its powerful eyesight, the eagle spots the high density from afar and responds to it. If this happens in view of the guns or the beaters they are likely to become annoyed because, despite all their planning and hard work, the shooting on a day or part of a day may be ruined. One has considerable sympathy for sportsmen paying large sums for the privilege of shooting grouse, either as owners or short-term tenants but this has to be seen as just one of those things that affect the size of a bag. They must take the disappointment on the chin. As apex predators, eagles tend to keep other birds of prey away; their presence tends to *reduce* the level of avian predation overall.

Eagles are early nesters. Although the young grow fast, they do not leave the nest until mid-July after which their parents teach them hunting skills within the natal area. By mid-autumn the parents are becoming aggressive towards their chicks and the young leave. Most fly to an area where there are no adult eagles but plentiful prey, notably hares, rabbits and grouse. Keepered grouse moors tend to supply the prey but often do not host adult eagles because typically they often lack suitable nesting trees or crags. In such circumstances, young eagles are usually left in peace and allowed to hunt over a large range that is unlikely to be defended by adults.

An owner whose moor hosts a breeding pair has to accept that nothing he does can stop the birds hunting over his ground and that they inevitably take some grouse. If he flouts the law and has the adults killed, he immediately creates a highly desirable area for non-territorial young; numbers of birds will appear and struggle for dominance. The criteria for breeding territories are the availability of food and suitable nesting sites; a place occupied by a breeding pair is, by definition, a prime site. If breeding birds are removed, the site becomes the subject of intense competition until a new pair beats off challengers and establishes undisputed ownership. Killing territorial birds merely *increases* the number of eagles using the territory in subsequent months.

Traditionally, Scotland's mountains have been widely used by the general public for recreation at all seasons of the year but winter use has probably increased faster than access at other seasons. There seems to be no let-up in the development of new rock- and ice-climbs, with some impinging on well-known eagle nesting crags. Climbers impact on golden eagles and ravens as both species nest early when the weather is often still wintery. Disturbance during incubation chills the eggs which may fail to hatch. There is some evidence that poor breeding success in well-used locations can be attributed to these changes.

There is also some evidence that the development of sections of the National Core Pathway Network is adversely affecting eagles. A sitting bird cannot tolerate being gawped at day after day and may well forsake a traditional nesting site and choose somewhere more secluded the following year but doing so does not guarantee either breeding success or protection from further disturbance. Zealous birdwatchers, finding the traditional site deserted, rake far and wide in search of the new one which will then be publicised and so become the subject of renewed inspection.[45]

Goshawks

The goshawk is a powerful, short-winged hawk similar to a sparrowhawk but over twice as large. It is adapted to hunt in and from thick cover and flies low and fast, pouncing on prey which it takes by surprise. Although it is a native and a forest dweller, recent increases are related to the creation of extensive blocks of commercial woodland. Augmented by occasional vagrants from Europe and topped up by falconers' escapes, goshawks are one of our rarer raptors but in the last fifty years they have returned to many parts of Britain as a regular breeding bird. They are known to breed in places as widely scattered as Speyside, the Peak District National Park, the southeast Borders and the North Tyne Valley in Northumberland. It is a bird of woodland edges though a goshawk kill has been observed over a mile (1.6 km) from woodland in the Derwent Valley in the eastern Peak District and one was seen to kill a grouse near Braemar well over a mile from the nearest woodland.[46] They often kill grouse living near woodland or plantation forest. As plantations approach the pole stage and undergo their first thinning or suffer local windthrow, extensive habitats are created that are ideal for goshawks. The afforestation of the past forty years has fragmented large areas of moorland; even in places where it is unknown at present, the goshawk is set to become an important predator of grouse in years to come.

Being forest based, they are easily overlooked. Naturalists who know of their presence as a breeding species do not draw attention to them by telling other people. Their depredations bring them into head-on conflict with keepers looking after pheasants as well as with the hen wife with a few free-range birds at her back door. According to eye-witnesses, their natural spread has been augmented by conservation bodies and raptor enthusiasts putting eggs from captive birds into the nests of sparrowhawks.[47] If this illegal practice is widespread, it is easy to see how large areas could be seeded with foundation stock and come to support flourishing populations. As it is, there are assumed to be one hundred and thirty breeding pairs in Scotland with 'possibly a further one hundred and sixty territories'.[48]

Hen harriers

It is usual for the majesty of the eagle to attract admiration if not affection from people of varied interests – including field sportsmen. No such tolerance is shown for the hen harrier which is particularly disliked by game managers, shooters and keepers alike. While their effect on grouse behaviour is broadly similar to that of eagles, in that their appearance sets off a panic response, they are also systematic and effective killers of territorial birds in winter and spring and so their effect on the bag may not be confined merely to their taking chicks in summer. Workers at Langholm found that twenty-four per cent of grouse chicks up to six weeks old were taken by one hen harrier from within that individual's hunting range. The study showed that grouse losses to harriers in particular could depress numbers to such a low level (less than sixty-five per sq km) that a driven shooting enterprise could become unviable. Because sportsmen maintain that harrier numbers on grouse moors have gone up in the last thirty years (although the national population appears to be stable), there is a tendency for high-profile conflict to develop between hen harriers and sporting interests.

Hen harriers are partial migrants so some moors in south Scotland and north England host resident birds all winter whereas the main Highland landmass has vagrants only at that season. Moors that have harriers as winter visitors often receive several at the one time. The well-known site at West Freuch in Wigtownshire (NX108544) may hold peak numbers of up to thirty individuals while a site in the Forest of Bowland may hold similar numbers.[49] From these roosts they fan out over the surrounding countryside and hunt for prey living there.

The status of the hen harrier has changed since the end of the 1950s when they were recorded as only occasional visitors over much of upland Scotland.[50] At that time, their breeding stronghold was Orkney where they were studied intensively.[51] Even today the heather moorlands of Orkney support many pairs which nest at a density bordering on the colonial. Despite the fact that there are few grouse, mountain hares or meadow pipits on this moorland, the harriers do well on the Orkney vole which is about twice the size of the short-tailed field vole of mainland Britain. Their breeding success is related to vole abundance.

The trigger to the harrier's rapid spread as a breeding bird beyond Orkney appears to have been increased afforestation providing extensive areas of undisturbed breeding ground and substantial prey populations of meadow pipits and small rodents. The fencing, ploughing, planting and fertilising of over-grazed sheep walks created improved conditions for voles and had a positive effect on hen harrier numbers. They nested discreetly in the young plantations which saw them largely undisturbed although, in the absence of gamekeepers, nest predation by foxes was often high.

It is the young dispersing in August from breeding foci that come in conflict with grouse shooters, paralleling the effect of eagles. Many keepers are afflicted by the problem with the flames of embarrassment fanned by derogatory remarks made by owners and sporting tenants; the atmosphere on many a shooting day can become toxic and spoiled by dislike for the bird should one show up. Whereas no definite orders are given to destroy hen harriers, many keepers see it as part of their duties to discourage them from breeding on 'their' moor though there is little they can do about immigrant birds which have been bred on neighbouring ground.

Whatever the rights and wrongs of the argument about the relationship between hen harriers and grouse and of the wider effects of predation on a grouse population which may be living on less-than-optimum habitat, it is hard to accept the allegation that hen harrier numbers are threatened nationally by keepers.[52] While data from RSPB show that hen harriers on keepered grouse moors are less likely to hatch their eggs and rear fewer young than they do on unkeepered ground, the hen harrier is holding its own as a breeding bird and would be in serious trouble were it not for the prey available on managed grouse moors. Something is driving the population's success though it is clearly not the SPAs designated to help them: some are spectacularly unsuccessful. The unkeepered heather moors of the western half of highland Scotland are virtually devoid of breeding harriers; the reason for this has nothing to do with the activities of keepers and everything to do with intraguild predation by eagles and goshawks and the poor prey base typical of these places.

In summary, heather moorland which is systematically over-browsed by sheep and deer changes to grassy heath (Rodwell M19) and consequently becomes more attractive to the male harrier seeking territory in the spring because it carries higher numbers of meadow pipits and voles. There is no doubt that heather/blaeberry/fine-leaved grass moors (Rodwell H9 to 13, 16, 18 and 19) are less attractive to harriers and, by their very nature, better for grouse. It stands to reason that owners can go some way to making their moorland unattractive to hen harriers by reducing the grass/sedge component of the M19 category. A benign view of resident eagles certainly helps to discourage harriers and offset the damage that they do.

When all is said and done, it is not a condition of moor ownership that an owner should be obliged to manage his hill for harriers. The law is there to be obeyed but it is seen by many to be unjust. For all too long, people with a general interest in birds have expected landowners to play host to species which work against their interests without ever suggesting that compensation be paid when substantial financial damage is incurred by them.

Kestrels

The kestrel's hunting method demonstrates that it is primarily a rodent and insect eater as examination of pellets confirms. However, there are always rogues; an individual occasionally becomes a specialist killer of a particular species. Goshawks are efficient killers of kestrels.[53]

Merlins

Few keepers today would lift a finger against the merlin although it was not always so. The 'lady's falcon' has declined over the western half of the country although numbers are stable nationally.[54] Merlins have suffered indirectly from over-grazing and over-burning as they nest mainly in beds of old heather which are often scarce on grassy sheep walks. They have suffered from the reduction in keepers because they are usually a ground-nesting bird and vulnerable to foxes. Often mistaken in flight for the larger kestrel, they have fledged their young and moved south by the time grouse shooting starts so few shooters ever see them. They are incapable of catching and killing adult birds. There is some anecdotal evidence that they are predators of grouse chicks. They possibly take a few tiny chicks from broods they encounter within hunting range of their nest. An eye witness reported that two grouse broods which were hatched within 500 m of a merlin's nest lost five out of nine and four out of nine chicks respectively before the chicks were four weeks old.[55] As other broods in the vicinity did not dwindle, it is plausible to suggest that they were lost to the merlin but the event has to be seen in context. Two broods of potentially eleven birds each were reduced to one of seven and one of six on a moor yielding one thousand five hundred brace in a season; one or two pairs of merlins is hardly likely to lighten the season's bag. Merlins are not implicated in any grouse declines.

Peregrines

Peregrines suffered badly from the indirect effects of organo-chlorine seed dressings in the 1950s. When their plight was recognised, steps were taken quickly to ban the products. The positive effect was seen almost immediately but it took several decades for the birds to re-colonise all their former haunts. The situation is now that saturation, or a point very close to it, has been reached; the national population is in a good state of health at some one thousand five hundred pairs with virtually every traditional territory occupied. Peregrines conflict with grouse interests in that they are highly effective predators which can depress populations particularly when grouse breeding numbers are low.[56] Their impact is much reduced when they have access to large sweeps of well-managed moorland which are home to higher densities of grouse as well as other moorland breeding birds. Over much of Britain away from moorland, their numbers are maintained by racing pigeons which crisscross the sky from spring until autumn with certain individuals becoming specialist pigeon killers. There is no doubt that plentiful supplies of this easily caught prey contribute greatly to high falcon numbers nationally.

Short-eared owls

Short-eared owls are sometimes numerous; their breeding density and success are closely related to vole abundance. They may be significant grouse chick predators in the year following a rodent peak as happens in Scandinavia in the year following a lemming crash.[57]

Short-eared owls are capable of taking prey up to three hundred grams and frequently do so.
(PHOTO: K. TAYLOR)

215

In early August 1987 at NN881361, a short-eared owl flushed during a grouse count flew some hundred metres before suddenly checking its flight and attacking a covey of well-grown grouse on the ground. The brood scattered and flew a few yards before the cock grouse attacked the owl, which backed off and flew away. Two months earlier, however, that bird would have taken a chick successfully. In July 1990, a nest from which the young owls had flown and which was located less than a km from where the attack on the grouse brood took place contained the remains of two black grouse poults about eight weeks old, the wings of an adult male grey partridge and fur from young mountain hares and rabbits, indicating that they take quite large prey considering their weight (320 g). Small rodents and meadow pipits are their main food items but they are certainly not blameless as predators of game birds.

Sparrowhawks

Sparrowhawks do not threaten moorland game interests significantly although they can devastate grey partridge populations on the moor edge. Like the goshawk, the sparrowhawk is a bird of cover: the woodland edge and hedgerow. They forage along scrub-covered stream sides running out into open moorland or in larger areas of natural regeneration and will also hunt up to half a mile (800 m) from trees. Occasionally they kill grouse out on the moor. As their main weapon is surprise, they gain little by hunting far from cover and their preferred hunting range includes only a small proportion of a moorland area. If the predicted increase in goshawks takes place, sparrowhawk numbers will fall because they will suffer intraguild predation: many will be eaten by their larger cousins.

A sparrowhawk at her nest. (PHOTO: K. TAYLOR)

Buffer feeding

Raptors are perceived as harming shooters' interests if they take game birds under any circumstances. The claim may not be borne out by the facts because there are surplus birds at the end of shooting in autumn on most moors in most years. Avian predation on a flourishing stock in winter is unlikely to reduce the breeding population the following spring materially if at all.[58] This is generally true of all avian predators but the principle only holds good if, at the end of the shooting season, the grouse population contains birds surplus to territorial requirements for the subsequent season. In the event of a territory owner dying, the loss will be made good by a surplus individual taking on the vacant space and so upgrading its status from being part of the doomed surplus to being a member of the breeding elite whose future is more assured.

In practice, there are moors in certain districts, and occasional years in many places, where a year's production allows little or no shooting and the pair density in October is not as high as the moor can hold. Under these circumstances, predation, particularly by peregrines, can account for a high percentage of the birds that are present especially where moorland is fragmented, though it is true of many such moors to a degree.

When cocks first attempt to take territories in autumn, they have to take risks. If they do not advertise themselves, they risk losing their proposed patch to a bolder bird – it pays to display widely and act aggressively even at the risk of attracting the attentions of a raptor. As long as there are some surplus birds to act as replacements, the population overall is unlikely to suffer whatever fate befalls the individual.

During the autumn, birds acquire detailed knowledge of their territory, its layout, feeding places, hiding places, lookout points and so on. Such knowledge takes time to acquire; during the learning process, a grouse needs to make only one mistake and it is dead. When grouse are scarce, losses of this kind can amount to a significant percentage of the pre-winter population and breeding numbers the following year can be lower than they would have been had the losses not occurred.

Buffering peregrine predation

Of all the raptors that may catch territorial birds in numbers, peregrines are the most likely to do so. On many blocks of moorland today, they are raising their young predominantly on passage or disoriented racing pigeons. With the end of the racing season in September, the supply of that luxury food dries up. Peregrines become short of pigeon prey just as grouse are starting to behave in a much more conspicuous manner, crowing and song-flighting. From October to the end of the year, peregrines may, and often do, cut quite a swathe through the potential breeding population. Buffer feeding aims to provide a population of free-living feral pigeons within the hunting range of resident peregrines. The technique involves creating a dovecote in an old farm building, redundant lunch hut or even in a purpose-built shed. The pigeons are confined to the house with its covered verandah for the whole summer during which time they rear their young on poultry food and grain. When October arrives, the front of the verandah is opened up and the pigeon population is set at liberty. The birds forage widely and the resident peregrines soon start to hunt them, taking two or three birds a week until the surviving pigeons become so educated to the threat that it is

A racing pigeon predated by a peregrine.

unprofitable for the peregrines to hunt them. By the time this happens, the end of the year is approaching and territorial grouse have familiarised themselves with their home ranges in enough detail to make them much less vulnerable to avian attack. By early spring, the pigeon house verandah is closed up, the survivors settle down and breed and the colony builds up again. It is important that the dovecote is run exactly as decribed. There is no point in making pigeons available to peregrines in the summer as it only helps to increase their breeding success and make the problem worse. Peregrines in summer have to take their chance with wild food. If it is scarce, they will rear fewer young.

A simple dovecote built inside a disused lunch-hut.

Buffering hen harrier predation

The technique has proved to work well for peregrines and would probably also work for goshawks; it appears to be more difficult for hen harriers although it should work in theory. Harriers conflict with grouse in the minds of keepers and moor owners because they catch hen grouse in late March and chicks in summer. (They are also perceived as a nuisance at shooting time because of disturbance but no amount of buffering can solve that.) Buffering by providing a pen of young poultry confined within the hunting range of a pair of nesting birds, the simplest and most effective way of doing it, is illegal because birds confined in this way constitute an 'animal experiment' and this needs a Home Office licence. Applying for a licence takes time as well as the signatures of two Fellows of the Royal Society who have to satisfy themselves of the legitimacy of the 'experiment'. The process is likely to be beyond the capabilities of the average factor or estate owner: the tactic appears to be impossible in practice however practical, elegant, inexpensive and desirable it may be in theory.

However, an alternative used by a number of owners is the subject of research by The Game Conservancy Trust and has been endorsed by RSPB.[59] It involves supplying dead rats, mice and poultry chicks on a regular basis on a feeding post within easy reach of a nest, starting when the harrier chicks are about two weeks old. Up to then, both cock and hen supply their young with meadow pipits and voles but, at about two weeks, the hen, who is capable of taking both chicks and adult grouse, starts to hunt larger prey. Providing alternative food appears to have promise as it seems that the harriers can be trained to take it in preference to grouse.

Predators in context

Of all contentious issues in field sports, the predator issue and the many conflicts surrounding it is the greatest, the most confusing for all parties and the most difficult to resolve. Many members of the general public are uncomfortable about game shooting and would rather that it did not happen. What is perceived as the wanton killing of birds and animals 'for fun' can be hard to justify to people who have an idealised view of nature and the countryside. High concentrations of small game do not sit easily with high concentrations of people using the countryside and the fact that game management in Britain involves intervention by man and the control of certain predatory species can be hard to accept. The fact that the higher end of the 'market' is enjoyed by an affluent minority also does it few favours.

Game management and shooting changed a great deal during the twentieth century, a point examined elsewhere in this book. The sport has become more egalitarian. Though more people go shooting than ever before and look to harvest wild creatures, which is an activity as old as history itself, there are also many who would be happy to see the whole activity outlawed. Into this mix comes the issue of the protection of species such as badgers, buzzards and sparrowhawks which are, on the one hand, perceived as pests and on the other as in need of protected status. The suburban dweller is as unhappy to see song birds snatched off the bird table by sparrowhawks as he is to see magpies and grey squirrels pulling baby birds to pieces. The predator/prey issue is not wholly a field sports one.

There have indeed been times when some predatory species needed assistance. It was provided and declining populations recovered. What has been lost is the ability to examine objectively the current status of these species and arrive at a consensus which allows the coexistence of reasonable numbers of game birds, the non-quarry species associated with them, such as farmland and suburban birds with equally reasonable numbers of predatory birds and mammals.[60] This is perhaps the big issue for the twenty-first century. Unless it is resolved urgently, game shooting as we know it is in jeopardy along with a number of treasured non-quarry bird species.

Wildlife tourism

Visiting the countryside, tourists and 'birders' like to see as many species as possible often with a strong preference for predators, avian or terrestrial. Whether the visitor is a holidaymaker or a resident, the number of boxes which can be ticked on the basis of spotting an individual creature is often regarded as the chief measure of the conservation worth of an area (see Chapter 4, *The value of 'bird reserves'*). The shooter's aspirations are different; the greater the *number* of quarry he sees, the higher he values a venue. Commentators stress the importance of 'wildlife tourism' but published figures quantifying the financial value to rural communities are hard to come by.[61] SNH did report in 2010 that 'visitors to the Galloway Kite Trail have spent... £21 million since 2004 with over £2.6 million spent by people who came specifically to see the (red) kites. The project has also supported some thirteen full-time jobs each year since the project started,' adding that the Trail was 'developed as a partnership between RSPB, Forestry Commission, local farmers and tourist businesses' and that 'it received initial funding... from SNH, VisitScotland and Leader+. Its further development has been funded through Sulworth Connections, a partnership part-funded by Heritage Lottery Fund, Dumfries & Galloway Council and SNH'.[62]

In short, the thirteen jobs have been paid for predominantly out of the public purse. Unlike employment in agriculture, shooting or forestry, they are not 'real' jobs in the eyes of many – the annual cost at, say, £20,000 per head per annum, is over half the *public* spend of £2.6 million over the six years 2004 to 2009 (£433,000 per annum). The project is not one that any hard-nosed private individual would ever contemplate undertaking.

Specialist naturalists and birdwatchers, whether singly or in groups, local or visiting from afar, are not big spenders and are unlikely to leave much money behind them. Without public funding, there is little or no impact on local employment levels. Indeed, knowledgeable 'birders' abhor crowds and are essentially solitary. Generally, they do not use high-level walk ways or specialist bird hides at visitor honeypots like Loch Garten and other Speyside locations catering for mass tourism. Their interests lie in the rare and the unusual which are most likely to be found off the beaten track.

Sporting tourism

Sporting tourists are often lavish spenders: their money keeps the roof on many a lodge or hotel and sees substantial council taxes being paid to the benefit of the wider community. Wages, permanent and seasonal, are obviously significant but money spent locally on clothes, sporting kit, souvenirs, food, drink and so on can easily equal that paid to proprietors for sporting rights. Grouse shooting is estimated to inject up to £30 million into Scotland in a good year, a sum that dwarfs that attributable to wildlife tourism. Sums of this order going into otherwise poor districts provide year-round jobs for estate staff and casual jobs during the season which keep the all-important infrastructure in place and primary schools and post offices open. Without these, the countryside dies. Income from moorland field sports owes *nothing* to public funding (except insofar as any estate income derived from associated sheep farming activities is subsidised).

End notes

1   ICI Game Services later became The Game Conservancy.
2   Malcolm & Maxwell, 1910.
3   Ellice, 1931.
4   McCreath & Murray, 1954.
5   Nethersole-Thompson & Watson, 1974.
6   Watson Lyall, 1912.
7   Ellice, 1931; Holloway, 1996; Moffat & McNally, pers comm.
8   St John, 1845; Birks, 1988; Moffat & McNally, pers comm.
9   Dougall, 1875.
10 Hudson, 1986.
11 The Heather Trust, 1995; Redpath & Thirgood, 1997.
12 Lovat, 1911.
13 Lindström et al, 1995; Watson & Moss, 2008.
14 Petty et al, 2003.
15 C. Graffius, pers comm.
16 Amar et al, 2007; Eaton et al, 2009; Barov et al, 2010.
17 Ash et al, 1967; C. Graffius, pers comm.
18 Macdonald, 1987.
19 Lloyd, 1980.
20 Kolb & Hewson, 1980.
21 Hudson, 1986.
22 Moss et al, 1990a.
23 Small & Rusch, 1985.
24 Erikstad, 1979; Jenkins et al, 1964b.
25 Lovat, 1911; Lindström et al, 1994; Smedshaug et al, 1999; Soulbury et al, 2007.
26 Speedy, 1886.
27 Rochlitz et al, 2010; Turner, 2010.
28 J. Grave, pers comm.
29 M. Naisby, pers comm.
30 Graham, 1908.
31 Warren & Baines, 2007; Thomas & McDiarmid, 2011.
32 B. Hendy & D. Macdougall, pers comm.

33 Author, pers obs.
34 Author, pers obs.
35 Author, pers obs.
36 Parker, 1985.
37 Picozzi, 1975.
38 Parr, 1990; Fletcher, 2006.
39 Author, pers obs.
40 The Institute later became the Centre for Ecology and Hydrology.
41 Speedy, 1886; Stuart-Wortley, 1895; Watson & Moss, 2008.
42 *Birdlife*, 2011.
43 Author, pers obs.
44 D. Phillips, pers comm. In October 2010, he reported having as many as nine birds at a time following the plough often within two metres of the machinery.
45 J. Lithgow, pers comm.
46 A. Watson, pers comm.
47 W. Veitch in Martin, 1990.
48 R. Cunningham, Scottish Parliament; written answer S3W-34186, 16 June 2010.
49 Watson, 1977.
50 Waddington, 1958.
51 Picozzi, 1980.
52 Etheridge et al, 1997; Thirgood et al, 2000; Galbraith et al, 2003.
53 Watson & Moss, 2008.
54 Orchel, 1992.
55 B. Mitchell, pers comm.
56 Lovat, 1911; Redpath & Thirgood, 1999.
57 Marcström et al, 1988; Steen et al, 1988.
58 Jenkins et al, 1964b.
59 Redpath et al, 2001; Scottish Natural Heritage, 2010b.
60 British Association for Shooting and Conservation, 2004.
61 Dickie et al, 2006; Watson & Moss, 2008.
62 Scottish Natural Heritage, 2010a.

Chapter 6
Parasites: ticks and worms

Sheep ticks

The relative sizes of stages in the sheep tick life cycle. Including: larvae, nymphs, adult male (bottom centre) and a fed adult female (the big one!). X6 natural size. (PHOTO: K. RYAN)

Sheep ticks *Ixodes ricinus* are localised in distribution but can, if numbers are high enough, significantly impact on land use practice. A member of the order *Arachnidae* (spiders), they depend on vertebrate blood and body fluids for survival and transmit a range of viral and bacterial conditions to the host while feeding. Their natural habitat is scrub woodland including heather moorland where they parasitise economically important species such as red and roe deer, the grouse family (red grouse, black grouse and capercaillie), mountain hares and, of course, domestic sheep and cattle.

Though farmers have recognised the impact of ticks for over two hundred years, Lovat devoted a mere 158 words to them, suggesting they were not seen as a problem for grouse a century ago though that does not mean they were not present. The Gaelic word *Mial* is a general term for biting insects such as lice, fleas and ticks. *Monadh nan Mial* (NN 866441) in Perthshire can be translated as the 'moor (or mountain) of the ticks'. It was ticky in the 1970s and it is reasonable to assume it was ticky when nineteenth century cartographers were drawing on the folk knowledge of local place names. A hundred miles (160 km) away, at the head of Loch Beannacharain, Ross-shire (NH212516), there is a shieling and settlement called Scardroy. A native of nearby Cannich born in 1899 reported that alluvial flats at the head of the loch which looked like good summer grazing for cattle were in fact dangerous because redwater fever (*babesiosis*) was endemic on them.[1] They were known in his youth as 'the place of the red skitter', a literal translation of the place name. It is likely that ticks had regularly infected cattle with the bacterium *Babesia divergens*, a blood parasite which causes the excretion of haemoglobin (which is red) in the urine.[2] These two anecdotes suggest that, though ticks were certainly recognised at that time, they had yet to be identified as the pests they are now. There is no evidence that louping ill virus – the scourge we associate with ticks today – was widespread in Highland Scotland until well into the nineteenth century.

Life cycle and biology

There are three instars in the tick's life cycle – larvum, nymph and adult; individuals are sexually differentiated during the last instar only. Its quest for a host is haphazard; when the larvum hatches from the egg, it crawls up vegetation and waits for a passing host with which to make contact. As it can live on any vertebrate during the first two instars, its hosts include reptiles as well as mammals and birds. Once on a host, it responds to the high levels of CO_2 which occur around the head, tends to make its way there and sink its mouthparts into the thin skin around the nose, mouth, eyes and ears though in mammals, the bare skin between the forelegs and the inguinal area can also be favoured. Feeding duration varies but is usually between five and eight days after which the tick disengages and falls off. It then crawls into moist litter close to the soil where it digests its blood meal and moults into a nymph, ready to crawl up vegetation and start questing again. The stage is likely to last a year but there are wide regional and temperature variations; under favourable conditions, a larvum fed in April can emerge as a questing nymph by the autumn or earlier. If the nymph finds a host, it feeds on it for up to ten days before falling back into the mat and moulting into a sexually differentiated adult, again usually a year later. The quest for a host is repeated except that, unlike the larvum and nymph, the adult tick requires a large mammal as host. They have been recorded on squirrels, foxes, wild cats, stoats, pine martens, brown and mountain hares, deer of all species, sheep and cattle. (Rabbits are rarely infested mainly because of the texture of their fur and because they tend to rest underground and feed on short herbage.)

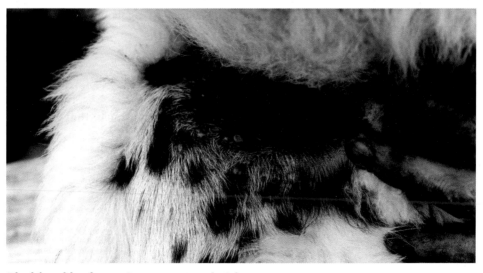

Blackfaced lamb carrying many nymph ticks.

223

The attachment period of the adult female may be as long as three weeks. She waits until a male attaches himself ventrally then drops off, lays some two thousand eggs in a cluster and dies. The male does not feed during the third stage but may attach to the female before the latter attaches to a host. The 'nest' of eggs lies dormant over winter and hatches when temperatures rise in the spring. A host that blunders into a concentration of eggs may end up carrying hundreds of larvae. Small numbers do not affect sheep or grouse but a burden of fifty or so on a week-old grouse chick usually kills it.

The influence of habitat

It takes only two of the couple of thousand eggs to attain maturity and breed to maintain a stable population. There is obviously considerable wastage; the vast majority of ticks die without ever finding a host. Optimum conditions for survival include a thick 'mat' layer on top of the soil for cover and re-hydration after questing, a plentiful supply of hosts and warm, humid weather to facilitate long questing periods. The converse – poor mat layers, few suitable hosts and hot, dry, windy summer weather – lowers the chances of survival which obviously depend also on chance events such as where a tick lands when it falls off the host. If the litter layer is too thin, it may not supply enough moisture or cover to allow it to complete the moult in security or re-hydrate properly following a spell of questing in bright sunshine and low humidity. Windy weather does not directly damage a tick but it does force frequent interruptions to questing because of the need to descend back into the mat to re-hydrate.

A tick-supporting mat about 12 cm (5 in) thick on top of mor humus.

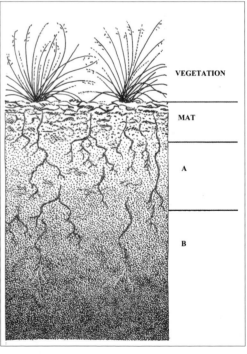

A drawing which further illustrates the photograph. (D. Pullan)

ABOVE: *Fine-leaved grasses on quality brown earth soil; almost no mat.*

RIGHT: *A drawing which further illustrates the photograph.* (D. PULLAN)

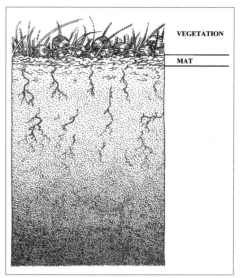

Optimum habitats

The best habitats for ticks are scrub woodland (with or without bracken and heather), bog flushes with plenty of mosses and lightly grazed grassland with an abundant mat of moist but well-drained litter originating from years of uneaten, partially decayed growth. Heavily grazed grasslands dominated by fine-leaved grasses and clovers make for poor habitat as does properly burned heather moorland with no deep layer of heather leaf litter and no fog (grass litter, mosses or lichens as understory). A heavy tick presence tends to be the result of inadequate hill management over a long period. Poor burning with cool fires on long rotations tends to increase mat levels. A large host population in summer (sheep, hares, roe and red deer, etc.) increases questing success rates and thus tick populations. Numbers increased in Moray and central Perthshire in the 1970s on moors where there were plenty of hares and where sheep farming based on low-input ranching was increasingly replacing cultivated crofts on the hill and the moor edge round about. Croft grasslands were losing much of the little fertility they had; there were no longer any cattle in summer and no general attempt was being made to maintain fertility with lime and phosphate. As the only grazing was by sheep and even that tended to be light and highly selective in favour of the fine-leaved grasses, coarse grasses increased at the expense of fine-leaved ones, causing the mat to thicken. As the coarse grass sites were still the best sources of food, herbivores concentrated on them. Ticks detach at random after feeding so the longer a host spends on a site, the greater the chance that ticks fall on it and, mat conditions being favourable, survive. They are thus more likely to find a host the following year. As the animals also browsed on moorland round about, ticks were shed in increasing numbers on the surrounding vegetation. Researchers in the 1970s found this to be a consistent pattern on neglected moors even though ticks were scarce on adjacent moors subject to different management regimes.[3] On the 'low tick' moors, heather was usually better burned with enough animals – including hill cattle in some cases – to keep 'greens' tightly cropped and dominated by productive grasses with

225

Blanket-dragging the line of a well-used path. (NJ126324)

little or no mat. However, they also tended to overlie base-rich rock that made for intrinsically more fertile soils – fertility that made for thinner mat layers where dead plants rotted more rapidly.

Hugh McBean was head keeper at Dunphail, near Forres, in the 1950s and 1960s. Towards the end of the 1960s, grouse declined generally throughout Moray and Nairn just as ticks on the hill started to become a problem for farm stock and grouse alike. An excellent observer, Hugh maintained that there were two reasons why grouse had declined as steeply as they had. The first was the abandonment of proper farming systems on the remoter crofts. He used to instance Bogeney on the southeast side of Cairn Eney (NJ024432) which was rotationally cropped by a crofter who grew oats, turnips and oats under-sown with grass seed for a four or five year ley and ran a herd of suckler cows and a flock of sheep. Mountain hares which used to draw in large numbers onto the arable land at night were snared and shot hard as farm pests. They carried no ticks. Hugh was of the opinion that conditions within the enclosed ground of the croft were inimical to ticks and he was clearly right. His other claim (which, though widely held, would probably not withstand scrutiny so readily) was that the packs of grouse that foraged on Bogeney's stooks, stubble fields and corn stacks in the autumn and winter were thereby fattened up for the winter and so survived better!

An eyewitness account of a morning driving grouse on Cairn Eney and the Moidach More in 1969 recorded ninety-two brace by lunch time. By 2000, the same part of the moor held fewer than fifty birds because there was no local recruitment due to ticks.

Juvenile grouse dying from louping-ill.

Although high-quality burning and adequate grazing pressure were identified by those studies in the 1970s as working against large tick populations, over-burned and over-grazed moors generally – common in parts of Argyll, the south-east Borders, Skye, Mull, Islay and Jura – were found to carry high tick numbers. The reason is essentially the same as it is on the drier, eastern moors. Where grazing and burning are excessive, dwarf shrub heath with heather as a main constituent changes into grass-dominated moorland in which substantial mats of dead white bent and purple moor grass litter build up over the years. Even where hares are scarce, there are usually sheep, sometimes at high densities, and, in some areas, red deer so there is no shortage of hosts. Most of the moorland provides good conditions for ticks to over-winter. As summers in the west tend to be moist, mild and humid, ticks can quest for long periods before having to descend back into the mat to re-hydrate; questing success rates tend to be high. The western moors may seem very different from the tick strongholds of Moray and Perthshire but the differences are superficial: the essential conditions for tick survival are present in abundance on both.

LEFT:
Wild goat, an important tick host in certain local areas.
(PHOTO: K. TAYLOR)

BELOW:
Pooly managed moorland can lead to bracken encroachment which creates suitable conditions for ticks – 2011. See p.47.

The moorland manager must understand how to mitigate the debilitating effect of large tick populations or, given that they are still virtually absent from many areas, avoid them altogether. It is difficult to attack them directly particularly when, as in Perthshire, Moray, Argyll and the islands, wild animals are the main hosts. Obviously, dipping hares, red and roe deer (all of major importance in these areas) is not possible. Recent informal proposals that red deer be fed systemic acaricides or provided with impregnated rubbing posts are equally far-fetched. True, sheep and cattle can be sprayed, dipped, smeared with anti-tick cream or treated with a systemic 'pour on' but they are very often not the chief hosts: if ticks are to be controlled, the whole suite of hosts has to be tackled. That means treating some and killing others.

Moorland in Moray typically comprises old croft greens, stream sides full of birch, pine scrub and long heather associated with fast-growing (and often under-burned) heather on the hill faces. Sheep at a density of perhaps a ewe to 3 ha (7 ac) summer graze on greens carrying fine-leaved grasses and browse heather on the open moor the rest of the year. Roe typically occur at densities of one to 5 ha (12 ac) of scrub and degenerate heather centred on gullies and stream sides. Itinerant red deer often descend onto the lower moors at night in summer. In peak years, hares reach densities of one per 0.8 ha (2 ac), particularly where there is plenty of young heather. In short, each moor type has a different combination of resident host species.

A charming tick host on a grouse moor in August. (PHOTO: N. MCINTYRE)

Roe are important hosts on heather moor. (PHOTO: D. KJAER)

Because of the way hares use the hill, their tick burdens can be high. A thousand and more specimens (including all three instars) have regularly been observed on an individual animal. If the ticks feed for ten days, tick recruitment over a four-month summer is at least twelve thousand per hare, equivalent to shedding 1.5 ticks per sq m. Hares tend to lie up in long heather and broken haggy ground during the day and graze rich greens and browse short heather at night. The vegetation around their resting places can be so infested with questing ticks that

Hares can carry as many as a thousand larvae, nymphs and adult ticks at one time.

A hare's form, a focus for ticks.

they can be seen with the naked eye.[4] If sheep are carrying thirty five to fifty adult ticks at a time (mean forty-three), their recruitment influence is 0.017 ticks per sq m. In crude terms, hares are a hundred times more effective than sheep at raising the density of questing ticks.

Studies on roe conducted in Perthshire in the 1970s showed that thirty-five to fifty adult females was a common summer burden – much the same as sheep though work in Angus showed that high numbers of adult females were carried by red deer.[5] Two marauding stags coming in from the open hill (NO575725) each evening and shot in June 1992 carried 235 and 142 adult ticks respectively as well as almost uncountable numbers of nymphs and larvae.

Any attempt to control ticks by dipping sheep on moorland which carries even a few wild animals is doomed to failure although the chances of success are substantially improved if the hill carries sheep as the only large host. In the Forest of Bowland and the North York Moors in the 1980s, systematic and thorough prophylactic treatment of the sheep stock reduced host potential to such low level that ticks ceased to be a problem. Excellent pioneering work on tick control began at Castle Grant (NJ041312) in 1991, Lochindorb (NJ969355) in the mid-1990s and, more recently at NJ009470 and other places. Implementing control policies based on the sound ecological principle of tackling host potential has resulted in grouse returning in some cases to the levels of fifty years ago.

Manipulation of vertebrates includes treatment of sheep to prevent them from hosting ticks through the summer and the reduction of hares to the lowest possible breeding numbers living on a low-altitude moor. It means tolerating no red deer on grouse ground and controlling roe deer living on the moor in summer. Allied to a proper burning regime and, if appropriate, upgrading of croft ground on the open hill, the situation can sort itself in the space of three to five years, depending on how rapidly all aspects of the problem can be addressed.

The use of sheep as 'tick mops' has been hailed as a magic bullet that would further the recovery of tick-affected moors. The concept involves running a flock on the moor during the active tick season that is dressed with anti-tick chemical every four to six weeks. In their normal diurnal travel around the moor, the sheep pick up questing ticks which are then killed by the chemical. The idea can only work if the sheep visit every corner of the hill on a frequent basis. Sheep do not do this naturally unless they are hungry. If they are that unsettled, they are being run at such a high density that they likely to over-graze and damage the hill. The sheep need a person dedicated to looking after them, which makes the programme expensive, and must be kept permanently toxic to ticks throughout the summer if they are to have a consistent effect. Heavy rain after treatment reduces the efficacy of the chemical so, unless treatments are applied systematically and maintained according to weather, money is wasted. Sheep are bare skinned after clipping in mid-July, cannot retain the acaricide and so have little impact on questing ticks during the critical six weeks running up to the end of August. Finally, if the sheep share the ground with wild animals, their effect potentially beneficial is limited still further.[6]

The need for hare control

Hares can be numerous but can be controlled with ease. High numbers of breeding hares living in summer below 300 metres (1000 ft) amplify the tick population. They do not themselves have levels of louping-ill virus in their blood that will infect every tick that feeds on them as is the case with sheep which become viraemic (actively producing the virus in its blood). Research in the 1990s showed that hares and small mammals facilitate the transmission of viruses from one tick to another during co-feeding.[7] As ticks feed on blood-rich skin, they locate where the host's skin is thinnest. When a tick starts feeding, the host's reaction is to increase the blood supply to the injured site, which raises a weal-like bump. A newly arrived tick seeks out just such a spot and inserts its mouth-parts. If the first tick has virus in its salivary glands, it injects the feeding site with them along with its saliva (which contains an anti-coagulant). The new arrival becomes infected not because the host passes it amplified virus but because the blood on which it is feeding is contaminated by its neighbour.

Managers should strive to bring hare populations under control by late spring – maintaining large summer populations is misguided as it puts grouse at long-

Mountain hares. Critical components of the moorland system. (PHOTO: NEIL MCINTYRE)

term risk on many a good moor in Scotland. People believe that high numbers of hares and big grouse populations go together and that large numbers of hares pre-date a surge in grouse populations.[8] It is indeed a regular occurence but not always at low altitudes of 580 m (1600 ft and below). Keepers in charge of ticky moors who pursue hares with vigour soon find themselves at loggerheads with conservation bodies that see large numbers of hares as a prey source for eagles, wild cats and pine martens.[9] Other workers argue that there is no link between hare control, the control of ticks and louping ill despite field experience from many owners and keepers to the contrary.[10] Proposals are afoot to outlaw snares for hare control altogether.[11] Were they to reach the statute book, the outcome would be a decline in the numbers of red grouse and waders breeding on grouse moors and the mountain hare would become a serious pest on agricultural land adjoining moors where heather management is good. If hares are to be an object of conservation concern, improved heather management is a first step to increasing their numbers – a point well understood by SNH.[12]

How to bring back grouse on tick-affected moors

The challenge for moor managers today is to re-create extensive swards of well-managed heather-dominated moorland interspersed with productive 'greens' characterised by the low hare and deer levels which prevailed fifty years ago. It is pointless to entertain the idea that a policy of diversity can be attained – leading to the creation of a sort of Noah's Ark on moorland – where significant populations of hares, roe and red deer can coexist happily with grouse on ground below 500 metres without affecting the grouse adversely.[13] Deer have no positive part to play in the management and maintenance of the best traditional grouse ground below 450 m (1500 ft).

By the 1970s, deer numbers on many Perthshire moors were increasing alongside grouse which were present in large numbers. Though ticks had ruined grouse shooting on some moors, notably Drumour, Grandtully and Moness, all of which ended up being planted with trees, they were absent or at low numbers on most others and not considered a threat by proprietors reluctant to exterminate the incoming deer. By the late 1980s, many had come to rely on the revenue that deer provided year on year for little effort. Even proprietors who might have preferred more grouse and fewer (if any) deer had to depend on the cooperation of neighbours to achieve their aim. The easy option was to go with the flow, quit open moorland sheep farming and grouse production and manage deer as the main wild game resource. Those who chose that option have found that they are unlikely to have grouse in any numbers though there is always the solace that, as less labour is needed, the business can often be run with part-time staff. The land use pattern of a whole district was influenced by circumstances which favoured ticks.

English moor owners do not have the option of letting deer stalking and so are not tempted to tackle cash flow problems at the risk of over-browsing and tick build-up. There are ticks on some English moors and they may sometimes be economically important but, since the hosts are usually sheep, a prophylactic approach involving a rigid programme of inoculation and dipping/pour-on treatment is feasible, if expensive. In the 1980s and 1990s, ticks became a problem over extensive areas in west Lancashire and the North York Moors and shooting results declined sharply. Following the introduction of tighter sheep management and first-class keepering, large grouse bags are being obtained again.

Problems posed by agricultural change

Over the last thirty years the number of shepherds has declined. Those shepherding out-of-the-way places generally prefer to live close to social facilities and travel to work; many farm and shepherd houses in remoter glens have ceased to be occupied by farm staff and are now often holiday houses used seasonally. Because there are fewer people about on a day-to-day basis, red deer, increasing in numbers following a lengthy run of mild winters and persistent under-shooting of hinds, have tended to colonise the lower, heathery slopes close to valley bottoms from where they descend to feed at night in the grassy straths. In times gone by, these summered herds of cattle and set stocked flocks of sheep; the green ground was kept short, fine-leaved grasses flourished and a mat of dead grass was largely absent. The extension of year-round red-deer range has led to a sharp increase in ticks particularly in the east Grampians including the Angus glens, parts of central and west Perthshire and as far north as east Sutherland. Places where ticks were unknown in the 1970s are now severely affected; grouse production is close to being abandoned because reduced revenues cannot cover the cost of keepering. Once the keepers go, burning standards fall, scrub woodland encroaches onto the heather ground, foxes and crows proliferate and moorland waders breed poorly and the economic life of a whole district declines rapidly.

It is a story as sad as it is avoidable. The decline in moorland birds generally is the product of vegetation change and predation. It may be the decline in wading birds that strikes the casual observer but passerines have suffered just as much.The scenario is all but universal over the sheep walks of Scotland but is less obvious on the base-rich rocks of the moorlands of the north of England that are now the national strongholds of grouse and moorland waders. The plight of waders such as lapwings, curlews and redshanks is a constant refrain of the conservation bodies (numbers of these species nationally are in free fall) as one half-baked idea after another is aired to encourage them to breed. Any wildlife manager appreciates that, for waders to rear young in big numbers, one must kill predators, however unpalatable this may be.[14] 'Habitat management' may be appropriate over extensive ranges in the US but, on its own at least, provides an incomplete solution for intensively managed habitats in Britain.

Changing land management options

Ticks have become an economic problem, red deer have increased and grouse are still considered a primary enterprise. Is there a practicable remedy? The answer is yes but not one for the faint hearted. Rectifying the agricultural aspect needs substantial investment allied to determined management though it should perhaps be regarded as 'delayed maintenance' – the result of under-investment in hill farm land over the past fifty years. The annual cost of prophylactic sheep treatments involving vaccination and dipping or pour ons for ewes and lambs to control ectoparasites can run to £20 per ewe. But farming is only part of the challenge – it is necessary also to take a hard look at land use and resources generally. If wild animal numbers are high, they may, unlike cyclic grouse populations, be providing a reliable income for the estate and a 'soft option' for owner and keeper alike. Many owners admit to being reluctant to forego the letting income obtained from shooting a score of stags even while acknowledging that they do not have a proper deer forest. Knowing they will have to do without it if they are to get grouse back, they console themselves by saying that they have deer every year without effort and that the sale of venison provides a limited but reliable cash flow. The temptation to abandon the grouse altogether is strong: it takes a confident man to go for the grouse and forego the deer.

On a site in Caithness (ND118226), grouse had declined from many hundreds of brace shot per annum to virtually nothing over forty-odd years. There had been a three-fold increase in red deer numbers over the same period and the tick burdens on them were high. The management decided to reduce deer numbers substantially; over four thousand were shot between 2000 and 2003, bringing the population down to fifteen hundred – enough to support an annual stag crop of a hundred or more. A tick control programme involving a herd of Luing cows to improve the grassland (the cattle were dressed with 'pour on' at regular intervals throughout the summer and carried almost no ticks) was coupled with bracken control and lime and ground rock phosphate applications on the 'greens' in the valley bottoms. A small hill flock was run under a stringent regime of dipping and vaccination.[15] Five years on, the grasslands had improved dramatically and the build-up of mat had ceased. Oystercatchers and lapwings had prospected the low ground despite having been absent for over twenty years and golden plover had returned and were breeding well; ticks on shot deer were in single figures and grouse were again producing young. In short, a programme that sought to rectify a gradual run down in fertility over decades was quickly delivering the predicted results.

Tick-transmitted disease and population dynamics

While the ex-sanguinating effect of a thousand ticks on a hare leads to its early demise, five hundred on a ewe, roe or red deer have less effect. Twenty to thirty nymphs on a grouse chick debilitate it and reduce mean chick survival rates but vigorous chicks survive provided they are not challenged by louping-ill virus. This sheep malady has been observed for centuries and was given the name of louping-ill or 'trembling' by shepherds. To 'loup' in Scots is to jump – affected sheep louped in an incoherent way before usually dying a few days later. The virus which caused the disease was discovered in the 1930s and that its spread from sheep to sheep was caused by ticks soon after. The disease had spread to

Nymph ticks localise around the beak and eyes of grouse chicks.

many parts of Scotland in the sheep boom of the nineteenth century. At that time many farmers, particularly from the Scottish borders, took the tenancies of thousands of acres of land vacated by crofter-clansmen following the failure of the 1745 Jacobite rebellion and the breakdown of a part-time farming system. The rate of spread of the disease increased sharply in the second half of the century as it became common practice to move sheep in large numbers over long distances by rail rather than by walking along drove roads. New grazing areas suitable for hogg wintering were opened up such as the extensive heather moors of northeast Scotland, Dava Moor being a prime example. They provided grazings which were neither rich enough nor diverse enough for ewe flocks but were adequate for young sheep which did not need as high a plane of nutrition over their first winter. The hogg loaded on a train on a September Monday morning at Lockerbie (complete with its cargo of ticks) was discharged at Grantown-on-Spey less than twenty-four hours later, turned out onto the hill and immediately started browsing heather and birch scrub. If the animal was viraemic, its ticks had the potential to be infective the following year. It was a very different beast to the sick animal that began the walk from Lockerbie to Moray but was dead by the time it got to Beattock. Even if it recovered sufficiently to complete the journey, any ticks it was carrying when it started out would have fallen off long before it got to the wintering ground. The disease could not have been spread by driven animals.

Although the louping-ill first emanated from the Scottish borders, once in Highland Scotland, it moved about at random. It was brought to Loch Tayside, for example, in the 1920s by ewe hoggs which came from the Oban area for wintering.[16] The random movement patterns ensured that the disease achieved total coverage of Scotland; by the mid-twentieth century, it had been recorded virtually everywhere. Once ticks carrying the virus arrived in a district, only creatures with a degree of immunity would survive. Hares and red deer showed no signs of the disease but sheep were severely affected and grouse catastrophically so.[17] It was by now obvious that the arrival of viraemic sheep was going to have serious consequences. Grouse have little or no resistance to the disease; the 1970s work showed how dramatic mortality could be. Modest quantities of ticks without the virus can be tolerated but ticks carrying it completely destroy the productivity of a grouse population. Chicks were dying at the rate of about twelve per cent per week; a brood of ten young hatching at the end of May was down to two poults or less by the time they were twelve weeks old. Where the death rate was twenty per cent per week, the whole brood would be dead before it was ten weeks old. The effect on commercial grouse moors was disastrous because there was no production and so no shootable surplus. When so few young are reared, there is little subsequent competition for space. Cocks holding territory that are among the fortunate few with an antibody response can live a long time. A marked cock, a second season bird when first caught, lived at NN831461 for nine seasons. Every March, he had a hen but every year she died; he raised no young during his life. Every mate was an incomer, an immigrant from perhaps several kilometres away where there were no ticks and no virus. She therefore possessed no antibodies and was doomed in her new home. As noted, cocks take territory close to where they were born but hens tend to disperse for a greater distance from their natal area and are more likely to end up in places where the chances of survival are poor.[18] They usually die during the summer – along with their chicks. In 1977, eighteen out of thirty-three (fifty-five per cent) of territorial cock grouse on a Perthshire moor (NN900460) carried antibody to louping ill, i.e. had been challenged and recovered. By 1983, none of thirty cocks sampled on the same area was affected.[19] The extinction of the virus in grouse coincided with a massive decline (eighty-five per cent) of ticks on the ground as measured by blanket dragging. The cycle was broken because the area was sold to a forestry company for planting. Sheep were cleared from the ground, resident roe were shot out and hares (the only other significant mammalian host) were tightly controlled. The host population became very low. Ticks were therefore bound to decline but it was agreeably surprising to find a result that was so unequivocal.

Transmission of disease

Given the absence of sheep as a potential virus pool, the only other reservoir was grouse. The data suggest that the louping-ill virus needs a certain size of tick population to perpetuate itself. If, for whatever reason, host numbers decline, fewer ticks emerge the following year. If the virus declines or dies out, the change has to be a function of the population dynamics of the hosts – grouse and sheep.The finding provides managers with a ray of hope provided that host population can be manipulated in the right way. Tick problems are management-related; those who face them must assess how and where the system has broken down. It is known that approximately ten per cent of nymph and adult ticks in an affected place carry louping-ill virus and that the virus is not egg-transmitted.[20] Ticks must therefore become infected when feeding on a viraemic animal. If the host is a grouse, the nymph has only a small time window to become infected before the bird dies. If the sheep stock is vaccinated against louping-ill, it does not amplify the virus. Even if challenged by a virus-carrying tick, a sheep has an antibody response and, though it may host many ticks and help to support the population, does not infect them. The incidence of disease falls in the following year. Of the adult female's two thousand eggs, about two hundred first generation larvae find a host, over-winter in the litter and emerge the following year as questing nymphs. If ten per cent of these succeed in obtaining a host and over-wintering, twenty adults emerge. In a stable population, only two (a male and a female) quest successfully; if three succeed, the population rises; if only one, it falls. This is, of course, an oversimplification but it shows how the dynamic of population change is a percentage game. If a few more ticks are successful, the population tends to rise; if a few more fail, it tends to decline though the dynamic in either direction is slow. Because not every tick carries virus, increases or decreases in the incidence of virus disease are also a matter of statistical probability. If one in two hundred ticks carries the virus and a susceptible sheep hosts ten ticks for a ten-day feeding period, the animal runs a one-in-twenty chance of infection making the chances of the 199 co-feeding ticks becoming infected low. If the sheep hosts not ten but two hundred ticks for ten days, the risk of viral infection is increased and the chance of the other 199 ticks becoming infected is very high. If ticks are reduced to low levels, virus transmission ceases.

Louping-ill in sheep

Louping-ill is deadly and greatly feared by the hill flockmaster. If the disease is present on a hill grazing, it can reduce profitability to such a low level that the farm is abandoned. Many otherwise sound hill grazings have been lost in this way. Adult ewes challenged by the disease have a lifelong active immunity. They pass on a passive immunity to their lambs in their colostrum or first milk that protects them for up to six weeks; the lambs develop their own lifelong active immunity if they are challenged by a virus-carrying tick during that period. If, as is normal for ewes mothering twins, the flock is lambed in an in-bye field and not turned out to tick-infected pastures during the summer, the passive immunity in the lambs wears off, no active immunity is obtained and the animals become at risk. In the autumn, when stock hoggs are selected for breeding, it is traditional for them to be wintered away on grass pastures on low ground before returning to their home range at the end of the following March. If they are then turned out onto the open hill (common practice as in-bye ground is used as lambing ground), they are immediately challenged by virus-carrying ticks and become infected. Losses occur. It is heartbreaking for the shepherd and a serious financial loss to the owner to see thriving hoggs suddenly keel over and die when turned away to the hill. A vaccine is available; inoculation protects the animal for life.[21]

Redwater fever in cattle

This is a bacterial disease (*babesiosis*) transmitted by ticks that only affects cattle. Ten to fourteen days after the tick bites the animal, clinical signs appear. The bacteria destroy red blood corpuscles, releasing haemoglobin which is passed in the urine, producing the characteristic 'red water'. Initially, the animal suffers from diarrhoea, fever and raised temperature. In-calf cows may abort and become unthrifty. Healthy animals rarely die but check and suffer a loss of condition and productivity. There is no vaccination against the disease but an injection of Imazol gives six to eight weeks protection.[22] Animals must respond to a direct bacterial challenge to achieve active immunity; they have to be turned out onto affected pastures immediately after treatment so there is time for them to be challenged before the effect of the inoculation wears off.

Strongyle worms and grouse disease – history

Though grouse disease was first described in 1838, it had been recorded earlier.[23] Grouse are commonly internally parasitised by tapeworms and the threadworms of which the most important economically is *Trichostrongylus tenuis*. The worm is a serious pest over much red grouse range (and is recorded in the closely allied willow ptarmigan).[24] It has received significant scientific attention on account of the mortality it tends to induce when grouse numbers are high. Considerable effort has gone into the development of prophylactic techniques aimed at reducing the pathogenic effects of high worm burdens. The biology of *T. tenuis* is well known – it is unnecessary to duplicate the many adequate descriptions here.[25] Of more direct concern are the courses of action open to the manager who recognises that worms may threaten his grouse. Adult worms are easily counted; many assess the worm burdens in their grouse as a matter of annual routine by sampling birds picked up dead as well as those shot in autumn. The birds, or at least their intestines, are sent to laboratories which report on the status of adult worms for a small fee. The technique is simple; many keepers carry out their own counts. People talk freely about worm burdens but adopt a simplistic 'low = no problem, high = disease next spring' interpretation.

Unfortunately, things are not so clear cut. Adult worms are present in grouse as a result of an earlier or *retrospective* event (the number of infective larvae ingested earlier in life) but what the manager and biologist need to know is the *prospective* effect of the worms, particularlty in the subsequent breeding season. This is known to correlate poorly with the number of adult worms present in autumn and only loosely with the number of worm eggs deposited on the ground.[26] The eggs require specific conditions to develop into infective larvae capable of damaging the host and appearing later as adults in the bird. Humidity and temperature have to be within narrow limits and, to be really harmful, the ingested larvae have to become *hypobiotic*, the term for a phase of arrested development in the host's gut wall. Larvae which develop 'normally' after ingestion and turn into the adult worms that are counted in autumn are unlikely to have a life-threatening effect the following spring because they are benign parasites. They float about in the lumen of the caecum adsorbing nutrient from the partially digested soup of grouse food which surrounds them. The hypobiotic worms that burrowed in the host's caecal wall for the winter pose a far greater threat to the host but are not included when adult worms are counted. Their number reflects not only past conditions for infection but also a prospective effect on the host's future. When the dormant larvae hatch synchronously in spring (timing probably controlled by changes in food quality on the onset of plant growth), they damage the caecum lining and impair its efficacy as a digestive organ. There may be a serious loss of body fluid and internal bleeding that lead to ineffective food absorption, debility and, in extreme cases, death. It is these worms that should be counted because they are the ones that cause 'disease'and because it is their numbers that relate closely to breeding success the following year.

Methods of control

Medicated grit

Towards the end of the 1980s, 'medicated' grit began to be promoted heavily by the GCT which had patented a process whereby grit particles were coated with fat containing a proprietory anthelmintic. Many keepers were encouraged to use the material on their moors though many often did so against their better judgement. Reports from sceptics at the time often indicated that the grit was not being consumed in significant quantities; for reasons already given (see Chapter 4, *Grit*) grouse can be reluctant to take artificial grit if plenty of natural material is available. The money spent was often wasted. Unfortunately, the technique was never rigorously evaluated by anyone other than the GCT at the outset nor has it been since. The bumpy ride it has been given in some quarters could have been avoided had independent trials been conducted.

In recent years, some owners, particularly in the north of England, have put out large tonnages of dressed grit containing Fenbendazole from the end of the shooting season in December until mid-July following.[27] At over £1000 a tonne, it is expensive, particularly if applied at a rate of a tonne per 400 ha (1000 ac). However, its effect has been remarkable on many moors. Routine counts of adult worms over whole districts have been low despite consistently high grouse numbers in recent years and the fall of worm eggs onto the ground is often all but non-existent. Moors which in earlier years could have expected a population crash triggered by disease have not had one. It appears that the material has effectively negated the effect of worms and the disease outbreaks which used to follow high numbers almost inevitably. However, even owners convinced that medicated grit delivers many more grouse for shooting would be unwise to use it at high levels all the time as the development of resistance is all but inevitable. Just because it has not occurred to date does not mean it never will.[28] The history of anthelmintics worldwide teaches us that resistance *always* appears in the target parasite sooner or later.

It seems that the chemist has achieved a partial solution to the problem of *T. tenuis*; the anthelmintic is undoubtedly contributing to the present run of good, even exceptional, bags particularly in the north of England. As this is unlikely to be permanent, new research and field trials should be being put in hand now so that, when the breakdown occurs, new anthelmintics will have already been developed, tried, tested and approved for general use.

'Catch and dose'

Twenty-five years ago, observant keepers lamping foxes at night noticed they could make a close approach to roosting grouse as they took little notice of them. It was a short step for keepers to start catching the grouse with a landing net and giving a dose of anthelmintic before releasing them with a marker on leg or wing.[29] It has been shown that direct dosing using a liquid worm drench (catch and dose) clears out adult worms in the same way as farmers control worms in domestic stock. The technique was reckoned to be better than putting wormer onto grit in the hope that birds somehow take the correct dose. Clearly, trying to get medication into a wild bird *via* grit is rather 'hit or miss'; some birds overdose, some take minimal amounts that may well have no effect and others pick up 'wild' grit exclusively. Dosing birds, on the other hand, was found to be highly effective in killing adult worms living in the lumen of the gut. The problem is that grouse tolerate these worms with no apparent ill effects; it is larvae encysted in the gut wall which do the damage.[30] As these are out of reach of the wormer and, because it has no systemic effect, are unaffected by it, it is questionable whether the treatment is as valuable as is made out to be. There is some evidence that, in some years, particularly the real 'disease' ones when the mortality of breeding birds is high, some treated birds may rear slightly more chicks and others, which would normally have died, do not succumb.[31] If this is the case, it is probably worthwhile to spend time catching up birds in certain years. In the summer following a wipe out, a chick is a chick and every bird is valuable; the more there are, the more rapidly the population recovers. Some people have become almost obsessed with 'catch and dose' with individual grouse being caught many times. There are cases where marked birds are five, six and seven years old with a series of patagial tags to prove it.[32] Regardless of how successful it is in controlling adult worms, it has not been shown to affect encysted L3 worm larvae. Whether the larvae are even susceptible to treatment administered by mouth while they are dormant is currently unknown. If encysted larvae are destroyed by 'catch and dose', it implies a systemic effect. Systemic chemicals should never be administered to vertebrates (particularly wild ones) because of the risk of their entering the food chain.

Worms as population limiters

Why do worms build up in grouse in the first place? The simple answer is that they are a symptom of over-population. Numbers rise, parasites build up and adversely affect the hosts; numbers decline. This is obviously an oversimplification but it is essentially correct. It is in the manager's power to limit population increases but, by the time he finds out his grouse are carrying large numbers of hypobiotic larvae, it is too late to prevent the forthcoming disease outbreak. He can at least reduce the financial impact.

Lovat described in 1911 what he considered to be 'safe' stocks on moors in various parts of the country and advocated hard shooting with good guns early in the season to crop the surplus. By 'safe' he meant breeding densities which should not suffer from disease. He was keen on assessing breeding success in advance so proprietors would not be taken aback by extra-large populations of birds when shooting started in August. In this, he was years ahead of his time; only in the last thirty years has grouse counting became general practice. As late as the 1960s, many keepers and owners would have nothing to do with counting either in July or March as they distrusted the data obtained. Having shot the desired amount of grouse by fair means, Lovat advocated 'cocking' (see Chapter 4, 'Cocking') in the belief, subsequently discredited, that two-year-old cocks were necessarily harmful to the future productivity of a moor. On a well-shot, well-burned moor, cocking brought the density of breeding pairs down to slightly below maximum, reduced inter-pair strife over territory and made slightly more space and food available to each pair. Others advocated it without understanding the mechanism.[33] Waddington recommended that shooting in good years should be hard during the first three weeks of the season (to crop a high proportion of old birds) and in late November and even into December (to reduce territory owners which would not necessarily be replaced). The effect is to keep the population slightly below maximum and so reduce the risk of a cyclic decline. Lovat had plenty of experience of disease and was nervous of its effect on shooting in subsequent years but was able to mitigate its effect, at least in major part, by applying the technique, which is entirely consistent with the 'hard shooting' school of thinking to which many managers subscribe today. 'Fire' and 'lead' together are important management tools. The aphorism hints at another truth – moors well burned with many small fires do not seem to suffer the devastating outbreaks of disease that less well-burned ones do. Once again, the old authorities identified the right result for the wrong reasons, their hypothesis being that intensive burning gets rid of 'unhealthy old heather' and replaces it with nutritious young growth which supports fewer parasites.

The real cause is probably different. Although there seems to be no record of pertinent fieldwork, casual observations record a seemingly disproportionately high frequency with which overnight roosting piles (with caecal droppings carrying worm eggs) occur on newly burned or pioneer heather or on bare ground such as short grass and hill tracks. Grouse seem to seek out the most open sites they can find on which to roost for the night, probably as an anti-predator device. As grouse do not feed on pioneer heather or bare ground, the effect is to distance dangerous caecal droppings from the birds' dining table. On the well-burned moor, the birds are implementing a natural sanitary strategy whereas on poorly burned moors carrying a high density of birds (and therefore a population at risk), there is, inevitably, an increased chance that surplus and territorial birds alike roost overnight on a 'dining table'. The hypothesis is untested but could explain why under-burned moors, especially those overlying rich rocks, can produce high breeding densities and enormous bags at the cyclic top and crash with cataclysmic suddenness while similar but better burned moors nearby may decline in sympathy but do not crash to the same low level.

Worms and cycles

There has been intense debate between GCT and the ITE's Banchory Grouse Unit in recent years as to the relative importance of intrinsic (behavioural) factors and extrinsic ones such as worms or weather in driving the grouse cycle. Without re-opening deep and only partially healed wounds, it should be noted that grouse cycles can be driven by intrinsic *and* extrinsic influences. Both viewpoints are tenable. Regional differences affect the likelihood of which hypothesis applies where.

Extrinsic factors

Over much of the north of England, grouse numbers fluctuate on a cycle of four to five years. The conditions which facilitate this are wet moors (leading to high rates of worm infection); moderate to poor standards of burning as assessed by Lovat particularly of wet heath; rich underlying rocks capable of supporting high breeding densities that result in large broods and, finally, under-shooting when populations are large in the years before a peak. In combination, they provide a predictable sequence of shooting results. The mechanism appears to be as follows.

Grouse choose to roost on bare ground.

Year 1: a thin stock of birds following a disease outbreak; excellent breeding on a low stock; the proprietor shoots lightly to conserve stock and encourage recovery and a rapid build-up of numbers.

Year 2: a fairly large but sub-maximal spring stock which breeds well; a large standing crop in August yields first class bags; the proprietor under-shoots either by accident or design though he recognises that the population is at risk; worm burdens are low but set to rise fast.

Year 3: a very large stock which breeds poorly because of disease and inter-pair strife; very large bags – the proprietor shoots and shoots but makes little or no serious inroad on the annual crop and so fails to get the stock down to Lovat's 'safe' level.

Year 4: a huge stock in spring; members of a high-density population have to expend more and more precious energy as the year goes on to maintain territory against neighbours. If a bird is carrying hypobiotic larvae which hatch synchronously in spring as territorial strife is at its height in late March, the energy demands become insupportable and death is likely. Such stock as survives rear scarcely any young because most adults die of disease; very few birds present in August, no shooting.

In this case, the cycle lasted for four years because of large brood production on a 'rich' moor largely free of fox and crows. It lasts longer in places where mean brood sizes are smaller and perhaps as long as fifteen years in places in Highland Scotland where three young per hen represents a good year. Population crashes in Scotland are just as often associated not with worms (extrinsic cause) but changes in behaviour (intrinsic ones). Both hypotheses are pertinent. Critics from the 'intrinsic' camp point to dry eastern moors, notably the Lammermuirs, North York Moors and low-lying ones on the eastern side of the Pennines which are not cyclic in the manner of moors elsewhere.[34] Numbers rise and fall but do so independently of worm burdens. Within each group of moors, there may be sub-populations cycling as a discrete entity out of phase with neighbouring ground. When one sub-population is in the increase phase – leading to a large-scale emigration a year or two later – losses may be made good by compensatory movement inwards from round about. Though unconfirmed, the notion is based on personal observation on a whole moor scale in areas like the Lammermuirs. Individual moors are burned in a variable manner; some are managed much better than others and support high spring breeding densities. Others yield bags that do not reflect their habitat management. Less-good neighbours with a few large shepherd fires enjoy good shooting days with bags of a size they do not deserve. A minority of poorly managed places may act as a safety valve for the better managed ones which surround them. They provide holiday camp accomodation for emigrant coveys from high-density breeding areas nearby. On shooting days in August and September, stretches of uniform-aged heather may hold plenty of grouse and provide bags which satisfy both owners and sportsmen. By mid-October, they hold a low density of territorial birds reflecting the poorer breeding territory though there may be plenty of birds in flocks floating around the moor and doomed to die over winter. This is another example of the 'sink effect' (Chapter 4).

Intrinsic factors

Grouse cycle without the stimulus of worms. Work carried out by the Banchory team at Kerloch (NO696879) and Riccarton (NO814893) over twenty years demonstrated that neither worms nor weather drove cycles on either place.[35] Numbers built up to a peak, breeding success declined, aggression levels changed and populations declined but the decline did not correlate with worm burdens. During the increase phase, territorial aggression was low. When a decline was triggered by overcrowding, aggression increased; during the decline phase, the spring population contained a number of unmated cocks. In biological terms, the presence of unmated, non-breeding birds is a sophisticated way of reducing recruitment. The Banchory work showed that, if a spring population was reduced by shooting cocks in the winter after the territorial fabric had been established, the population did not decline whereas, on a neighbouring site where manipulation did not take place, it did.[36] A trigger mechanism rooted in over-population set off the decline sequence that could be overridden if management intervened. The finding was important in that it bridged a bitter gap which had opened up between scientific workers and provided convincing evidence that a tactic of hard shooting can prevent cyclic fluctuation. It also lent credibility to the mantra so often heard in shooting circles that 'such-and-such a moor crashed or got disease because they did not shoot enough'. Of course, it also provided indirect evidence of the efficacy of cocking.

Tapeworms – effect and importance

If a covey of grouse is flushed in summer, it is common to see a cream-coloured streamer of worm segments spurt out of a bird's bottom. As the worm's other host is reckoned to be cranefly, the chicks probably become infected when they are very young and feeding in damp flushes. The worm *Paroniella urogalli* is often accompanied by a microscopic tapeworm *Hymenolepsis microps* but both appear to be well tolerated by grouse as affected birds are rarely in poor condition.

Bacterial disease

A century ago, coccidiosis was regarded as a serious scourge of grouse and a killer of chicks.[37] It was recorded in blackgame in the 1930s.[38] The causal agent is a bacterium *Eimeria avium* which attacks the gut lining and causes the bird to contract an enteritis which is often fatal. It was clearly important a century ago as in *The Grouse in Health and in Disease* it attracted thirty-seven pages compared with forty-nine for strongylosis. We know that grouse densities were high over extensive areas at that time. They tend to be generally lower today which probably explains why the disease is now all but unknown. Even in the north of England, where as many as two birds per hectare can be shot annually, the disease is not seen as a problem for managers although it is thought to occur on some moors in some years.[39] There is no suggestion that shortages of grouse today can be attributed to coccidiosis outbreaks though it is common in free-range poultry kept on 'dirty' ground, particularly in warm, damp summers. The author's view is that there is a distinct possibility that, with serious strongyle worm problems possibly a thing of the past, the factor that will limit grouse populations in years to come could be the coccidiosis that constituted such a widespread problem in grey partridges a century ago.

In summary, there is always a catch when the performance of any animal is pushed to the limit. As the damaging effect of one pathogen is eliminated, another invariably arises; truly can it be said that 'No tree ever grows to heaven'.

End notes

1   W. Chisholm, pers comm.
2   *Sgard* is the Gaelic for to squirt or sprinkle, *sgaird* is diarrhoea or loose stools and *ruaidh* means red.
3   Duncan *et al*, 1978; Reid *et al*, 1978.
4   The same can be true of vegetation on the side of sheep and deer paths in woodland or on the open hill.
5   Author, pers obs.
6   Smith, 2005.
7   Jones *et al*, 1997; Gilbert *et al*, 2000; Laurenson *et al*, 2000; Laurenson *et al*, 2003.
8   Watson *et al*, 1973.
9   Newey *et al*, 2008.
10   Harrison *et al*, 2010.
11   Rochlitz *et al*, 2010; Turner, 2010.
12   Newey *et al*, 2008.
13   Smith, 2005.
14   Merricks, 2010.
15   This was a strictly commercial decision, not one driven by the concept of 'tick mops'.
16   M. K. Browne, pers comm.
17   Reid, 1975; Reid & Moss, 1980.
18   Warren & Baines, 2007.
19   Phillips & Duncan, unpubl.
20   Hudson *et al*, 1995.
21   In 2011, £6.00 per dose.

22   Active ingredient Imidocarb; cost about £16 per adult animal in 2011.
23   Macdonald, 1883; Speedy, 1886.
24   Holstad *et al*, 1994.
25   Hudson, 1986.
26   Moss, 1989; Delahaye & Moss, 1994.
27   There is an obligatory withdrawal period of four weeks between ending medication and the birds being regarded as fit for human consumption.
28   Schoenian, 2008; Cox *et al*, 2010.
29   Levamisole formulation.
30   Delahaye, 1992; Delahaye & Moss, 1994.
31   Newborn & Foster, 2002; Newborn, 2004.
32   In wildlife management terms, the best thing the keeper can do when he encounters such aged birds, be they cock or hen, is to knock them on the head because, wormy or not, they are occupying space likely to be better used by juveniles.
33   Waddington, 1958.
34   Moss *et al*, 1990a; Moss *et al* 1990b.
35   Moss *et al* 1993.
36   Moss *et al* 1996; Hudson et al, 1998.
37   Fantham, 1911; Thomson, 1921; Cazenove, 1936.
38   Scott, 1937.
39   J. Grave, pers comm.

Chapter 7
Afforestation, fragmentation and pollution

Proximity to woodland – the problem

Our farming and sporting forebears of a hundred years ago would scarcely recognise large tracts of Scotland and Wales today, particularly where national forest policy has had its chief effect. Since the formation of the Forestry Commission in 1919, the Scottish national forest estate has jumped from five per cent (385,000 ha) to seventeen per cent (1.3 million ha) of Scotland's land surface in 2010. In 1998, the Westminster government introduced a presumption against large-scale planting on unenclosed upland in England, a measure which has safeguarded large areas of open space and prevented ugly forest sprawl. No such constraints apply in Scotland where substantial sums of public money are even now spent on schemes designed to establish forest, particularly native woodland, on any available stretch of sheep walk or deer forest.

Red deer populations will have to undergo reductions if the present forest targets are to be attained.
(Photo: N. McIntyre)

The Scottish Government proposes that forest cover be increased from seventeen to twenty-five per cent by 2020, an increase of *c*500,000 hectares or more than 60,000 ha per year, much of which will be at the expense of heather moorland. It would confer modest conservation advantages locally but would otherwise be useless as anything other than amenity woodlands. In a resource-poor country like Scotland, public investment on this scale is quite an indulgence.

The bald figures conceal the alarming fact that most twentieth-century afforestation has been on unenclosed upland; Scotland holds seventy-five per cent of the world moorland resource, twenty-four per cent of which (1034 million ha) has been lost in the past sixty years, eighteen per cent of it between 1940 and 1980. Dumfries & Galloway has lost a staggering sixty-three per cent of its moorland over this period (moorland managed for shooting lost twenty-four per cent and non-grouse moors forty-one). Though the FC planted the majority of new acreage in its first fifty years, private sector acreage had overtaken the state by the 1970s. The programme was driven by a tax and grant regime under which trees were planted using pre-tax income but treated as capital once the trees were established. Some proprietors boasted that the entire cost of forest establishment had been met by planting grants. Though it undoubtedly developed the national forest at small cost to the exchequer, such fiscal largesse was always open to abuse. Trees were planted on unsuitable sites and occasionally on places of international conservation interest such as the Caithness Flow. The entire head-water catchments of some extensive river systems were planted, ugly blocks of exotic conifers were scattered across parts of the country with scant regard for their effect on landscape or adjacent land users and many others were placed randomly on open hill ground.

Though the tax loophole was closed in the 1988 Budget and the scheme phased out by 1993, inappropriate afforestation remains a problem. However, as forests planted by both private and state operators in the 1960s and 70s come to the end of their rotation, managers have the opportunity to re-design and re-shape the hard-edged and extensive blocks and are encouraged to do so under conditions of grant. In some cases, replanted areas contain a significant propor-tion of native woodland species, a substantial improvement on the extensive sweeps of exotic Sitka spruce and lodgepole pine they replace.

Heavily subsidised but essentially uncontrolled tree planting had run amok for seventy years; *bona fide* farmers could no longer match land prices inflated by the forestry boom. Sound hill farms stretching from Ross-shire (NC424101) to south Scotland (NS746972) were bought up for trees and good stretches of productive heather moorland in Kincardineshire (NO665835) were converted to forest despite intense local opposition. The investment forest industry had become a force of some power with criticism of it often muted. Farmers muttered imprecations among themselves but were conscious of the financial safety net forestry offered if, for example, they wanted to quit farming. The landowning community, seemingly slightly in awe of forestry investors from outside, were equally silent, not least because they exploited the same tax breaks as the incomers though on a smaller scale. Conservation bodies were overwhelmed by public relations campaigns of the forestry companies pouring out streams of massaged facts about discounted net revenue, employment opportunities, balance of payments deficits and conservation advantages resulting from the creation of the new forests. The claims were almost entirely spurious and largely over the heads of country people who nonetheless knew a scam when they saw one.

As the forests grew, so did incidental but largely unforeseen problems. The forester typically had neither hill farming nor moorland conservation skills; in farming districts, he was a 'foreigner' to be greeted with suspicion and handled with care who rarely lived on site and worked out of an office many miles away. Many of the job opportunities in the new forests were of limited duration, required specialist skills and rarely went to dispossessed local farm people. Once the initial works of planting, fencing, road construction and the like had been completed, the forest was effectively a dead place where one could walk for days and see no one. As spring approached and heather burning and lambing time drew near, officious correspondence from the forester stressed the need for fire precautions but the farmer who sought help from forestry staff with routine moor burning was often obstructed by bureaucratic administration and clock-watching staff. That said, prior to 1988, individual foresters, some of whom were formerly employed on the farms which the forests replaced were often helpful to farming neighbours with fox control in particular. Many would run snares in the forest, shoot foxes on sight, help at fox dens during lambing and encourage keepers to 'lamp' foxes at night. Later, as forestry margins tightened, the obvious savings to make were ranger and warden staff; the cutbacks changed everything. Remaining staff were instructed to shoot deer to protect trees and offset their wages with the sale of venison. Wardens told neighbours they no longer had time for systematic fox control and nodded at the 19 81 Wildlife and Countryside Act which made it compulsory to inspect snares every twenty-four hours. As they were on a five-day week, they argued, they caught little or nothing if snares were set up on a Monday morning and 'pulled' on a Friday. When farmers offered to 'look' the snares themselves, wardens claimed that higher authority forbad it and suggested that foxes were in any case not pests in the forest because they kept down capercaillie, black grouse, rabbits, hares, mice and voles and so prevented tree damage. Relations between forester and farmers, shepherds and shooters soon deteriorated to their present low level.

As local authorities and other bodies opposed inappropriate planting applications in Scotland, the FC came to accept that safeguarding landscape required exchequer pressure. Grant approval conditions for planting schemes currently stipulate that twenty per cent of a proposed forest area is left unplanted and that the species mix includes trees of conservation merit but little commercial value. This is a step in the right direction but forestry nevertheless remains distorted by a grant system which leaves landowners with substantial profits as soon as they complete planting and establishment – long before any harvesting revenue comes on stream. As a well-known Galloway landowner observed 'currently, grant is being used to create thickets of rubbish on otherwise useful land. In fifty years' time, when the country is land-hungry again, the mess will be considered offensive. Clearing it up and turning it back into something useful will become a priority and grant aid will be available to do so.'[1]

Habitat fragmentation and the effect on game and other wildlife

Where game occupy large areas of contiguous good habitat, populations remain high and resistant to natural disasters but, as soon as habitat becomes fragmented, weak points in the population are created (see Chapter 4, *The sink effect*). The subsequent deterioration may be due to overgrazing, land use changes (afforestation, land reclamation) or plain bad management and neglect of the heather moorland. Initially, good habitat vastly outweighs the bad and the effect is scarcely noticeable – a block of 400 ha (1000 ac) of spruce forest or bad burning in 10,000 ha (25,000 ac) of well-managed moorland is not a problem if keepers round about are competent. But when tree plantations and/or neglected heather moor make up a significant percentage of an area, the negative effects start to appear. At whatever level, the process is commonplace on heather moorland in Scotland and England; only in isolated pockets (notably in the Lammermuirs in southeast Scotland and in the English dales) does one find blocks of land in excess of 20,000 ha (50,000 ac) without serious problems stemming from land use changes or heather mismanagement. It should not surprise us to find that the grouse bags obtained in the more intensively managed areas rival those of a hundred years ago. Particularly in the north of England and some Angus glens, dynamic modern moorland management allied to heavy investment has dramatically increased bags in recent years.

Fragmentation takes many forms. The effect of poor standards of burning, grazing and tree planting are obvious but it can also be a reflection of district-wide factors such as the levels of professional interest and investment and the competence of owners, managers and keepers in executing management plans.

Varying standards of moor management. (1) Moor reclaimed by farm tenant in 1970s; (2) Heather badly under-burned; fires too small and rotation too long; (3) Well-burned moorland currently yielding days of a hundred brace; (4) Over-grazed/poorly burned yielding very little. (NN900367 – Settlement, 1997).

Pollution

We live in a pollution conscious age; *Homo sapiens* faces serious environmental problems due largely to the waste products that he generates. We are all familiar with the gross effects of pollution – atmospheric CO_2 rising relentlessly, oil tankers running aground and destroying marine life, power station emissions falling as acid rain and killing fish in Scandinavia, SO_2, nitrous and nitric oxides (NOX) causing the chronic ill health of German forests and power stations disseminating radioactivity over wide areas – incidents highlighted by environmental pressure groups who point accusing fingers at mankind, the wanton consumer.

Much less widely known is that minute quantities of cosmic dust arrive constantly on earth, a form of natural pollution occuring every hour of every day – when it rains, materials from space are trapped or dissolved in water droplets. When the wind blows strongly, tiny droplets of sea water are picked up and form an aerosol which ultimately falls to earth carrying the salts. None of these nutrient sources is large, especially when compared with the volume of artificial fertilisation, but they have landed indiscriminately, winter and summer, spring and autumn since the beginning of time. Though it is correct to describe them as pollutants, they arrive in minute quantities and typically have a beneficial fertilising effect.

Problems arise when dissolved salts increase to the levels at which certain organisms become attenuated or die. One man's meat is truly another man's poison: NOX falling on a neutral or base-rich soil can increase nutrients to the advantage of plants and the organisms which feed on them but the same deposit falling on sphagnum moss growing on blanket bog kills it. Sphagnums are the building blocks of peat formation; the climate is too hot and dry for large-scale peat formation over most of Britain today though there are local exceptions in west and north Scotland. Peat is lost naturally to oxidation and erosion but the process is accelerated by the historical loss of the sphagnum understory in heather moorland particularly over much of the Pennine chain. Even if the climate becomes cooler and moister in the future and so again favours peat formation, it will not form as long as sphagnum is absent. The loss is the direct result of one hundred and fifty years of industrial fallout from the north Midlands industrial conurbations. To many, this probably seems unimportant; who cares if eight per cent of the Peak District National Park is covered in bare peat?[2] A few far-sighted ecologists and conservationists may worry, a hill walker or two may recognise eroding peat hags, water utilities may incur costs de-colouring water stained by humic acids dissolved and washed out of peatlands but nobody else expresses much concern. The economic value of the land is small and few livelihoods are threatened as the process continues apace. People get excited about wrecked Alaskan oil tankers, destroyed Amazon forests and poisoned European forests but the lingering death of moorland on our very doorsteps never makes the TV news, carries little weight and attracts few votes. Conservation groups and the wider eco-lobby are as silent as any. Their silence on the conservation of peat was deafening as a majority abandoned high-profile campaigns to defend peatlands once the power supply sector developed an interest in developing peatland as wind farms. Though the peatlands of Britain are vulnerable to pollution, moorland positively benefits from air-borne inputs – the nutrients otherwise available to plants on a northern heath are limited. Nitrogen, calcium and phosphorus are particularly scarce, shortages which limit the performance of grazing animals. Pollution may explain at least in part why many of the grouse moors overlying nutrient-poor millstone grit in northern England produce large bags. The classic example is Broomhead moor in Yorkshire. Its 1600 ha (4000 ac) lie at the eastern end of a spur of moorland which peters out northwest of Sheffield above Stocksbridge. At the turn of the nineteenth century, Stocksbridge was a grimy place indeed; it had been the first site in Britain to use the Bessemer steel-making process which involved 'blowing' molten iron to oxidise impurities and produced significant base-rich airborne effluent high in phosphorus. Some would have fallen on the moorland and been taken up by the heather. Eyewitnesses working as pickers up on shooting days a century ago describe the grouse as covered with grime and the sheep as a dingy grey-black because of the fallout. In 1913, a record number of 1421.5 brace of grouse were shot in one day by nine guns, though the record was to stand for a mere two years before being broken by Lord Sefton and his party who shot 1464.5 brace to eight guns at Abbeystead, Lancashire on 12 August 1915. Broomhead overlies infertile millstone grit which had been constantly enriched by the steelworks, Abbeystead overlies limestone.[3] In 1916, the Bessemer process was superseded by the Siemens-Martin open-hearth furnace which produced less effluent and hence had less fertilising effect. Although big bags were obtained at Broomhead in the 1930s and even after the Second World War, they never approached those of the early years of the century. Bags in the same region of the West Riding but away from any Bessemer influence were more or less steady over the same period.

Until the end of the 1970s, many distilleries in northeast Scotland routinely discharged the lees from their pot stills (which had to be disposed of after distillation) onto moorland nearby. As the material was toxic to heather, significant areas were changed from good-quality dry and accessible moorland to rushes and acid grassland. Like the effluent which seeps from a poorly designed silage pit, it was also damaging to fish stocks. Heather killed by these pollutants was often upgraded into improved grass pasture by re-seeding in subsequent years. Happily, the practice has been abolished.

End notes

1   W. D. Kennedy, pers comm.
2   Phillips *et al*, 1981.
3   Though Abbeystead is three times the size of
    Broomhead, approximately the same acreage was
    shot on each day; the Broomhead bag is broadly
    comparable with the Abbeystead one despite
    Broomhead having less natural advantage from
    its soil.

Chapter 8
Moorland regions – a summary

The north of England

The only part of Britain where grouse populations bear comparison with those of a hundred years ago is the north of England. In many districts, numbers are in fact substantially higher due to a better understanding of heather burning and management coupled with an end to the 'industrial' poaching to which almost all writers of the mid-nineteenth century refer. Though problems persist throughout the region and some, particularly those associated with commons and stints are more severe than they are in Scotland, managers deliver consistently high numbers. Some are producing bags of 5 birds per ha (2 per ac), many times that of nearly every moor in Scotland. How are these results being achieved?

Underlying geology

A crucial feature of the scene in the north of England is that a majority of its moors overlie good-quality, base-rich rock. A glance at the solid geology map covering the Pennines shows extensive areas of carboniferous limestone and of millstone grit which, though not as naturally productive as limestone, still supports high grouse breeding populations. They do so partly because of good heather management but also because wind-borne fertilising particulates probably serve to make moors on Category 1 rock (granites and grits) more akin to the richer Category 4s and 5s (basalts, diorites and limestones (see Chapter 7, *Pollution*).

Fragmentation

English moorland tends to occur in extensive, relatively unfragmented blocks. The significant areas of forest that encroach onto sweeps of open moor can be counted on the fingers of one hand. Apart from the vast Kielder complex to the north and the North Riding Forest Park at the eastern side of the North York Moors, few moors adjoin forest. Slaney forest lies against Blanchland moor but there is nothing else of significance for some 32 km (20 miles) south to Hamsterley Forest between Weardale and Teesdale. It is another 19 km (12 miles) south to the Stang between Teesdale and Swaledale and 32 more to Langstrothdale at the head of Wharfedale. Further south by 22 km (14 miles) lies the Stocks Reservoir forest in Ribblesdale. Patches of woodland, some large, between forest blocks provide havens for foxes and crows but the pattern overall is very different from the core grouse areas in the Southern Uplands and highland Scotland.

Fragmentation of heather ground by stretches of 'white' ground can cause problems. Heather has been lost as a dominant plant in many places in Cumbria, the Howgill fells and the western dales because of over-browsing by sheep in winter. The resulting white savannahs support low or non-existent breeding grouse populations but may locally contain immigrant broods seeking space in summer and autumn. White ground can provide excellent shooting but it lives off more productive heather/blaeberry ground nearby (see Chapter 4, *The sink effect*).

Keeper density

The density of gamekeepers is a great deal higher in the north of England than in Scotland. Over ninety per cent (350,000 ha) of England's heather moorland (over twenty per cent of its uplands) is considered as grouse moor and owned by members of the Moorland Association who employ between them three hundred and fifty full-time keepers, an average of one per 1000 ha (2500 ac).[1] It is not uncommon to find estates of 2000 ha (5000 ac) with two or more full-time keepers and seasonal help from a lad or pensioner. Many small estates of under 600 ha (1500 ac) employ a man whose main, perhaps sole, duty is to look after the moor. As a result, common predators such as foxes, crows and stoats are kept at low levels over thousands of hectares. Because large areas are also devoid of woodland and isolated trees are scarce, the suite of predators is much reduced. With little predator pressure, intensive grouse cultivation is significantly easier.

Base-rich soils grow heather plants that are higher in nutrients than those growing on acidic granites or deep peats. Good food creates large broods of young raised on small breeding territories, resulting in big standing crops of birds at the start of the shooting season. It is not uncommon to find areas carrying over 7.5 birds per ha (3 per ac) in August, leading to quite modest-sized drives that contain literally thousands of birds. Between half and three-quarters of these can be cropped without reducing the subsequent breeding density meaning that an area of only 400 ha (1000 ac) can yield an annual bag of seven hundred and fifty brace or more. In contrast, the best-managed Scottish moors overlying the richest rock would be yielding exceptionally good bags if four hundred birds per 400 ha were shot on a ten-year average. Even in the good years of the 1960s and 1970s, only a handful of moors in Perthshire and Aberdeenshire achieved those levels; it is questionable if any are doing so now.

English moor owners are thus better insulated against the financial problems commonly experienced in Scotland due to a shortage of grouse. A two-day driving moor in the north of England has the prospect of shooting four thousand brace in a season; unless the owner is leisured and has hosts of friends, the likelihood is that a portion of that shooting will be let. The manager can market six or ten two-day packages at £150 per brace or more and gross close on £200,000. That goes a long way toward providing for the general upkeep of the moor, the cost of two keepers and a handsome profit.

English moors readily forgive poor management even where there are extensive patches of over-browsed heather. They may suffer local damage because they are used as fothering sites and trampled by many wintering sheep but they still carry territorial pairs of grouse in the spring. Essential cover is supplied by straggly stick heather, soft rush and tufts of cotton sedge although it may be sparse. The moor may appear to be bare from a distance but avian predators are scarce and territorial birds tend to over-winter successfully. In spring, the few heather and blaeberry plants which have escaped the attention of the sheep provide food nutritious enough (particularly if some cotton sedge catkins survive) to bring the birds into good breeding condition. It is normal to find a dozen or more young in each covey even though broods do not occur at the same densities as on the main heather moorland.

Some moors have the disadvantage of having been designated as common grazings, many of which accommodate far too many sheep, particularly in winter.[2] Tackling the over-use is difficult because it is the result of economic pressures. A moor may be subject to stints, an essential part of the sheep system that can be traded between farmers. Typically, farmers in hill districts use small areas of in-bye ground to the maximum at tupping and lambing as well as for growing hay or wrapped silage for winter forage. As soon as possible after lambing, they turn the ewes and lambs out to the hill where they stay through summer and autumn into early winter. As stocking densities of four or more ewes per ha are common, over-browsing inevitably puts heather into regression. The problem is essentially social in origin. Modest-sized farms carrying two to three hundred ewes are simply not big enough to provide a family with a decent living – they need three or four times as many animals to be viable. The only 'solution' is to reduce the number of farmers in remote places but it is *not* advanced here – the areas concerned already have enough depopulation issues. The problem is intractable; it has been around for many years and will be with us for many more.

In the north of England the grouse cycle is normally short; peak populations are usually followed by devastating crashes. The declines are usually driven by worms in England whereas in Scotland they are more commonly caused by behavioural changes unconnected with worm numbers (see Chapter 6, *Worms and cycles*).

Keeper activity and effectiveness

Grouse keepering is a different activity in England and Scotland. A Yorkshire keeper is likely to have charge of 1000 ha (2500 ac) of hill, his opposite number in Inverness-shire up to 4000 ha (10,000 ac). The Yorkshire man can get all round his beat every day and micro-manage it. As his annual burning programme amounts to a modest 80 ha (200 ac), he invariably gets enough suitable weather to fulfil it so his rotations are short and as a result his hill tends to be dominated by young heather. He has few predators on his ground and the time to run literally dozens of tunnel traps for small ground vermin, as do his neighbours. If he gets an influx of crows or foxes in late summer, he has plenty of neighbours to help get on top of them. He can put metal plates on any stock fences that threaten to kill his grouse, can ensure his springs always run clear and has time to re-seed places where bracken or acid grassland encroach on the heather. Often overlying limestone, his ground tends to be naturally productive and 'grouse friendly'. Compared with the man looking after a moor on schist or granite, he is fortunate indeed. His butts are solid and need little maintenance on account of the fine craftsmanship that was the norm when they were built many decades ago. He probably works for someone with plenty of money and so is never short of seasonal help or proper equipment for burning; his quad bike and Land Rover are replaced regularly.

If he has low-ground game to look after, it is probably based on wild birds with any rearing programme being of limited extent. Because there are grouse to shoot most years, his employer is less tempted to diversify into shooting reared partridges on the moorland fringe. His moor is a haven for rare and endangered wading birds. Last but by no means least, he is not isolated socially but is respected as the professional he is and usually enjoys good relations with neighbouring farmers. It is not surprising that the grouse yield from his land is so large and that average bags may even exceed those of a century ago. All in all, his future is secure, even rosy.

Scotland

Underlying geology

The main massif in Scotland is dominated not by limestone but by granites, schists, slates and quartzites interspersed with small pockets of better-quality fertile metamorphic rocks, foliated granites and limestone. To the south of the Highland Boundary Fault, andesite lava and igneous basalt make up the majority of the Sidlaws, Pentlands, Moorfoots, Gargunnock and North Ayrshire hills. These weather into fertile soils. Further south, the productive coal measures of South Ayrshire give way to the broad band of Silurian shale which runs west to east from Girvan to Dunbar, yielding a free-draining soil where heather is productive if it is well managed. Compared with the north England moors, managing vegetation to produce grouse in quantity has to be more skilled and painstaking if farming, conservation and shooting interests are to be reconciled.

Fragmentation

Fragmentation is the curse of Scotland's moorland and has had serious negative impacts on wide areas. We think of plantation forest as the major intruder but, though 'management by indifference' is less easy to identify, it is just as common and just as insidious. A high proportion of what was considered prime grouse land a century ago has since been broken up and now comprises a mixture of overgrazed sheep walk, land fenced for natural woodland regeneration, forest blocks established since the end of the Second World War and land which is barely managed at all because those employed to do so have to look after areas that are far too extensive for their work to be effective. Many forest blocks are approaching the end of their first rotation or have been harvested and are now being replanted. (In the north of England, plantation forest is often not re-planted and is instead returned to heather.[3]) Only at altitude in the Central Highlands can extensive tracts of moor be found without areas of negative influence within a mile or so of their boundaries. Elsewhere, such quality moorland as remains usually occurs in medium-sized pockets mostly in the Lammermuirs, Moorfoots, Pentlands and the Buccleuch/ Linlithgow moors on the Ayrshire/ Dumfriesshire/Lanarkshire border.

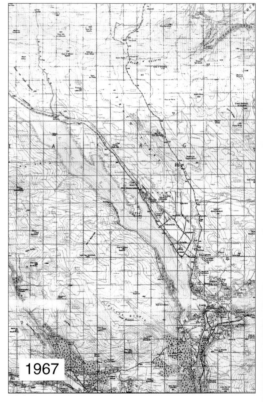

1967

Increased afforestation fragments moorland areas and leads to increases in crow and fox populations.

ABOVE: *The situation in 1967.*

OPPOSITE: *1989 and 2008; illustrates the process which is far from unusual.*

Scotland is rapidly ceasing to be home to the extensive, good-quality grouse shooting it was as recently as fifty years ago. Some places still yield bags comparable with or, in isolated cases, in excess of those of former days but they become fewer as the years go by.[4] Under the fiscal regime current in Scotland, the decline is set to continue and may well accelerate.

Keeper density

Keepers in Scotland are expected to look after 'beats' that are not only much larger than their English counterparts but lie on ground that tends to be less productive for grouse because of its mineral and vegetation composition. It is not unusual to find one man single-handedly looking after a moor of 3000 ha (7500) ac without seasonal help and counting himself lucky he has not got more to look after. Many often have responsibility for large tracts of so-called grouse ground which includes a substantial deer interest. With so many demands on

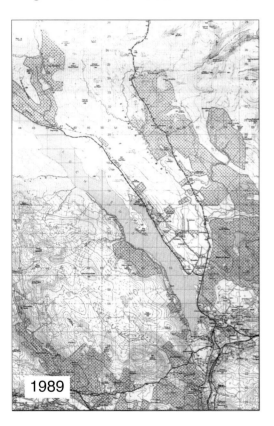

1989

2008

his time, his keepering efforts often become too diluted to exert much positive influence on small-game performance.

Granites may support extensive sweeps of purple heather which gladden the eye of the tourist in August and give the appearance of being wonderful for shooting but the swards tend to conceal fewer birds than mottled, grassy mosaics on more fertile rock. No amount of keeper effort can change that: under a given heather management regime, ten-year-average bags are *always* lower on moors overlying

Roe stalking can make strong demands on a keeper's time and compete with essential fox control in the spring. (Photo: D. Kjaer)

granite than on rocks that weather into more fertile soils. It does not follow that very heathery moors can never yield large bags some years but, to achieve them, management has to be first rate. If it is not, the yield will disappoint.

Another problem in Scotland which lowers the effectiveness of keepers is the demand on their time made by the stalking of roe, sika and fallow deer. Even where the 'big' deer are absent, there is hardly a moor in Scotland that does not carry a stock of roe. As the April and May buck shooting season coincides with the peak time for fox control, many keepers become tired to the point of exhaustion. On estates which were pure grouse moors in former times but which now carry a stock of red deer, the development of stalking enterprises over the past half century has added an economic string to the owner's bow but grouse numbers have fallen as deer numbers have risen partly because of the dilution of keeper effort. The impact is both direct (over-browsing and tick build-up) and indirect (a reduction in that detailed management without which grouse tend to be scarce). It is not the keeper that is at fault; the change is driven by market forces.

All that said, some estates, particularly in the Angus glens and south Scotland, have defied the trend. Resuscitation has come about because determined owners with deep pockets have taken on moorland with potential and provided the investment essential for good grouse management. The rewards have been substantial as 'covey' birds (grouse, partridges and ptarmigan) respond to positive management with astonishing rapidity; numbers increase very quickly if the habitat is to their liking. It is no coincidence that some moors have bucked the decline and yielded outstanding, even record, bags in recent years. The vision of their owners highlights the failure of others who have been less successful in raising decent numbers of birds. That said, while it is undoubtedly quicker and easier to resuscitate a derelict moor if there is plenty of money to invest, knowledge and hard work are just as important. Moorland can be successfully restored by those of modest means if correct management techniques are used (see Chapter 1, *The repair and restoration of moorland*).

Keeper activity and effectiveness

It is difficult to assess how many grouse keepers there are in Scotland. GWCT say that two-thirds of the one hundred and forty 'core' grouse moors (the Moorland Association puts the figure at one hundred and fifty) are located in Highland, Grampian and Tayside regions. Each moor employs at least one full-time grouse keeper, some more. If the average is assumed to be between 1.25 and 1.5 per moor, the number of full-time employees dedicated to grouse keepering can be estimated to be between 192 and 206. Given that the 'core' moors average about 4000 ha (10,000 ac), the typical Scottish keeper is looking after four times as much ground as his English counterpart.[5] However Herculean he may be (and many are indeed Herculean), he cannot micro-manage in the way of his English counterpart, cannot burn heather to the highest standards because of the shortage of casual help and the money to pay it, cannot pressure stoats and weasels properly because he cannot get round his traps every day as the law dictates. When he does produce some grouse, it often comes as a surprise. If his neighbours are short of stock, innate caution tends to make him under-crop the enhanced population because he fears stock losses from emigration. If he also has a stalking responsibility, he has to stop shooting grouse in mid-September to start on his stags while his English colleagues are free to hammer away, bird numbers permitting, through October and November. Fear of over-shooting is so widespread that good stock management is often compromised and short-term tenants restricted to shooting a certain number of brace. In walked up situations, this is seldom justified because the birds soon learn to look after themselves and rise out of shot if they are harried. In good years on driving moors subject to day-letting, it is not unknown for grouse to be 'spilled' from an afternoon drive if the bag is deemed to be approaching the contractual figure. In other cases, a bag-filling return drive off a piece of ground where many birds have been gathered together and 'bottled up' is neglected in favour of more modest prospects elsewhere. The tactic may safeguard cash flow but it is at the expense of stock quality – the performance of birds the following season may well be compromised. Some owners persist in charging 'overage' if the expected bag for the day is exceeded, arguing that, as they have to give money back when bags fall short of 'reasonable expectation', it is fair to ask for more when expectations are exceeded. Quite apart from the lack of professionalism and the guns' understandable reluctance to pay a levy, the notion of 'overage' works against proper stock control and management of the moor. If there are grouse in quantity, they need to be shot and that's that.

One has great sympathy for grouse keepers in Scotland. Every aspect of their work is hedged with regulations, many of them trivial; many work for cash-starved employers who cannot support them with the proper equipment and seasonal help. Lovat defined keepers as follows:

> A keeper must be a good trapper, an observer of natural conditions and a man with the interests of the moor at heart; above all things, he must be a worker; not a gentleman who goes to the hill with a gun.

Only moors blessed with such paragons can deliver grouse in consistent quantities today.

Wales

The middle of this century will see the Welsh grouse shooting flame finally flicker and go out. There are still small pockets of moorland on which a few pairs cling on precariously from year to year but their prospects are not good. Scotland's problems of fragmentation are as nothing compared with Wales where land holdings tend to be smaller, over-browsing almost universal and predator numbers high. It is hard to imagine the glory days of 1904 when 781 brace were shot in one day at Ruabon in north Wales. In years to come, grouse will still be recorded as breeding birds but it is unrealistic to think of them as anything more than a tick on a Red List of endangered species. It would surprise no-one if the Welsh Assembly afforded the red grouse Schedule 1 protection. When that day dawns, its days are truly numbered because it will become the responsibility of nobody. It will assume the status enjoyed by grouse in western Ireland where they cling on against all the odds at a density of a breeding pair to sixty to a hundred ha. Farmers, on the hill every day, hardly ever encounter them and sportsmen have no interest in going in pursuit of them.

Northern Ireland and the Republic of Ireland

There are many physical and social differences between the north and the south of Ireland but moorlands on both sides of the border suffer from much the same problems. Chief among these is that there are no core areas of good moorland management from whence best practice can be taught and surplus birds diffused into surrounding areas. The result is that heathlands are generally in poor condition, being either over-browsed and over-burned or neither burned nor browsed at all. Broom and whin are common and colonise extensively; piecemeal land use changes have resulted in widespread fragmentation; foxes are ubiquitous, sustained by large numbers of casualty livestock which are not removed from farmland and disposed of securely.[6] Even modest levels of crow control are rarely practised. Generally speaking, sportsmen in both countries spend little time on DIY keepering but are keen to go shooting. When the opening day comes and sportsmen find no moor game, they complain to whoever they think should lend an ear that 'Something Must Be Done' while hinting that government should do it. Changing such habits may prove impossible because there are few, if any professional heather managers of knowledge and competence such as are scattered over England and Scotland. Apart from the odd 'Trophy Bird', the days of even modest bags of grouse in Ireland, north or south, seem numbered.

End notes
1   A. Anderson, Moorland Association, pers comm.
2   Denman *et al*, 1967.
3   A. Anderson, Moorland Association, pers comm.
4   Waddington, 1958.
5   Author's estimate.
6   Watson & O'Hare, 1980.

Chapter 9
General matters

The point of no return

If the 'gentle reader' has got this far, he or she may be suffering from mild depression and confusion fatigue. It is clearly difficult to manage a stretch of moorland with the objective of having some grouse to shoot; the reader might be thinking it is all too difficult and that dreams of improvements are just that – DREAMS.

It is important to be confident and ensure that both enthusiasm and investment are directed in the right direction; this chapter is designed to help the would-be owner decide on what is or is not worth tackling. There are unscrupulous people marketing Scottish property who try to sell a place on its past history knowing full well that it has been changed irrevocably and that there is little or no likelihood of a return to the glory days of the 1970s, let alone those of the 1930s or earlier.

Though the reasoning that should govern a decision to purchase and become involved in the re-creation of productive moorland is easy to identify, it is still possible to get carried away. A lovely house with a good view or a nice garden can influence a female's decision; her male companion may be more pragmatic and have doubts about the neighbours, the state of the hill and so on. A place can be bought for all the wrong reasons. As a general rule, potential owners should start by looking at two key factors. The first is the ground. They should answer the following questions, rated + (yes) or - (no). Is it very heathery? (+). If it has not got extensive swards of heather, has the hill plenty of blaeberry on it? (+). Has it got a large block (> 200 hectares) of plantation forest within 3 km (2 miles)? (-). Is there any keepered ground adjacent? (+). Is it subject to crofting tenure? (-). Has it got red deer resident below 500 m on the property all year? (-). Is the farming in hand? (+). Do the sheep suffer from louping-ill? (-). Is the property located in the western half of the country? (-).

The second factor is the geology. What is lying underneath? If the estate overlies base-rich rock, it is potentially good whatever it looks like superficially; if it overlies basalt, coal measures, diorite or carboniferous limestone then (++). Is it on the Silurian series, slates and phyllites? (+). Even if it is quite grassy, heather and blaeberry can be encouraged to greater dominance by management. It need not be expensive; it is worth undertaking a rehabilitation programme in most grass-dominated situations.

Almost every other factor is secondary to these two. Burning standards can be improved, over-browsing of heather in winter can be rectified, bracken can be destroyed and heather be grown in its place, 'white' grass dominance can be reduced, red deer and the ticks associated with them can be eradicated.

What cannot be put right is if the moor itself is heavily degraded, overlies deep peat, poor-quality granite or quartzite, is surrounded by forestry, has no keepers on neighbouring ground and so on. These factors make it a poorer prospect. All this is common sense. You cannot make a silk purse out of a sow's ear but there are still opportunities out there to purchase silk purses if you know what you are looking for. All they may need is to be dry-cleaned, have a seam or two re-sewn and some new buttons stitched on and they'd be as good as new.

Over a decade, such places would provide their owner with excellent sport and a substantial capital gain into the bargain.

Politics

In the coming years, the two disciplines that will be of overriding importance for the future of moorlands and their management throughout the UK are politics and economics. The former is usually influenced by the latter.

In Scotland, the political climate in 2011 is unfavourable for the future of moorland. The Scottish Executive Government is made up of 127 members of whom 105 (eighty-three per cent) are socialist (either sourced from the Labour or Scottish National Party (SNP)) and currently fewer than twenty members (sixteen per cent) have an affiliation with, grasp of or well-focussed understanding of country matters. In consequence, informed country voices are seldom heard and their views are casualties of party political positions. Policy is driven largely by the wishes of the urban majority and takes scant account of either the views or the requirements of land managers. The Scottish Government takes its instructions from the groups which are strong numerically and which shout loudest. As a result, policy is formulated under the influence of RSPB, Royal Society for the Prevention of Cruelty to Animals (RSPCA), NTS, SNH, MoD, JMT, SWT and the FC, all of which are quasi-government, government or organisations receiving significant government support. The SRPBA and National Farmers Union (NFU) in Scotland who represent the owners of most of the heritable property, are minority voices and their views often run contra to the left-of-centre consensus displayed by policymakers. The people who actually own and manage the land are therefore seen as some sort of obstruction to the further democratisation of the countryside. Despite the 'Right to Roam' legislation passed in 2003 which reinforced the long-standing right of unfettered access on foot everywhere in Scotland, there is still a body of opinion which seeks more control over how the countryside is managed. This comes in many guises – viz. from Brussels in the form of cross-compliance – 'we give you bountiful farm subsidies from the public purse and in return you will deliver... ' 'You want to plant trees? You will do so in this way with these species or you will not get grant aid' (essential to cover establishment costs). SNH is never slow in getting in on the act. 'You bought a piece of a National Nature Reserve which currently carries too many red deer (in our opinion). We want fewer deer on this ground in order to satisfy our "conservation objectives". If you do not comply with our wishes, we will come in and kill the deer ourselves – no compensation will be paid.' The fact that this high-handed action destroys a privately owned asset (a valuable deer forest) for objectives desired by an un-elected coterie appointed by government is the action of a Marxist state.

Moor owners and their managers are obliged to function under edicts such as 'You want to kill crows to improve your game shooting (and incidentally benefit many non-quarry species of widespread conservation interest)? Your traps must be labelled with a Personal Identification Code-number so that some "expert" can ensure that the law is being complied with.' 'You want to control foxes? If you use snares, you must use the most inefficient design and employ methods which make it *less* rather than *more* likely that you will achieve satisfactory results.' 'You have a rarity of scientific interest on your land? We – the Government – zone the area as a SSSI/SPA without either consultation or appeal and the value of your land is reduced greatly. No compensation is payable.' 'You have a golden eagle nesting on your land. Disturb the eagle and we will fine you £5000."You want to burn heather within half a mile of an eagle's eyrie? We "recommend" that you do not do so after end of February.'

No such sanctions are applied to relays of ice or rock climbers who choose to explore a challenging pitch within the same distance of an occupied eagle nest – thus keeping the female off her eggs so that the embryos die. In consequence, no young are reared, the eyrie is classed as 'unoccupied' and the landowner usually gets the blame. It is unsurprising that landowners and farmers as a group feel that their way of life is being destroyed by people who at best, demonstrate low levels of comprehension about country matters and at worst, are often positively vindictive.

These problems have become substantially worse since devolution in 1999. Scotland is governed in crucial areas (some powers were devolved by a unionist Labour regime) by a regional government which is demonstrably hostile to land ownership and field sports. The executive pays lip service to conservation and formulates grand policies for the enhancement of 'wildlife interest' but it is never proactive and interventionist in helping to create favourable conditions for the desired species. Generally speaking, to improve breeding densities and success of scarce species (such as greenshank and capercaillie), the twin prongs of habitat and predation have to be addressed. Habitat improvement is fashionable and 'showy'. A group of mixed-age volunteers portrayed in the act of carrying out positive conservation work planting trees or creating wetland provides good journalistic copy. A gamekeeper destroying a crow's nest or bolting a vixen from a breeding earth is probably exerting a greater positive influence on the target species but his actions attract hostile criticism and he ends up being almost universally reviled.

Urban man likes his conservation sanitised and SNH sees that he gets it by publishing a quarterly populist journal, *The Nature of Scotland*, the content of which is mostly emotive claptrap. No country people advocate or condone cruelty in any shape or form but there is a general failure to accept that if people want enhanced numbers of one species, there usually has to be a reduction in the numbers of some others. The issue of intervention is always side-stepped for fear of offending people and the current debate about how to deal with illegal beaver releases in Scotland provides a good example. Game and wildlife management depends on setting clear objectives which are then delivered by implementing a programme of robust management. Habitats in the United Kingdom are too small and mostly too fragmented to respond to the broad brush management designed to favour wildfowl or prairie grouse in north America where single blocks of prime habitat for a target species (for example sharp-tailed grouse and prairie chickens) can extend to 19,000 square miles (12 million ac) in one block. Just one example is the sandhill prairie in Nebraska – an area as big as the whole of the Scottish moorlands and heaths. In Britain, micro-management has to be the basis for the enhancement of desired species – whether it is a quarry one or not. This means a commitment to intervention.

It is impossible to be optimistic about the influence of politics on traditional moorland management in the future. The year 2011 is merely a snapshot year in a long sequence of curtailments and changes. We are well rid of many of these: a hundred years ago, a farm tenant could be thrown out of his holding at the whim of factor or landowner. Strychnine was used indiscriminately and in large amounts until the end of the 1930s. Rabbits were trapped for market using the inhumane gin trap and their maimed limbs were a reproach to civilised people.

The Agriculture Act of 1946 opened the door to an increase in tree planting and the reclamation of much moorland to grassland of varying quality and productivity. Organochlorine-based insecticides which were invented during the Second World War and used as cereal seed dressings did immense damage to wildlife until they were withdrawn in the 1960s. When the persistent chemical dieldrin was withdrawn as a sheep dip, it heralded the start of the progressive decline in the standards of sheep husbandry which has culminated in today's ranch culture and the development of easy-care systems. The UK's entry into the Common Market in 1973 has proved a mixed blessing. It has resulted in many highly damaging management decisions concerning moorland. The long-term degradation of the poorer hill soils in the wetter western half of the country has gathered pace because of a combination of high numbers of subsidised sheep; poor range management and modern-day small-minded and short-term political decisions continue to exert their baleful influence widely.

Change is inevitable. The trick is to manage it – in both its rate and detail. Fifty years ago, national forest policy led to the destruction of vast swathes of high-quality moorland. Public opinion brought the process to a halt in the 1990s. Currently, The Scottish Executive is reactivating this discredited policy and together with its obsession with wind-power, both will have a long-term negative effect on landscape quality.

Although the size of a so-called 'wind farm' may be limited in area, the development of what is essentially an industrial site in wilderness areas creates an infrastructure which by its very nature, destroys for ever that very wilderness as well as affecting the hydrology of peat and associated drainage channels. Construction of the turbine bases involves disturbance of peat as well as deep excavation to bed-rock in order to facilitate the construction of concrete foundations. At a time when other countries in Europe are backing away from this expensive and unreliable power source, the SNP is hell bent on achieving its target of 6000 towers nationwide associated with self-sufficiency in so-called 'green' energy by 2020.

The future

The Grouse in Health and in Disease was published a century ago in response to a decline in grouse numbers, a decline that threatened the prosperity of large areas of the UK, particularly Scotland. The people in power recognised that, if grouse numbers declined, revenue streams in remote parts of the country where there were few alternative economic opportunities would decline also. At that time, Scotland was the place that people chose to go in order to shoot grouse and deer, catch salmon, see and be seen socially and so on. Large areas of the country and the people who lived there were dependent upon an influx of wealth which created a great deal of employment in an otherwise resource-poor region. There were grouse in good numbers on the softer hills of the north of England and in Wales but pursuit of them did not attract the attention of the rich Victorian industrialist in the same manner as *Caledonia Stern and Wild*. The book was made up with compilations written by acknowledged experts of that epoch and much of its content is as sound today as it was when it was written. It represented the epitome of scientific thought and applied management at that time; nobody questioned that grouse and grouse shooting were anything other than wholly desirable and good. Every effort had to be devoted to making sure that the torrent of English gold continued to flow north unabated. That was why the book was written.

A mere three years later came the Great War, the skies fell in and things were never the same again. After the Armistice in 1919, new land use issues became important and single-species management of the kind that prevailed before 1914 became less possible to achieve as well as less desirable.

For the next thirty years, matters moved along quietly. The fourth decade of the century was a good one for grouse but the Second World War put a stop to the recovery. By the 1950s, many people were hoping to see grouse numbers improve; the Scottish Landowners' Federation commissioned Professor Wynne-Edwards of Aberdeen University to suggest ways of restoring populations. Work began in Glen Esk in 1957 and metamorphosed into the Nature Conservancy Council Grouse Research Unit at Blackhall, Banchory under the direction first of Dr David Jenkins and then of Dr Adam Watson. Robert Moss joined the team as a PhD student in 1963. The unit became part of ITE in 1973. The Reconciliation Project, which later became the Heather Trust, was formed in 1984 and the GCT started its Upland Research Unit at Newtonmore in 1988.

Grouse, particularly in Scotland, were now under scientific and applied management focus as never before. The effort yielded new and rigorous scientific knowledge which was gradually converted into useable management techniques. The knowledge was disseminated to owners, factors, farmers and keepers by a combination of open days, Game Fair exhibitions, articles in the popular press, training courses and, finally, demonstration moors where best practice was in place. The investment by both private individuals and public bodies started to fill the void in information on moorland management in a way that had been lacking since Lovat edited the *Grouse in Health and in Disease* all those years ago.

In the 1960s, grandee politicians became associated with grouse shooting in the public mind and the 'Grouse Moor Image' which stemmed from Harold MacMillan, later the Earl of Stockton, and Sir Alec Douglas Home's interest in shooting became by association something of a poisoned chalice as they were lampooned and the smear filtered down to landowners in general. It was a short step from there to the seasonal barrage of cheap shots from tabloid journalists aimed at those who shoot grouse. The topic became hijacked and used as a stick with which to beat political right-wing dogs, and those of a left-wing persuasion were quick to capitalise on this advantage. In consequence, the beneficial influence on the maintenance and conservation of the national moorland by the landowner – particularly in Scotland – has declined in step with the public perception that grouse shooting is seen to be an elitist, cruel and offensive activity and because of this, has forfeited political and official support. RSPB/SNH have responded by joining forces on all matters to do with moorland management in an attempt to satisfy the 'demands' of the general public despite the fact that they themselves have neither the land nor the expertise to deliver many of their visions.

By about 2000, the tide of research effort started to ebb. There were good reasons for this. The acquisition of knowledge relating to 'why' and 'how to' had peaked. The Banchory Grouse Unit ceased work on the moorland at Riccarton in 1989. Government cut-backs led to the closure of that research unit in 2007 and official objectives included greater emphasis on general conservation and biodiversity. This meant that, although the grouse unit was no more, the knowledge base was still there for anyone prepared to learn. Those who had both the will and the money could grow grouse. Those who lacked one or the other generally could not. The result was a further fragmentation of grouse ground; pockets of best practice were often surrounded by areas where less effort was applied. This in turn tended to weaken the efforts of those who were trying the hardest. 261

A centre-left Westminster government was elected in 1997 and began to implement a manifesto promise to give Scotland a measure of devolved power. Sections of population living in the country started to feel disenfranchised as it became more and more official policy to open up the countryside as a play park for town folk. There is nothing wrong with that in principle but it has had unfortunate consequences for people who live and work in rural areas.

The Right to Roam, enacted in 2003, has provided an official charter and excuse for those who want to interfere and poke their noses into other people's business. Farmers are under constant scrutiny on welfare, conservation and animal health issues. As part of their SFP and attendant cross-compliance, they have to conform to minutiae of apparently ever-increasing irrelevance which is sometimes sourced from the comments of the weekend visitor to the countryside. Short-staffed and over-worked at lambing time, they are pilloried for the incorrect disposal of a casualty ewe. If they cut a hedge after the end of February or a meadow before mid-July, their support payments are cut. Rules are in place which control when stubble fields can be ploughed; if these are broken, grant money is withheld. And all of this takes place against a background of escalating global population and a sixty per cent increase in the price of wheat in one year (2010–11). Realism is lacking. Without farmers, conservationists would starve.

Gamekeepers are just as vulnerable. Those who have nothing to hide and who carry out their work wholly within the law are harassed. They are assumed to be incapable of using a crow cage without having first obtained a training certificate. It is illegal for them to set such a trap without it having a tag or label on it which identifies them personally as responsible for it. Even although they conform to these rules and observe the regulations to the letter, it does not stop mindless and random vandalism. Placing a notice on a trap stating its purpose and describing the benefits which accrue to wildlife from crow control provides no protection. Often a live decoy will be let out and the trap smashed. These criminal actions are anti-conservation and have little or nothing to do with animal welfare. They are carried out by people driven by a toxic mixture of political motivation, ignorance and sheer malice. They come from the same family as hunt and shoot saboteurs. Complete with binoculars and anoraks, they arrive with the swallows in spring and prospect every corner of wild country as if they were some sort of self-appointed cadre of special constables.

Snares are now in the sights of these activists. They are anxious to see their use banned; there is a body of opinion that would like to see all snaring made illegal without being in the least bit interested in the effects on wildlife of such a ban. It is now mandatory that all snares (including those set for rabbits) are required to have an identity tag attached. Keepers are encouraged to attend so-called accreditation courses run by GWCT and BASC which explain what can and cannot be regarded as acceptable methods of use. In one recent example, the authority organising such a course was the Cairngorms National Park Authority – a government-funded quango located in the middle of some of the best-managed heather upland in Scotland. It was attended by keepers, most of whom had a lifetime's knowledge and experience. They found that they were lectured and hectored on the subject by people who had no practical knowledge whatever.

Restrictions on snare use are petty; for example, you cannot attach a snare to a drag nor can you attach a snare to a fence. Quite why these conditions are needed and what benefits they provide for animal welfare are unclear. The infringement of an impractical regulation then lays the operator open to a charge.

The keeper or his employer is expected to pay to get this so-called qualification; it is but a short step to specifying that certain tasks cannot be carried out unless the person doing them is accredited. That would make it illegal, for example, for a keeper's unqualified teenaged children to stand in for him temporarily and perform mandatory daily inspections while he was ill or absent.

The 2011 WANE act included populist measures such as abolishing the need to hold a Game Licence in Scotland stemming from the Game Licences Act, 1860 which all who shot game were obliged to possess and other, more controversial ones such as shooting tests for deer stalkers and the establishment of a register of those who have demonstrated certain levels of skill and knowledge. In fairness, the Act contains some good clauses. Though it is to be welcomed that managers now have the option to burn heather in September, this is discretionary as SNH is the arbiter as to whether or not a licence is granted. Controversial matters such as permitting the control of ravens, sparrowhawks and buzzards under licence are ignored altogether. The photogenic badger and bright-eyed hedgehog – beloved as subjects for greetings cards and tablemats – are forgiven all their many sins as nest and chick predators because they are popular with the urban population. In response to public demand, it will not be long before hedgehogs are given the same sort of protection as certain birds receive under Schedule 1 as they are in steep decline nationally. When that happens, it will be impossible for keepers to run a network of tunnel traps for controlling mustelids because they will risk killing hedgehogs as a by-catch. A combination of predation by badgers and widespread road kill has depressed hedgehogs to the point of extinction over swathes of Britain, particularly in southern England and giving them the protection of the law will not save them.

The increasingly widespread use of volunteers in conservation has created cohorts of people who go to the country at weekends and public holidays both to work and enjoy themselves. Some obtain temporary jobs as wardens or in working groups on nature reserves. Many are good-hearted people who readily admit to having only limited knowledge and field experience but a proportion are busybodies pursuing a political agenda to disadvantage the gamekeeper and discredit the 'rich' landowner.

It is not surprising that keepers become dispirited. They feel sometimes that everyone is against them, the world is their enemy and that they are misunderstood and unappreciated. In consequence, many good men give up the unequal struggle and tend to be replaced by people of lesser calibre and knowledge, some of whom are appointed to posts of high influence merely because they have the necessary papers.

If the present trend continues (it began in the mid-1970s) further large stretches of country will be looked after by people who are not game creators, but will have more in common with the European 'guard de la chasse' who is essentially a policeman. As this happens, game populations will decline further and in some cases to levels close to extinction. If there is no game, there is no revenue and if there is no revenue there is no investment and if the investment ceases, predators run riot and there is nothing for anybody. If present policies are continued, it is not hard to envisage a Scotland in twenty years time in which there is no grouse driving of any kind because numbers will be too low. The only grouse shooting will be confined to occasional days walking in line or hunting over bird dogs. Even this activity will be confined to land above the tree line or on the extensive bogs of the Flow country of Caithness and Sutherland. What will have come of the rest of our traditional grouse moor? Much will have been fenced against farm livestock in order to return it to natural scrub or, if the Scottish Government has its way, will be the subject of extensive plantation forest if its target of twenty-five per cent land cover by 2020 is achieved.

Does the reader think this is far-fetched and fanciful and that such a thing will never happen? Do not believe it; for a good example, look at Wales. On 15 August 1913, 1540 birds were shot in one day on 'Ruabon Mountain' in Denbighshire (now Clwyd). Over the past fifty years, proper heather management and rotational burning has virtually ceased; farmers control the management of the majority of moorland throughout the Principality and run sheep where they like and in numbers over which there is little or no control. Ticks and tick-borne diseases are widespread. Predators are rife; land holdings are small and becoming smaller. Despite current estimates, it is possible that as little as 35,000 ha of heather remain; it is doubtful whether there are fifteen hundred red grouse in the whole of the Principality today. Black grouse are reduced to a handful of individuals; lapwings and curlews attempting to nest on the open hill are matters of remark and almost always fail to breed. As a productive callunetum, the moorland of Wales no longer exists.

In 1894 was written 'the thoughtless persecution of birds of prey, or of any other feathered fowl, is culpable in the extreme. But the interests of the sportsman and of the naturalist are closely allied and the one ought to help the other in the wise management of the grouse moor'.[1]

The Victorians and their game management are condemned by many people today because they are perceived as wanton killers of rare birds and mammals. To an extent, this is a valid comment but at that time, moderates among the shooting fraternity were voicing the need for a balanced view. What landowners and managers did was accepted almost universally at that time. If there was dissent from any section of the general population, it was muted. The Great War changed all that and marked the start of a gradual decline both in the influence of the landowner and the dominance of shooting as a primary land use. During the twentieth century, national priorities changed and shoot managers had to learn that the joint management of grouse and grazing animals was the only way forward. Although the BASC, (formerly the Wildfowlers Association of Great Britain and Ireland (WAGBI) which was founded in 1909) was able to boast a membership of over a hundred and thirty thousand and an income of about £3 million by 2010, in the same year, the RSPB (founded in 1889) laid claim to nearly ten times as many members and boasted an annual income of over £120 million. Acknowledging that these organisations are the two Big Beasts in terms of *realpolitik*, it has to be recognised that the shooting fraternity is in a minority and its interests are vulnerable to laws enacted by people who have a voting majority but little grasp of the issues involved. The shooting of grouse is the only thing which puts an economic value on heather. Without it, we would have far less of this already scarce and threatened habitat and far fewer of the interesting birds which are dependent upon it and which are esteemed by naturalists.

Unfortunately the two sides – shooters and naturalists – are further apart now than they have ever been and the gulf widens constantly because of often outrageous claims and accusations from both sides. The intellectual calibre of the debate is rarely high.

Shooters have got to accept that certain activities are unacceptable and naturalists have got to accept that some of their 'demands' are undeliverable in practice. Warfare benefits nobody and compromise is the only way forward.

To revert to the Victorians, when it came to the management of game, they had things all their own way and could do what they liked. Naturalists were in a minority and had no say whatever when it came to decisions affecting land and its uses because they did not own any. A century and a half later, the boot is on the other foot. When it comes to the formulation of policy, private landowners and managers act in a reactive way when in fact, their salvation long term lies in being proactive. They have to recognise that while naturalists hold most of the aces because they outnumber shooters and landowners, the one thing they do not own is the majority of the country's land, so their interests can only be furthered by allying themselves to those that do. Both SNH and the RSPB would be well advised to take note of this indisputable truism and stop being so confrontational. They would gain so much more for both their conservation objectives and the interests of their members and the birds whose interests they espouse if they adopted a more conciliatory stance and demonstrated a greater willingness to enter into joint management projects with those people who do own the land.

Wildlife policy is no longer made by the people who have a majority interest in it – namely the landowner or occupier as was the case when Lovat published in 1911. Power to make decisions has slipped away from these people over the last century in direct proportion to the rise of an articulate and affluent urban middle class and concurrent with them, a group of self-appointed spokespeople in the conservation movement led by the RSPB whose corporate thinking dominates the government-appointed SNH. The RSPB has gathered numerical and financial strength so that it is now the largest and most powerful conservation organisation in Western Europe. Nothing is done by executives in Brussels, Westminster or Edinburgh without an opinion being sought and approval given from the RSPB's Bedfordshire headquarters. Because it is so well funded, it can afford a large public relations section which is forever ready with an opinion, soundbite or scrap of journalistic copy which can fill a few column inches. When the Scottish Executive was created in 1999 to form the devolved parliament in Edinburgh, the Scientific Director of RSPB Scotland was appointed as adviser on wildlife matters to the relevant ministers. The influence of SNH followed close behind that placement (a senior member of the Uplands and Peatlands Department was runner-up for that post when it was set up) and conservation organisations like the JMT, NTS, SWT and WT soon slotted in as satellite bodies. Anxious to be seen to be supportive of the thoughts of the new administration, it was not long before they all recognised on which side their bread was buttered. Ten years on, there is never a word of dissent against anything proposed by the Scottish administration from any of these bodies – however ill-considered and impractical their comments may be. With the executive being of a left-leaning stripe for the whole of its existence, it is easy to see how these organisations dominate any debate on country issues and leave the opposition – BASC (membership 130,000), GWCT (< 20,000), SGA (5000) and SRPBA (52,000) seriously outnumbered. In terms of *realpolitik*, policies which impact on the wider countryside are wholly republican in flavour and private interests are given scant attention if they conflict with the one perceived as being supported by the urban majority.

The situation has also changed in so far as the RDC and SNH have been amalgamated (2010) in the name of administrative efficiency and the chief executive in charge of this grouping owes his appointment to a decision made by the Scottish Government and clearly has to serve that master. It is possible to envisage further 'consolidations' in the name of 'efficiency' as time moves forward.

The concentration of so much power into so few hands tasked with making decisions which affect the countryside is sinister and distinctly unhealthy. The situation is becoming Cromwellian. The Roundheads stifle debate, freedom of choice is outlawed, and the diverse views held by the Royalists are ruthlessly suppressed. The administrative changes which have been wrought over the past decade have created the foundations for changes of positively draconian proportions, and this does not augur well for the long-term interests of the landowner, the farmer or the field sportsman. The prospect of any sort of Restoration would appear to be many years away.

This dominance results in some unsatisfactory outcomes when it comes to the interests of people who live by and in the country. Decisions based on anthropomorphism attract majority support – irrespective of the facts. Legislation affecting tail docking of working dogs was enacted, mainly on the basis that the animals' ability to self-express would be compromised if they did not have tails to wag. The outlawing of fox hunting by separate statutes in Scotland and England was motivated by class hatred and had little to do with animal welfare or fox preservation. Indeed, foxes have a much tougher time of it now than they did before the legislation was enacted. The conservation bloc relish the changes; they are part of the theme 'we know best how to run rural Scotland' and this pressure extends to the dastardly people who shoot 'for sport'. A mechanism which is in widespread use is the obsessive attitude towards the preservation of birds of prey. So thorough and deep-rooted is this indoctrination that no politician dares to question any figures on raptor deaths from any source whatever.[2]

Misinformation plays its part as well. A good example is the progress made by red kite releases in different regions. The Black Isle in Easter Ross and the Chilterns in Buckinghamshire were locations chosen as sites for releasing birds and the programmes started at the same time (1989). The Chiltern sub-population is doing well while the Black Isle one is making only modest progress. The bird lobby explains this by blaming gamekeepers for the widespread use of traps and poisons in Scotland.

A more likely cause is the variation in prey base between the two sites. The Black Isle contains some of the most intensively managed arable land in the UK and the woods are predominantly coniferous. Consequently the country supports only a modest prey base so road kills are rare and carrion scarce. In the Chilterns, a district which is predominantly urban, people put food out in their gardens specifically for kites,[3] the woods are deciduous and there is a substantial base-load of road kill carrion. The situation is close to replicating the habitat available to kites two hundred years ago when they throve by scavenging the refuse thrown onto the midden at the back door of every house.[4]

The deification of raptors is a crucial, clever but cynical policy of RSPB. It is based on considering 'these beautiful birds' with the same sort of reverence as was applied to 'the noble savage' in Africa and North America in the eighteenth and nineteenth centuries. Most rational people would consider all bird species to be aesthetically pleasing particularly when involved in courtship or the raising of young. Prey species are just as worthy of admiration as some predators. It is a poor heart that cannot rejoice at the spring displays of great crested grebes, black-game, a hen harrier food pass, or the tender 'necking' that goes on between a pair of fulmars on a sea cliff.

The powerful forces which have been marshalled in support of raptors have been successful in increasing numbers throughout the United Kingdom over the last fifty years. Only the kestrel – a bird which has never been at odds with game interests – has declined. Most species have held their own or have shown modest increases and the buzzard and peregrine are now both at saturation point. Because of the scarcity of suitable nesting sites, the latter are often forced to nest on the ground on scree or boulder fields where they are at risk of being predated by foxes. Natural forces are now controlling their numbers but not before both species have contributed greatly to the decline of other species such as moorland waders. Peregrines kill many adults; buzzards prey on the chicks. Does RSPB care? Not a whit because wader declines provide it with yet another *cause célèbre*. The wicked farmer with his silage cutter and acres of winter corn is then castigated as being the proximate reason – despite the fact that over extensive areas of the country, his field practices have hardly changed at all in the past three decades.

It is not difficult to see just what is happening on a decade timescale. It is loss by attrition; it is erosion of traditional practices and their replacement by *laissez faire*. Whatever wildlife is the subject of personal interest, it will only thrive if it is the subject of clear management objectives. In PR terms, RSPB have been highly successful selling their raptor management policy to their public despite the fact that these same policies elsewhere are catastrophic and have swept up many hearts and minds in the process.[5] Those that live by the land and its products have been left standing and the future looks bleak.

Demonstration moors

John Maynard Keynes (1883–1946) said 'The difficulty lies not in the new ideas, but in escaping from the old ones.' Academics and theorists can write all they like and publish their thoughts and experiences in a variety of media but it is common for them as a group to fail – often completely – to reach people on the ground. Academics reach each other by the learned journal route but landowners and farmers are influenced by word of mouth and keepers tend to look to magazines such as the BASC's quarterly *Shooting & Conservation* or IPC's *The Shooting Times & Country Magazine*. People from all walks of life are influenced by what they see and hear at country game fairs and agricultural shows.

Sixty years ago, the HFRO was formed and three separate farms – Glensaugh, Kincardineshire (NO670782) bought in 1943, Lephinmore, Argyll (NM987926), purchased 1949 and Sourhope, Roxburghshire (NT845203), leased 1946 – were acquired and run as experimental and demonstration farm units. New hill farming and management techniques were developed and the pioneer work was widely published and put on show. Farmers from all over the country would attend open days in order to see the improvements at first hand. They looked, learned and returned home carrying the new ideas which they then put in hand on their own farms. Technical back-up was supplied by the three Scottish Agricultural Colleges (West, East and North) who, because of the location of the three farms, each had a demonstration site nearby. In terms of improving agricultural performance, the recipe was a winner and farmers have good reason to be grateful to the government of the day for setting up these facilities. In farming terms, what was a matter of wonder in 1950 is so mainstream today that it is unremarkable.

In the 1950s, the ICI Game Services, later the Eley Game Advisory Station (EGAS) leased the shooting on 1600 ha (4000 ac) of downland on the Hampshire/Dorset border at Damerham and ran it as a test-bed site for the management of grey partridges. Best practice was put in place with four full-time keepers employed. | 267

The work, warts and all, was published annually in *Your Year's Work* – a magazine which described all aspects of the shoot management. EGAS hosted a series of partridge shoots each year and invited 'the great and the good' to participate before they returned to their manors to put in hand the techniques they had discussed during their visit. They also ran training courses.

When the grouse unit moved from Glen Esk to Banchory, The Nature Conservancy Council later ITE, leased the Kerloch moor from the Gladstone family and used the site to carry out their ground-breaking work on grouse ecology. The unit provided the science which supported the thoughts of many practitioners and became highly regarded for the quality of its work. When Kerloch was sold, field studies continued at Riccarton, near Stonehaven and it was there that the all-important work on population cycles and behaviour-induced grouse declines took place.

On 30th March 1983, the Game Conservancy held a conference in Edinburgh on moorland-related themes at which the author presented a paper called *Upland Neighbours; Grouse and Sheep – Friend or Foe?* As a direct result and on that very day, proprietors began offering sites for experiment: almost overnight, the *Reconciliation Project* was born and the first tentative steps taken towards demonstrating best practice. Sir Joseph Nickerson provided much-needed funding at a critical point and the project evolved into the Heather Trust – a registered charity. In the years that followed, many facets of best practice were developed and demonstrated to interested parties all over Britain. Demonstrations and training courses are an essential part of disseminating good practice.

In 1992, the GWCT was bequeathed Loddington farm in Leicestershire by Lord and Lady Allerton. This 300 ha (800 ac) arable farm became a test bed for all matters to do with the impact of modern arable farming on game and wildlife. Experiments carried out and data published provided sound evidence to support

A Heather Trust demonstration day in the mid-1990s.

the notion that interventionist management benefits many wildlife species. Current studies continue to do so. The concept of the test-bed farm lends itself to imparting knowledge through open days, popular articles in the press and peer-reviewed scientific papers.

At the start of the second decade of the twenty-first century, fundamental moorland research has all but ceased. Indeed there is little need to put money in that direction. The disbanding of the ITE's grouse unit in 2007 marked the end of an epoch. That team of able and knowledgeable scientists retired which created a vacuum; the shooting public was left largely to its own devices. There was, and still is, a marked scarcity of sound, practical independent hands-on advice in the market place coming from people with a multi-disciplined approach. Single-issue conservationists started to have things all their own way as no competent, well-reasoned and science-based case was being presented regularly by hardline shooting interests in an attempt to offset the dead hand of official thinking. The interests of the private landowner have been eroded. Traditional moorland management was and still is dying a death by a thousand cuts. Landowners have lost confidence in the face of high-level harassment by SNH, RSPB and politicians generally. Many people have given up proper moor management, often because they have serious cash flow problems. It is almost a part of the Highland laird's genetic make-up to go to ground when times are hard, spend nothing and hope for the tide to turn.[6] We should not blame him. By that route have many families held on to a high proportion of their ancestral acres for many decades and even centuries. So, short of selling up and retiring to Brighton, Bournemouth, Bognor or Budleigh Salterton, what is his solution?

The situation requires the adoption of proactive policies. The initiatives with regard to land management and particularly hill land require the accumulation of funds to take a long lease of the farming and the shooting on a substantial stretch – at least 3000 ha (7500 ac) – of moorland in a couple of problem areas such as west Perthshire and the lower Spey valley. It would then be necessary to put active young keepers on and set about managing these moors properly. The project requires the farmland to be in hand so that agronomists can become involved and play an active part in partnership development. The management strategy would be based on traditional best practice for grouse, grazing animals and general wildlife diversity. All aspects of the effects of this could be measured, recorded and written up. There would be scope for bolting on additional studies in association and collaboration with other bodies – even SNH and RSPB – so that some of the conflicts which sour relations between shooters, farmers and bird people could be addressed. If it was properly structured with a multi-disciplined approach **and dominated by the needs of the private sector**, the enterprise could be self-financing after the initial start-up expenditure. It could become a Mecca for best practice and the management would host open days for interest groups from every different discipline. Should people argue that it cannot be afforded, it should be noted that The Scottish Committee of the GCT oversaw the expenditure of £1 million on grouse research over the twenty-five years of their existence to 2011 (£40,000 per annum) sourced from private subscription. There are plenty of people today who would question whether that research expenditure has resulted in a single extra grouse flying in Scotland. Many would argue that the widespread implementation of best practice from a century ago would have achieved more. Following the talisman of grouse research based on the three Ps (predators, protein and parasites) has proved to be an expensive route to go and the landowning fraternity should put all that work in context and concentrate on putting best practice in place. The concept of Demonstration Sites whether for agricultural, conservation or sporting purposes is a proven route by which best practice is implemented widely. It is essential that the issue is addressed with urgency and a site or sites set up in the years ahead.

Epilogue

Benjamin Franklin (1706–1790) wrote that 'idleness and pride tax with a heavier hand than kings and parliaments. If we can get rid of the former, we may easily bear the latter. We must indeed all hang together or assuredly we will all hang separately.' If ever there was a sentiment apposite to the problems faced by moor owners in Scotland at the present time, it is spelled out in that passage.

In the previous section, there is a plea for people in the private sector to put scarce resources towards demonstrating best practice, tackle perceived problems and go onto the front foot and justify their positions. It is just as valid to propose that the same disciplines be adopted by SNH and RSPB. The RSPB's Geltsdale reserve in Cumbria provides no answer to this criticism; hen harriers (for which the site is designated) struggle and foxes provide the management with a permanent problem. Management intervenes with reluctance and at a level which is often too little and too late. SNH, with an annual budget of £64 million and nearly seven hundred and fifty staff on its payroll has no apparent wish, capability or ability to take on a piece of moorland and manage it for its self-imposed and desired objectives, preferring to pressure and cajole the private owner to do its bidding. It is not difficult to see why they funk such a challenge. The level of in-house expertise in hands-on management that the organisation can call upon is not great; where is the cattle person who could set up a large-scale hill cow unit in a deer forest and demonstrate its conservation advantages? Where is the person who understands the management of hefted sheep flocks? If SNH's various publications on moorland management accurately reflect its viewpoint, any averagely competent moor manager who describes the organisation as 'not fit for purpose' can do so with justification. More than a decade ago, the then Head of Peatlands and Uplands was strongly in favour of Heather Trust proposals for September burning and agreed to put forward proposals to the legislature under an Agricultural (Miscellaneous Provisions) Order. Nothing was achieved at that time but September burning under licence was enshrined in the 2011 legislation.[7] The proviso that it can only take place under SNH licence largely negates the change. In 2011, a Perthshire proprietor applied in late August for a consent to burn heather in September and received a reply refusing permission in a letter dated 30 September!

The situation is that most of the land in Scotland – increasingly regarded by authority as the disputed territories – is still in private hands. A fifth of the nation's moorland is keepered grouse moor on which management methods can conflict with predatory species like hen harrier. Rather than go on harrying moor owners allegedly killing harriers and so limiting the spread of the national population, why do SNH or RSPB not take on the lease of a former grouse moor in Argyll or Ayrshire and manage it with their own conservation objectives in mind? Are they frightened to do so because they feel they lack the on-ground expertise to achieve their ends? If this is the case, then a spot of humility in dealing with those who own land would not go wrong; £3 million being spent on "Langholm 2" over ten years is already on course to achieve little; the money could be spent to much better effect. The investment (£300,000 per annum) should be compared with GCT's recent annual expenditure on research of £40,000 mentioned above. (See p.269.)

Public bodies seek to impose their ideas upon the heritable owners. In the main, owners have failed to appreciate what has been happening and have acceded to apparently benign suggestions for SSSIs, NNRs and SPAs on their land, all of which apply management constraints. By doing so, they have taken the Queen's Shilling when, for example, generous offers have been made to clear sheep and grow native woodland. Subesquently, many find that their land diminished in value but that no compensation is on offer for the devaluation.

It is time for proprietors to assert their rights and begin to be more proactive in defence of their interests and start to fight back. They should take comfort from moor owners in the north of England who have shown the way and achieved much by setting up the Moorland Association to look after their rights and interests. It has been successful because it has been the forum for publicising the remarkable achievements of its members. Natural England now fall over themselves to be nice to Association members and extol what they have done in the name of conservation. In contrast, in Scotland, the daily fare of landowners is brickbats. It is time for them to get organised and take the fight to the opposition. If they do not recognise this and act accordingly, their end is nigh.

Heather burning for grouse and sheep: It comes no better than this. (Whitfield, Northumberland)

End notes

1 Stuart-Wortley, 1895
2 One such organisation (Partnership for Action against Wildlife Crime, PAW) [*sic!*] wrung its collective hands because thirteen buzzards had (allegedly) been poisoned throughout Scotland in 2010 out of a total national spring breeding population of between 31,000 and 44,000 (0.035 per cent loss). On a 2.8 ha field of winter wheat (NN749013), seven buzzards picked up in one

day in January 2007 had starved to death – a pattern of events which would have been repeated many times over elsewhere in Scotland.
3 Nature Notes, *Daily Telegraph*, 31 May 2011.
4 *Daily Telegraph*, August 2011: comment by station commander at RAF Benson.
5 See Appendix 2, *Capercaillie*.
6 Grey Graham, 1937.
7 Wildlife and Natural Environment (Scotland) Act, 2011.

Chapter 10
Appendices

Table 9 shows the mean number of predators killed annually in Glengarry, Inverness-shire from 1837 to 1840 and the mean number of predators killed and (below) the mean numbers of game shot in Glenquoich from 1843 to 1862.

The estate at that time included what we now know as Aberchalder, Glenquoich and Ardachy and totalled over 57,000 ha. These data are discussed in Chapter 5.

The 1931 edition of *Place-names in Glengarry and Glenquoich*[1] featured a list of every bird recorded as nesting in or visiting Glengarry at that time as compiled by Murdoch Matheson, a keeper/stalker who, born in 1870, lived in the glen all his life.

In 2011, Derek Moffat, an amateur ornithologist and frequent visitor to Glengarry, and Lea McNally, a resident in the glen for over forty years, checked the 1931 list and compiled a contemporary record of their own which they passed to the author.

Alongside mid-nineteenth century predator numbers so high as to be almost inconceivable today, the 1931 and 2011 data summarised in Table 10 provide a striking measure of the scale of change in the wildlife of the district over the last eighty years.

Table 9. Glengarry, 1837–62

Predators recorded as killed	Glengarry,1837–40 (57,000 ha)	Glenquoich 1843–62 (16,000 ha)
Badgers	17	1
Buzzards, common	71	
Buzzards, honey	1	
Buzzards, rough-legged	93	
Cats (house or feral)	19	5
Eagles, golden	4	2
Eagles, sea	7	
Foxes	3	5
Gerfalcons	2	
Goshawks	16	
'Hawks'	39	
Hen harriers	16	
Hobbys	3	
Hooded crows	358	58
Kestrels	115	
Kites	69	
Magpies	2	
Marsh harriers	1	
Merlins	20	
Ospreys	5	
Otters	12	3
Owls, barn	3	
Owls, long-eared	9	
Owls, short-eared	18	
Peregrines	25	
Pine martens	62	9
Pole cats	27	11
Ravens	119	13
Stoats & weasels	75	43
Wild cats	50	10

Game shot, Glenquoich, 1843–62	
Grouse	851
Ptarmigan	64
Blackgame	100
Woodcock	49
Snipe	44
Hares	105
Stags	16

Table 10. Glengarry bird numbers, 1931 and 2011

	1931 Breeding	1931 Visitor	2011 Breeding	2011 Visitor		1931 Breeding	1931 Visitor	2011 Breeding	2011 Visitor
X = Present									
XR = Present but rare									
Peregrine	X			X	Redstart	X		XR	
Merlin	X			X	Spotted flycatcher	X		X	
Golden eagle	X			X	Grasshopper warbler	X		X	
Kestrel	X			X	Willow warbler	X		X	
Buzzard	X		X		Chiff chaff	X		XR	
Sparrow hawk	X		X		Wood warbler	X		X	
Tawny owl	X		X		Whinchat	X		X	
Long-eared owl	X				Stonechat	X		X	
Shelduck	X				Wheatear	X		X	
Wigeon	X		X		Goldcrest	X		X	
Teal	X		X		Great tit	X		X	
Shoveller	X				Blue tit	X		X	
Mallard	X		X		Coal tit	X		X	
Eider duck	X		X		Marsh tit	X			
Goosander	X		X		Long-tailed tit	X		X	
Merganser	X		X		Crested tit	X			
Heron	X		X		Common wren	X		X	
Black-throated diver	X		X		Tree creeper	X		X	
Red-throated diver	XR		XR		Raven	XR		XR	
Great crested grebe	X				Hooded crow	X		X	
Little grebe	X			X	Jackdaw	X		X	
Waterhen	XR		XR		Starling	X		XR	
Corncrake	XR				Crossbill	X		X	
Coot	XR				Bullfinch	X		X	
Black-headed gull	X		X		House sparrow	X		XR	
Common gull	X		X		Greenfinch	X		X	
Herring gull	X		X		Hedge accentor	X		X	
Kittiwake	X				Siskin	X		X	
Oystercatcher	X		X		Chaffinch	X		X	
Golden plover	X		X		Linnet	X			
Lapwing	X		X		Twite	X		X	
Curlew	X		X		Lesser redpoll	X			
Common sandpiper	X		X		Reed bunting	X		X	
Redshank	X		X		Yellow hammer	X		X	
Greenshank	X		X		Skylark	X		X	
Dunlin	X		X		Pied wagtail	X		X	
Common snipe	X		X		Grey wagtail	X		X	
Woodcock	X		X		Yellow wagtail	X			
Woodpigeon	X		X		Tree pipit	X		X	
Nightjar	X				Meadow pipit	X		X	
Great spotted woodpecker	X		X		Rock pipit	X		X	
Cuckoo	X		X		Swallow	X		X	
Dipper	X		X		House martin	X		X	
Missel thrush	X		X		Swift	X		X	
Blackbird	X		X		Sand martin	X			
Song thrush	X		X		Ptarmigan	X		X	
Ringouzel	X		X		Red grouse	X		X	
Robin	X		X		Black grouse	X		X	

Table 10 cont.

X = Present
XR = Present but rare

Species	1931 Breeding	1931 Visitor	2011 Breeding	2011 Visitor
Pheasant	X		X	
Grey partridge	X			
Short-eared owl	XR			
Barn owl	XR		X	
Greylag goose	XR		X	
Gadwall	XR			
Pintail	XR			
Pochard	XR			
Scaup	XR	X		
Tufted	XR			
Black-necked grebe	XR			
Waternail	XR			
Cormorant	XR		XR	
Greater black-backed gull	XR			X
Common tern	XR		XR	
Dotterel	XR		XR	
Whimbrel	XR			
Red-necked phalarope	XR			
Stock dove	XR			
Rock dove	XR			
Blackcap	XR		XR	
Whitethroat	XR		XR	
Tree sparrow	XR		XR	
Snow bunting	XR		XR	
Corn bunting	XR			
Goldfinch	XR		XR	
Carrion crow	XR		XR	
Greylag goose			X	X
Shoveller			X	
Goldeneye	XR		X	
Shag			X	X
Gannet			X	

Species	1931 Breeding	1931 Visitor	2011 Breeding	2011 Visitor
Jacksnipe		X		X
Fieldfare		X		X
Canada goose				X
Redwing		X		X
Rook		X		
Osprey		X		X
Kite		X		X
Hobby		X		X
Hen harrier		X		X
Goshawk		X		
Snowy owl		X		
Whooper swan		X		X
Black scoter		X		X
Great northern diver		X		
Puffin		X		
Little auk		X		
Richardson's skua		X		
Great skua		X		
Great snipe		X		
Kingfisher		X		
Green woodpecker		X		
Grey shrike		X		
Sedge warbler		X		
Jay		X		X
Brambling		X		X
Reed bunting		X		X

	1931	2011	Change
Breeding birds	93	75	-19%
Rare breeding birds	31	16	-48%
Visitors	30	20	-33%

Other grouse

In a book which contains information about red grouse in some depth, readers might expect at least a passing comment on the management of our other three grouse species. All of them use heath land at some point in their annual cycle, and this is the justification for including them here.

Black grouse

Over a century, blackgame have shown a marked decline. The numbers shot in Dumfriesshire a hundred and more years ago are remarkable.[2] In the nine years 1864 to 1872, an average of 2369 blackgame were shot on the Buccleuch Estates, a figure exceeded in the run-up to the Great War when over 4000 were shot in 1910. The biggest year of all was 1915, when 4309 were shot. The bird, which was almost as plentiful over extensive swathes of Britain at the time, throve on the edge of woodland, copse and forest where there was bosky scrub, long heather, rushy fields and arable cropping at the moor edge with a field or two of oats, a corn stackyard and a patch of weedy turnips.

Several reasons for the decline of the bags (huge by our reckoning) were advanced in 1937 by 'an owner of one large estate in southwest Scotland' (probably Lord George Scott) who noted that:

- In the 1914–18 war, blackgame were shot on sight as 'food was scarce';

Blackcock lekking. (Photo: N. McIntyre)

- During the 1930s, many birds died from disease (one bird was identified as having died of coccidiosis in October 1936);

- The fragmentation of estates had led to reduced keepering with smaller estate units being let to syndicates who took all the game they could;

- Large Forestry Commission plantations created since the Great War led to more vermin;

- Following releases, wild pheasants increased on ground formerly good for blackgame which possibly led to a cross-infection with *Eimeria* (coccidiosis) and histomoniasis (blackhead);

- Many shoots, even 'preserved' ones, shot blackgame harder than hitherto;

- There had unquestionably been a large increase in vermin in south Scotland since the Great War.

His recommendations were to stop shooting greyhens; to shoot *all* pheasants on blackgame ground; to shoot no blackcock before 1 October; to make efforts to rear and release and to supply winter food and grit and 'change the blood' as formerly done. He makes no mention of habitat destruction which, during the 1930s, would probably not have been a significant factor; the big changes in agricultural practice came with the appointment of the regional War Agricultural Committees in 1940.

Another commentator wrote that

in autumn, blackgame congregate in valleys …and the vital question of winter feeding for blackgame has seldom, if ever, been seriously considered…[3] When blackgame congregate in a stormy autumn, nearly the entire stock of a district can be seen collected together. Under these conditions, any reserves left behind on the hill are probably small [implying that hard shooting of these concentrations could over-crop them]… Greyhens are not first-rate mothers…and it seems probable that where blackgame are decreasing in numbers, the three main causes are over-shooting, increase of vermin and insufficient food.

275

In the four years 1972 to 1975 in north Ayrshire (NS280550) and on a scale much smaller than Buccleuch estates, there was an area of low hill of some 80 ha (200 ac) made up of heather, blaeberry and cotton sedge on the drier ground and dense beds of soft rush in the hollows. There were no woods within a kilometre and the area was summer-grazed by cattle only. There were no trees on the area bar a small strip of about thirty straggly hawthorn bushes three metres high, the remains of an old hedge that ran through the middle on a raised bank. From mid-September onwards, it was the winter home for between eighty and a hundred blackgame as well as around fifty grouse. It used to be driven to eight guns three or four times a season. All the surrounding sheep walks were grassy, wet heath where frequent broods of blackgame were to be found in August. Throughout the district, blackgame were more common than grouse. The terrain bore no resemblance to what is perceived today as good black grouse habitat but it was typical of large areas of west and south Scotland and consistent with the Dumfriesshire description of fifty years earlier. Those halcyon sporting days came to an abrupt end when keepering ceased and the land was sold for planting with Sitka spruce. Within three years, the birds had gone. Thirty years later, they are extinct in the district.[4] The decline has come about for a number of reasons: the specialised habitat no longer exists, the upland farmer grows no crops, sprays both rushes and bracken, cuts silage and over-browses such heather as there is. Not only is the birds' house in ruins but the district is alive with crows and foxes. They are also shy breeders compared with red grouse – if a greyhen fledges a brood of five young, she has done exceptionally well.

It is easy to see why black grouse have declined so dramatically and continue to do so all over Britain. The ninth Duke of Buccleuch summed matters up well in an informal conversation in the mid-1990s when he opined that 'blackgame were not good republicans' by which he meant that they did not thrive if they had to compete with the interests of groups such as progressive farmers and the public seeking access in large numbers. That wise man also felt that the widespread over-browsing at the moor edge – a factor usually allocated pole position by conservation bodies in the blame game of the bird's decline – was probably over-emphasised and that their reduced fortunes (particularly in the Dumfriesshire strongholds with which he was familiar) was associated with the development of pheasant shoots that relied on releasing hand-reared birds. He did not speculate whether it was competition for food and nesting sites or a cross-infection such as coccidiosis or blackhead that brought this about but who is to say he was wrong?

The current thrust by the conservation movement to restore the bird's fortunes owes more to European knowledge (where they are almost wholly forest-based) than to an appreciation of the circumstances which prevailed traditionally over most of Britain from Exmoor through Staffordshire to the Borders and beyond. Public money spent bribing farmers to reduce sheep together with the creation of 'little woods here and little woods there' is, by itself, a misplaced tactic; the lessons of *regional* history must be learned. If blackgame, whose daily range in adulthood is up to 3 km (2 miles) are to thrive, the clock has to be turned back: extensive contiguous areas of land must be devoted to growing crops and to allowing the moor edge and upland fields to revert to something like what they were a hundred or even fifty years ago. Unless this is done extensively and allied to crow and fox control, the black grouse will be familiar to future generations only in a glass case and certainly not as an admired and treasured constituent of a mixed bag obtained on an October day along the moor edge.

Capercaillie

Capercaillie lekking. (Photo: N. McIntyre)

Our largest grouse became extinct at the beginning of the nineteenth century almost certainly because of habitat loss.[5] In the middle of the century, the forest of Darnaway in Moray was described thus:

> Few know what [D]arnaway was in those days – almost untrodden except by the deer, the roe, the foxes, and the pine martens, its green dells filled with Lilies of the Valley, its banks covered with wild hyacinths, primroses and pyrolas and its deep thickets clothed with every species of woodland luxuriance, its blossoms, grass, moss and timber of every kind, growing with the magnificence and solitude of an aboriginal wilderness, a world of unknown beauty and silent loneliness, broken only by the sough of pines, the hum of the water, and hoarse bell of the [roe] buck, the long wild cry of the fox, the shriek of the heron or the strange mysterious tap of the northern woodpecker.[6]

Forests fitting this description once extended over huge areas of Scotland, dominating the valleys of Findhorn, Spey, Dee, the Inverness-shire Garry and elsewhere but were logged systematically from the first quarter of the eighteenth century onwards. The logging destroyed the bird's habitat. Only when more settled times came in the early nineteenth century did afforestation re-create a suitable niche for it. An odd pair may have survived here and there but it was popularly held at the time that the bird had gone.

Re-introduced at Taymouth Castle in 1837, it flourished during Victorian times and re-established itself throughout Scotland from east Sutherland to Argyll. In suitable woodland sites north of the Highland Boundary fault, it withstood extensive habitat destruction caused by clear felling to meet the demand for timber during the First World War just as it withstood deer fencing and re-planting in later times. By the mid 1920s, it was even classed as a forest pest on account of its habit of browsing the shoots of young trees, particularly Scots pine and larch and this view was held by many foresters even as late as the

1960s. Notwithstanding the attendant persecution, it remained esteemed as a game bird. The author's father was one of a team of eight guns who shot ninety-eight birds over four days in December 1936 (two of the days being curtailed because there was a foot of snow on the ground). The venue? Abernethy, in Speyside, now owned by RSPB and, seventy-five years on, generally regarded as the core capercaillie area in Scotland. In April 2011, it boasted about forty calling cocks.[7] As recently as the early 1970s, forty-nine birds were shot in one day on the old Taymouth estate on Drummond hill (NN755464) over some 900 ha (2200 ac) of mixed-age conifers and broadleaves. Nowadays, apart from small areas in Deeside, Badenoch and Upper Speyside, the bird is all but extinct in Scotland. What caused this catastrophic contraction in range in less than forty years?

In the 1960s and early 1970s, capercaillie occurred in shootable numbers in many places. Populations of varying size were to found from Novar in Easter Ross through Culbin Forest on the Moray coast, the great woods of Upper Deeside in Aberdeenshire, in Angus, Atholl, the estates of South Perthshire and west to the islands of Loch Lomond, the north foothills of Ben Lui at Tyndrum and the Black Wood of Rannoch and most of them were lightly shot almost every year. At that time, estates in Upper Speyside yielded many days of around twenty birds: there was no sign of a fall in numbers attributable to shooting. By the late 1970s however, concern was being expressed that numbers were declining. After scientists identified one reason as adult birds colliding with deer fences, pressure on the forest industry from conservation bodies saw many fences dismantled and those deemed essential as a barrier against deer encroachment 'flagged' with plastic bunting and cleft chestnut fencing. The initiative doubtless saved birds but numbers continued to decline; some other cause had to be identified. Specialists began to blame poor chick survival rates on cold, wet weather at hatching (capercaillie chicks are susceptible to wet and cold, grow slowly and are vulnerable to disturbance when small) coupled with poor supplies of the sawfly larvae which depend on blaeberry swards and are important food for chicks.[8] In Speyside, swards were being over-browsed by red deer or over-topped and suppressed by heather. The hypothesis fitted nicely with the anti-deer sentiment of the time but, despite various initiatives, numbers continued to decline. Gamekeepers were demonised for setting snares to catch foxes because capercaillie cocks were recorded occasionally as a by-catch.[9]

Although landowners put in place a voluntary ban on shooting them, it too had no effect on numbers. In 2005, capercaillie were taken off the quarry list and put onto Schedule 1 where they receive the same level of protection as raptors and corncrakes, stone curlews and choughs. As a quarry species, it is unlikely ever to return to the game bag: the one group that could have saved it (because it could put a price on its head for the trophy sportsman) was sidelined and their views ignored.

Some in the conservation lobby seem surprised that numbers continue to plummet. If the decline persists at the present rate, we are likely to lose the bird as a breeding species for the second time.[10] What makes the situation lamentable is that plenty of keepers, factors and landowners know full well why numbers have collapsed and why the bird is now entering the terminal zone of 'Red Book' status. Of course, we may have had a spell of predominantly cold and wet springs unfavourable for chick survival but there have been similar spells in the last hundred and seventy-five years – the birds have weathered that kind of adversity before. The big change in the last thirty-five years has been the reduction in keepering and the substantial increase in predator pressure.[11] In 1970, buzzards were rarely seen in upper Speyside but are now commonplace. Pine martens used

Hinds on heather moorland in winter. A widespread problem on grouse ground.
(PHOTO: N. McINTYRE)

to be scarce but, following intensive protection and a series of (illegal) translocations, are now everywhere. Goshawks occur widely in areas where, in former times, they were unknown. Though foxes remain relatively scarce and controlled in the great forests that were capercaillie strongholds, badgers are everywhere and a major ground predator. Carrion crows occur widely in good numbers, their populations supported by insanitary farming and tourist waste. The recent founding of the Cairngorm National Park has seen an increase in public access and general disturbance, particularly in spring.[12]

When alarm bells really began to ring in the 1990s, one key worker told the author that the bird's salvation depended on applying 'The Precautionary Principle'.[13] If the capercaillie is to be retained as a breeding bird, the Principle has to be observed immediately – there is no time to be lost. The bird's plight is now desperate; it will be saved only if every measure available to managers is taken. The five main points which need to be addressed are :

- Identifying one or more core areas such as upper Speyside or Deeside where the bird has traditionally thriven and where its needs can be prioritised;

- Controlling foxes, pine martens and badgers as tightly as possible. If that results in a by-catch of wild cats, so be it – at this eleventh hour, management must take tough decisions;

- Preventing crows and buzzards from nesting throughout the area;

- Closing forests to public access for the crucial eight weeks from end-April to end-June to avoid brood disturbance;

- Extending the use of fire under the canopy of mature pines to increase blaeberry and improve conditions for chicks. The practice, widely used in the southeastern United States to enhance quail performance, would have the same effect on the shrub zone in Scotland. There is nothing difficult about the work and it does not lead to extensive forest fires.

The blinkered approach of SNH and RSPB is particularly depressing. The former persists in throwing good money after bad; in 2011 it employed two rangers to 'step up conservation of threatened capercaillie' on the islands in Loch Lomond[14] despite the virtual certainty that capercaillie are no longer present on them.[15] A spokesman noted that 'capercaillie are under great threat due to a range of factors including climate change [*questionable at best*], low habitat quality [*certain*], predation and disturbance of breeding birds and young chicks [*doubly certain*]'.

SNH seems unable even to learn from the experience of another gallinule with problems of terminal decline: the grey partridge. These have declined nation-wide by ninety per cent since the 1950s and are now extinct over large areas of the country[16] with the reasons for the decline identified.[17] However, on three flagship sites in Northumberland, in Cambridgeshire, Hertfordshire and in Sussex, dedicated, knowledgeable managers reversed the trend and brought them back to populations comparable with those of the 1960s in the space of three seasons. On one site, pair numbers rose from seventeen to over two hundred in four years following the simple expedient of creating conditions suitable for their needs.[18] The initiative involved significant investment in improving habitat for nesting and brood cover and for areas for chicks to forage, in providing ample winter food, in reducing fungicide, herbicide and pesticide programmes (at the expense of farm profitability) and in *thorough predator control.*

It was a project driven by people interested in shooting but, as the capercaillie is now listed, it is a route to salvation now unavailable despite proven success. For all practical purposes, landowners have been stripped of responsibility for the capercaillie – the species has effectively been nationalised. The endgame is now clear: unless its problems are tackled in the dynamic way they were for partridges, it will become extinct within ten years however many 'Precautionary Principles' are observed. The blame for that will lie unequivocally at the door of the conservation organisations. It is just not possible to cherry pick survival strategies and expect substantially improved numbers quickly or even at all; the problem has to be considered in the round. Some may find the essential remedies unpalatable but, unless they are administered, those now solely responsible for the bird's survival will have only themselves to blame when they wake up one April morning to discover that the *leksang* of our great wood grouse can be heard no more.

Ptarmigan

Ptarmigan habitat is less vulnerable to anthropogenic change than moorland at lower altitudes and is less exposed to over-browsing damage although high concentrations of deer in summer can leave their mark. In core areas in the high mountains, there is little hard evidence that numbers have declined over the past fifty years although they have become scarcer in more marginal massifs and those small enough to be classed as isolated islands.

Numbers have held up well over much of their range because they live in relative equilibrium with predators which occur by and large at 'natural' densities. The vegetation on which they subsist is not burned and is little used by concentrations of sheep. Shooting some surplus birds in August is, of course, a form of predation but there is no suggestion that populations are over-shot. Nor are they likely to be as only young, fit shooters are willing to exert themselves to obtain a yield of a few brace from the high tops and there are probably fewer of those

A pair of ptarmigan in late winter. (PHOTO: N. MCINTYRE)

than there were fifty years ago. One indirect impact is increased public access at high altitude. Walkers often leave behind food scraps and excrement, luxury foods for crows and foxes alike. If there is enough, as there is over parts of the Cairngorms, scavengers over-winter in artificially high numbers and turn to becoming nest and chick predators in the summer.[19] Some harbingers of doom forecast that ptarmigan will decline as 'global warming' increases but this can safely be ignored. Ptarmigan occur close to sea level in northwest Sutherland where they are subject to a climate a great deal milder and more oceanic than the colder boreal conditions they experience on Scotland's highest mountains and have maintained numbers in these places without problem for decades.

Perhaps the biggest threat comes from the anti-field sports lobby who might attempt to put ptarmigan onto Schedule 1 on the grounds that the bird is 'rare', 'in decline' and in need of 'protection'. If it were to be listed, the result would be the same as happened when grey partridges in Ireland[20] and capercaillie in Scotland were removed from the quarry list.

Moorland waders

Apart from snipe, moorland waders have no commercial value. They occur in high densities where there is no predator pressure.[21] The plight of waders in the uplands attracts the attention of conservationists throughout Britain. Curlews and lapwings have declined to low levels over wide areas of moorland, bog and marsh; snipe and redshank are now scarce in places where they used to be common. It is clear that habitat loss, predation and land drainage have all contributed to this state of affairs but there is robust debate between interested parties as to proximal cause.[22] Travelling in Sutherland in 1848, Charles St John remarked on large concentrations of greenshanks, redshanks, curlews, snipe and wigeon in the marshes at Aultna-harrow (Altnaharra). He must have been referring to the Mudale valley (NC533359) which runs west from the village for four miles or so.[23] Waders are all but unknown there today. Elsewhere in Britain, it is obvious that there are plenty of waders on well-keepered moorland particularly where it overlies base-rich rocks but that they are largely absent from the areas of granite and undifferentiated schistose grits which dominate highland Scotland. This was not always so: much of the loss has taken place within living memory and is related to increased predator pressure. Organisations such as RSPB and SNH, though they pay lip service to the benefits of keepering, evade the contentious issue of raptor management.

Moorland waders exist on invertebrates that are at their most numerous in fertile, moist places. In the nineteenth and early twentieth centuries, there was still residual fertility from the high density human settlements that had occurred in every upland valley. Once these were abandoned, the fertility gradually leached away until the sites ceased to support invertebrates in numbers high enough to supply the waders' needs, so they in turn disappeared.

To illustrate the point, old-fashioned septic tanks did not have an integral subterranean soakaway; the supernatant liquid overflowed, seeped over a down-slope area and created a fan of enriched soil below the tank (NC306563). Farm middens operated in the same way. The ground so fertilised, which could extend over several acres next to a group of houses, provided an extensive area in which breeding birds could nest and feed in summer and a largely frost-free feeding place in winter. The author knows of many places which sixty years ago yielded common snipe and teal over guns during a grouse drive where the birds are unknown today (NS293643, 231678). Even small, flushed sites below an issue from base-rich rock which always held a snipe in frosty weather in winter seldom do so now (NS214589, 231585). It can be argued that any local decline in numbers is simply a reflection of the decline the national snipe population but this is incorrect: it reflects a general degradation in habitat caused by a loss of fertility and many decades of over-grazing by sheep. Artificial enrichment of wet areas using farmyard manure or slaughterhouse blood can produce immense concentrations of snipe, particularly in winter. On one such area in southwest Scotland (NY003855) in the 1990s, it was normal during a day's shooting to flush two thousand snipe from a forty-acre bog also heavily used by lapwings, curlews and snipe as a breeding area. Twenty years after artificial enrichment (and, to be fair, keepering) ended, there are no snipe in winter and almost no waders. Though modern regulation of the disposal of animal by-products forbids enrichment with fresh blood, there is no doubt that if the practice were to be applied widely to suitable areas of upland marsh, the positive effect on the numbers of a whole suite of moorland waders would be dramatic.

Salmon and trout

Drainage water is by definition an integral part of moorland. The path by which it reaches the ocean consists of a network of marshes, bogs, streams, tarns, lochs and larger water bodies which, under the influence of gravity, combine to make up both our great and lesser rivers. The Victorians pursued all sporting opportunities with vigour. They sought 'the most', 'the biggest', 'the most difficult' prizes with great dedication and enthusiasm. In the second half of the nineteenth century, fishers were catching large numbers of what would be referred to today as 'trophy fish'. Many hotels and highland lodges display upon their walls casts of huge salmon, pike and brown trout in glass cases that date predominately from the last quarter of the nineteenth century. Precious few trophy fish have been taken after 1950 and preserved for posterity to admire.

Atlantic salmon

Upland management influences the salmon population nationally.

Before leaving the topic of fertility decline and its effect on both numbers and productivity, there is a suggestion (but no hard data to support the idea) that changes in moorland vegetation may have contributed to this state of affairs by reducing stream flows, particularly in summer. The author knows of areas of formerly heather moorland where purple moor grass and cotton sedge have taken over from the ericaceous plants and become the dominant species. Many shooters have noted the change from heather moor to grass heath over a fifty-year time span. The process has been driven partly by fire and over-use by grazing animals in winter but also by misguided attempts to foster heather by removing sheep. All that this achieves is an *increase* in dominance by grasses and sedges.[24]

The effect of this change is likely to be negative in its effect on fisheries on account of differences in transpiration rates. A diverse sward of grasses and sedges has a greater leaf surface area and so exerts a larger water demand on the soil than if the same area were covered with heather. Under a solid sward of monocotyledonous plants, there is likely to be less water available to drain out of the moorland and maintain water levels and stream flows in nursery streams. In effect, the grasses tend to dry up the moor. In hot and dry spells, the streamlets shrink and may become too warm to support fish life: a proportion cease to function as nursery streams. Invertebrate life suffers and small salmonids have to try to establish territory further downstream. The vegetation changes may also contribute to slower growth in juvenile fish and to a reduction in the number of young fish a river system can support. (It is, alas, not difficult to imagine some eager scientist noticing a reduced flow and attributing the phenomenon to 'global warming'.)

Brown trout

Fertility in a drainage basin translates into base-richness in the water which in turn supports a larger and more diverse river flora and an enhanced suite of invertebrates. As extensive areas of moorland have clearly declined in fertility over the past century, so have average weights of fish, particularly brown trout.

In the twenty-first century, only lochs which receive water from base-rich basins and maintain a pH of at least 6.0 deliver individuals and baskets of rod-caught fish consistently which are comparable with former times. The hill loch which yields an occasional trophy fish is doing so either because that individual achieved dominance over its peers and grew large by eating the smaller ones or because limited recruitment (poor spawning conditions) resulted in a small resident stock. The biomass of fish which any loch can support is related to the food supply contained in it – a given body of water can either support a few large fish, many small ones or a mixture of the two.

Five trout weighing a total of 10.5 lbs caught on fly from a Sutherland hill loch – 2011.

End notes

1 Ellice, 1931.
2 Gladstone, 1922; Scott, 1937.
3 Scott, 1937.
4 Mary Craig (born *c.* 1890) was the farm tenant of this ground all her life and told the author in 1974 that before the 1939–1945 war, blackgame were so common on the farm that they arrived in September 'like a flock of crows' and pulled the frandies to pieces in the cornfields along the moor edge.
5 Hesketh Pritchard, 1921.
6 Haldane, 1962.
7 D. Dugan, pers comm.
8 Baines *et al*, 1994; Baines *et al*, 2011.
9 Watson & Moss, 2008.
10 Moss, 2001.
11 Baines *et al*, 2004; Haysom, 2011.
12 Watson & Moss, 2008.
13 R. Moss, pers comm.
14 Walters, 2011b.
15 D. Dugan, pers comm.
16 Potts, 1986.
17 Chaimberlain *et al*, 2001.
18 Phillips, 2011.
19 Watson, 1996.
20 Phillips, 2011.
21 Merricks, 2010.
22 Merricks, 2010.
23 St John, 1882.
24 Tucker, 2003; Walters, 2011b.

Glossary

Acclimatisation	Adaptation by a sheep stock to a home range, its endemic diseases and parasites.
Allotment	Fenced plot of land – often on moor edge – subject to farm tenancy.
Anthropogenic	Originated by man.
Anthropomorphism	Attribution of human form or personality to animals.
Babesiosis	Redwater fever in cattle; usually tick-transmitted.
Beck	Hill stream (North England)
Besom	1. Long-handled fire-beater made of birch twigs.
	2. Mischievous woman.
Birn	Spiky stick heather following fire. See Scrogg.
Blackhead	Disease of gallinaceous birds – Histomoniasis
Boulder clay (Till)	Glacial deposit, a mix of clay, pebbles, boulders and rock fragments.
Breek, breeking	Sacking 'knickers' stitched to hip-wool of female sheep to prevent insemination.
Brim, Brimming	Mammal in oestrus. See Clicket
BSE	Bovine spongiform encephalitis. A disease of cattle. See Scrapie.
Burn	1. Hill stream (Scotland)
	2. Fire (misuse)
Callunetum	Association of moorland plants with ling heather as a main constituent.
	Cf. Pinetum, arboretum.
Cast	Draft ewes when sold off the hill, typically after four crops of lambs. Also known as crogs.
Cheeper	Immature grouse or ptarmigan.
Cleg	Biting fly – summer pest of hill-living mammals. See Horse-fly.
Clicket(ing)	Mating season – Foxes
Clit-ill	Interdigital infection causing lameness in sheep and cattle. See Scad.
Coupie	Sheep lying on its back and unable to right itself.
Crane-fly, tipulid	Long-legged slender fly resembling a mosquito but which does not bite.
Crepuscular	Active in twilight at dawn and dusk.
Cut	Alternative name for heft.
Dicotyledon	Flowering plant with two embryonic seed leaves.
Dinmont	Wether between its first and second clipping. See Gimmer, Teg.
Dyke	Dry-stone wall.
Ewe	Adult female sheep.
Flow	Low-lying wet heath usually with many bog-pools, north Scotland.
Fluke, liver	Flat-worm parasite of sheep, common on wet pastures.
Flush	1. Area of wet ground caused typically by a spring, often base-rich.
	2. Rapid increase in the nutritive level of hill sheep.
	3. (out) Cause animal to get out of hiding-place.
Fog(g)	1. Understory of mosses, lichens and litter below shrub canopy.
	2. Flammable blanket of dead vegetation overlying soil after fire.
	3. Grass – Yorkshire fog – Holcus lanatus.
Foggage	Grass aftermath following a hay or silage crop.
Fother(ing)	Act of feeding sheep or cattle. Same as fodder(ing).
Frandie(s)	Small rick(s) of sheaves in a cornfield. Also hutts, hand-ricks.
Gait	See Stint.
Gibble	A device made of cloth and wire for lighting a fire.
Gimmer	Female sheep between first and second clipping. See Dinmont, Teg.
Grip	1. Open sheep-drain on moorland.
	2. Collie that bites sheep when working with them.
	3. Dung-channel in byre behind cattle tied in stalls.
Hag	Peat bog dissected by gullies and water-courses.
Hain	To exclude grazing from a pasture, either for the making of hay or silage later or for saving foggage.
Hart	Red deer stag six years or more. In common use in Atholl until the 1970s.
Heft	Home range of a collection of sheep, both family members and wider kinship groups.
Hirsel	Collection of hefts which collectively make up the responsibility of one shepherd – traditionally 500 ewes and gimmers.
Histomoniasis	Blackhead
Hogg	Sheep of either sex until first clipping.
Horse-fly	see Cleg.
In-bye	Enclosed ground.
Instar	Stage of an insect or other arthropod between moults.

Joint-ill	Pus-filled joints and septicaemia caused by tick-bites.
Ked	Sheep louse.
Keel	Colour-mark put on sheep fleece for flock identification. Marking-fluid.
Lagomorph(a)	Order which includes hares and rabbits.
Law	A roundish or conical hill.
Leveret	Immature hare.
Louping ill	Viral encaphalitis of tick-borne origin. See Trembling.
Lug-mark	Cut, hole or tattoo in the ear of sheep and cattle to aid farm identification. See Nip (or nick).
Lunkie-hole	Gate-way in dry-stone wall allowing passage of sheep but not cattle. Also creep.
Machair	Herb-rich coastal grassland, usually overlying shell-sand.
March	Boundary of farm or estate. Also mearn, meving.
Mawks	Greenbottle maggots on sheep.
Meristem	Growing-point in grass and sedge species.
Microtin(e)	Voles and lemmings.
Monocotyledon	Flowering plant having a single embryonic seed leaf (typically grasses and sedges).
Monogamous	Having one mate at a time.
Montane	Growing in or inhabiting mountain areas.
Nematode	Slender worm affecting sheep and grouse. Of significant economic importance.
Nip (or Nick)	See Lug-mark.
Philopatry	Dispersal mechanism in which individuals attempt to remain close to their birth-place.
Poach	1. To cut up a sward into a morass with hoof-prints. 2. To take game illegally.
Polygamous	Having more than one mate at a time.
Prothallus	Germinating spore of bracken.
Raddle	Colour-marker pad in a harness between the fore-legs of a tup to indicate when he has served a female.
Rake	Daily movement of sheep over their home range.
Redwater fever	See Babesiosis.
Rigg (Ridgeling)	Entire male sheep where only one testicle (or none) has dropped into scrotum. See Tup.
Rumevite	Urea-based feed block supplement for ruminants.
Rylock	Robust square-mesh wire netting for stock fences.
Scab (sheep)	Highly infectious condition caused by two species of mites.
Scad (Scald)	Interdigital infection in sheep and cattle causing lameness. See Clit-ill.
Scrapie	Encephalitis in sheep closely related to BSE in cattle.
Scrogg	Stunted deciduous bushes. Dead heather sticks after burning.
Set-stocking	Confining the same number of animals to an area year-round.
Shan-gabbit	Overshot lower jaw in sheep and cattle.
Shearling	Tup between its first and second clipping.
Sheiling	Settlement on upland pastures used in summer.
Soo-mooth	Undershot lower jaw in sheep and cattle. Pig-jawed.
Souming	Same as 'stint' but term confined to crofting townships.
Spean	Act of weaning calves or lambs.
Stell	Dry-stone circular storm-shelter, normally one per heft.
Stint	Measure of an individual farm's grazing rights for sheep and cattle on a common grazing. Various permutations of livestock according to district.
Stook	Six or eight sheaves of cut corn, usually oats, set up against each other for drying in the field. In windy districts, four sheaves tied together with a straw strap called a sugan.
Stot	1. Bullock. 2. To jump – stiff-legged.
Tapeworm	Segmented worm. Different species parasitise sheep and grouse; of limited economic importance.
Teg	Two-year-old sheep – South Pennines. See Dinmont, Gimmer.
Tipulid	See Crane-fly.
Trembling	Louping ill disease in sheep.
Tup	An entire male sheep. See Rigg.
Two-pasture system	Husbandry practice whereby hill sheep are put on improved pasture at critical times of year, notably pre-tupping and prior to lambing.
Weed	Mastitis in lactating ewe. Also udder-clap.
Wether	Emasculated male sheep.

Latin names

Birds

Auk, little	*Alle alle*
Blackbird	*Turdus merula*
Blackcap	*Sylvia atricapilla*
Brambling	*Fringilla montifringilla*
Bullfinch	*Pyrrhula pyrrhula*
Bunting, corn	*Emberiza calandra*
Bunting, reed	*Emberiza schoeniclus*
Bunting, snow	*Plectrophenax nivalis*
Buzzard, common	*Buteo buteo*
Capercaillie	*Tetrao urogallus*
Chaffinch	*Fringilla coelebs*
Chiff chaff	*Phylloscopus collybita*
Chough	*Pyrrhocorax pyrrhocorax*
Coot	*Fulica atra*
Corncrake	*Crex crex*
Crossbill	*Loxia curvirostra*
Crow, carrion	*Corvus corone*
Crow, hooded	*Corvus cornix*
Cuckoo	*Cuculus canorus*
Curlew	*Numenius arquata*
Dipper	*Cinclus cinclus*
Diver, black-throated	*Gavia arctica*
Diver, great northern	*Gavia immer*
Diver, red-throated	*Gavia stellata*
Dotterel	*Eudromias morinellus*
Dove, rock	*Columba livia*
Dove, stock	*Columba oenas*
Duck, black scoter	*Melanitta nigra*
Duck, eider	*Somateria mollissima*
Duck, gadwall	*Anas strepera*
Duck, golden-eye	*Bucephala clangula*
Duck, goosander	*Mergus merganser*
Duck, mallard	*Anas platrynchos*
Duck, pintail	*Anas acuta*
Duck, pochard	*Aythya ferina*
Duck, red breasted merganser	*Mergus serrator*
Duck, scaup	*Aythya marila*
Duck, shelduck	*Tadorna tadorna*
Duck, shoveller	*Anas clypeata*
Duck, teal	*Anas crecca*
Duck, tufted	*Aythya fuligula*
Duck, wigeon	*Anas penelope*
Dunlin	*Calidris alpina*
Eagle, golden	*Aquila chrysaëtos*
Eagle, white-tailed (sea)	*Haliaeëtus albicilla*
Fieldfare	*Turdus pilaris*
Fly-catcher, spotted	*Muscicapa striata*
Gannet	*Sula bassana*
Goldcrest	*Regulus regulus*
Goldfinch	*Carduelis carduelis*
Goose, Canada	*Branta canadensis*
Goose, greylag	*Anser anser*
Goshawk	*Accipiter gentilis*
Great skua	*Stercorarius skua*
Grebe, black-necked	*Podiceps nigricollis*
Grebe, great crested	*Podiceps cristatus*
Grebe, little	*Tachybaptus ruficollis*
Grebe, Slavonian	*Podiceps auritus*
Greenfinch	*Carduelis chloris*
Greenshank	*Tringa nebularia*
Grouse, black	*Lyrurus tetrix*
Grouse, red	*Lagopus lagopus scoticus*
Grouse, willow (willow ptarmigan)	*Lagopus lagopus lagopus*
Guillemot, black	*Cepphus grylle*
Gull, black-headed	*Larus ridibundus*
Gull, common	*Larus canus*
Gull, greater black-backed	*Larus marinus*
Gull, herring	*Larus argentus*
Gull, kittiwake	*Rissa tridactyla*
Gull, lesser black-backed	*Larus fuscus*
Gyrfalcon	*Falco rusticolus*
Hedge accentor	*Prunella modularis*
Hen harrier	*Circus cyaneus*
Heron	*Ardea cinerea*
Hobby	*Falco subbuteo*
Jackdaw	*Corvus monedula*
Jay	*Garrulus glandarius*
Kestrel	*Falco tinnunculus*
Kingfisher	*Alcedo atthis*
Kite, red	*Milvus milvus*
Linnet	*Acanthis cannabina*
Magpie	*Pica pica*
Martin, house	*Delichon urbicum*
Martin, sand	*Riparia riparia*
Merlin	*Falco columbarius*
Nightjar	*Caprimulgus europaeus*
Osprey	*Pandion haliaetus*
Owl, long-eared	*Asio otus*
Owl, short-eared	*Asio flammeus*
Owl, snowy	*Nyctea scandiaca*
Owl, tawny	*Strix aluco*
Oyster catcher	*Haematopus ostralegus*
Partridge, grey	*Perdix perdix*
Partridge, red-legged	*Alectoris rufa*
Peregrine	*Falco peregrinus*
Phalarope, red-necked	*Phalaropus lobatus*
Pheasant	*Phasianus colchicus*
Pipit, meadow	*Anthus pratensis*
Pipit, rock	*Anthus spinoletta*
Plover, golden	*Pluvialis apricaria*
Plover, green (lapwing)	*Vanellus vanellus*
Prairie chicken, greater	*Tymphanuchus cupido*
Ptarmigan	*Lagopus mutus*
Puffin	*Fratercula arctica*
Quail, bobwhite	*Colinus virginianus*
Raven	*Corvus corax*
Redpoll, lesser	*Acanthis cabaret*
Redshank	*Tringa totanus*
Redstart, common	*Phoenicurus phoenicurus*
Redwing	*Turdus iliacus*
Ring ouzel	*Turdus torquatus*
Robin	*Erithacus rubecula*
Rook	*Corvus frugilegus*

Sandpiper, common	*Tringa hypoleucos*
Shag	*Phalacrocorax aristotelis*
Sharp-tailed grouse	*Tymphanuchus phasianellus*
Shelduck	*Tadorna tadorna*
Shrike, grey	*Lanius excubitor*
Siskin	*Carduelis spinus*
Skua, great	*Stercoriarus skua*
Skua, Richardson's (Arctic)	*Stercorarius crepidatus*
Sky-lark	*Alauda arvensis*
Snipe, common	*Gallinago gallinago*
Snipe, great	*Gallinago media*
Snipe, jack	*Lymnocryptes minimus*
Sparrowhawk	*Accipiter nisus*
Sparrow, house	*Passer domesticus*
Sparrow, tree	*Passer montanus*
Starling	*Sturnus vulgaris*
Stonechat	*Saxicola torquata*
Stone curlew	*Burhinus oedicnemus*
Swallow	*Hirundo rustica*
Swan, whooper	*Cygnus cygnus*
Swift	*Apus apus*
Tern, common	*Sterna hirundo*
Thrush, missel	*Turdus viscivorus*
Thrush, song	*Turdus philomelos*
Tit, blue	*Parus caeruleus*
Tit, coal (cole)	*Parus ater*
Tit, crested	*Parus cristatus*
Tit, great	*Parus major*
Tit, long-tailed	*Aegithalos caudatus*
Tit, marsh	*Parus palustris*
Tree-creeper	*Certhia familiaris*
Twite	*Acanthis flavirostris*
Wagtail, grey	*Motacilla cinerea*
Wagtail, pied	*Motacilla alba*
Wagtail, yellow	*Motacilla flava*
Warbler, grasshopper	*Locustella naevia*
Warbler, sedge	*Acrocephalus schoenobaenus*
Warbler, willow	*Phylloscopus trochilus*
Warbler, wood	*Phylloscopus sibilatrix*
Waterhen (Moorhen)	*Gallinula chloropus*
Water-rail	*Rallus aquaticus*
Wheatear	*Oenanthe oenanthe*
Whimbrel	*Numenius phaeopus*
Whinchat	*Saxicola rubetra*
Whitethroat	*Sylvia communis*
Wigeon	*Anas penelope*
Woodcock	*Scolopax rusticola*
Woodpecker, greater spotted	*Dendrocopos major*
Woodpecker, green	*Picus viridis*
Woodpigeon	*Columba palumbus*
Wren, common	*Troglodytes troglodytes*
Yellow-hammer	*Emberiza citrinella*

Mammals

Badger	*Meles meles*
Beaver, European	*Castor fiber*
Deer, fallow	*Dama dama*
Deer, red	*Cervus elaphus*
Deer, roe	*Capreolus capreolus*
Deer, Sika	*Cervus nippon*
Fox, red	*Vulpes vulpes*
Hare, brown	*Lepus europaeus*
Hare, mountain	*Lepus timidus*
Hedgehog	*Erinaceus europaeus*
Marten, pine	*Martes martes*
Mink, American	*Mustela vison*
Mole	*Talpa europaea*
Moose, European	*Alces alces*
Mouse, wood	*Apodemus sylvaticus*
Otter	*Lutra lutra*
Polecat	*Mustela putorius*
Reindeer	*Rangifer tarandus*
Stoat	*Mustela erminea*
Vole, Orkney	*Microtus arvalis orcadensis*
Vole, short-tailed field	*Microtus agrestis*
Weasel	*Mustela nivalis*
Wild cat	*Felis sylvestris*
Wolverine	*Gulo gulo*

Plants

Alder	*Alnus glutinosa*
Ash	*Fraxinus excelsior*
Aspen	*Populus tremula*
Asphodel, bog	*Narthecium ossifragum*
Avens, mountain	*Dryas octopetala*
Barley	*Hordeum vulgare*
Bearberry	*Arctostaphylos uva-ursi*
Bedstraw, heath	*Galium saxatile*
Birch, brown	*Betula pubescens*
Birch, dwarf	*Betula nana*
Birch, silver	*Betula pendula*
Blaeberry	*Vaccinium myrtillus*
Bracken	*Pterideum aquilinum*
Broom	*Sarothamnus scoparius*
Cloudberry	*Rubus chamaemorus*
Clover, wild white	*Trifolium repens*
Cotton-sedge – common	*Eriophorum angustifolium*
Cotton-sedge – hare's tail	*Eriophorum vaginatum*
Cowberry	*Vaccinium vitis-idaea*
Cranberry	*Vaccinium oxycoccus*
Crowberry	*Empetrum nigrum*
Daisy	*Bellis perennis*
Deer sedge	*Trichophorum cespitosum*
Dock, broad-leaved	*Rumex obtusifolius*
Dock, curled	*Rumex crispus*
Eyebright	*Euphrasia spp*
Foxglove	*Digitalis purpurea*
Fir, Douglas	*Pseudotsuga menziesii*
Fir, Scotch/Scots pine	*Pinus sylvestris*
Gorse, see Whin	*Ulex europaeus*
Grass, bent	*Agrostis* spp
Grass, brome	*Bromus* spp
Grass, creeping red fescue	*Festuca rubra*
Grass, crested dogstail	*Cynosurus cristatus*
Grass, fescue	*Festuca* spp

Grass, foxtail	*Alopecurus* spp
Grass, meadow	*Poa spp*
Grass, purple moor	*Molinia caerulea*
Grass, rye	*Lolium perenne*
Grass, sweet vernal	*Anthoxanthum odoratum*
Grass, wavy hair-	*Deschampsia flexuosa*
Grass, white bent	*Nardus stricta*
Hazel	*Corylus avellana*
Heath, cross-leaved	*Erica tetralix*
Heather, bell	*Erica cinerea*
Heather, ling	*Calluna vulgaris*
Holly	*Ilex aquifolium*
Juniper	*Juniperus communis*
Lady's mantle	*Alchemilla mollis*
Myrtle, bog	*Myrica gale*
Oak	*Quercus spp*
Oat	*Avena sativa*
Pine, lodge-pole	*Pinus contorta*
Pine, Scots	*Pinus sylvestris*
Moss	*Polytrichum* spp
Moss	*Racomitrium* spp
Moss	*Sphagnum* spp
Nettle, stinging	*Urtica dioica*
Pansy	*Viola spp*
Rose bay willow herb	*Epilobium angustifolium*
Rowan (mountain ash)	*Sorbus aucuparia*
Rush, hard	*Juncus inflexus*
Rush, heath	*Juncus squarrosus*
Rush, soft	*Juncus effusus*
Saxifrage, yellow mountain	*Saxifraga aizoides*
Saxifrage, purple	*Saxifraga oppositifolia*
Sorrel, sheeps'	*Rumex acetosella*
Spruce, Norway	*Picea abies*
Spruce, Sitka	*Picea sitchensis*
Thyme, wild	*Thymus serpyllum*
Tormentil	*Potentilla erecta*
Violet, bog	*Viola palustris*

Whin	*Ulex europaeus*
Whortleberry, bog	*Vaccinium uliginosum*
Willow, creeping	*Salix repens*
Willow, downy	*Salix lapponum*
Willow, eared	*Salix aurita*
Willow, woolly	*Salix lanata*
Yorkshire fog	*Holcus lanatus*

Amphibians

Frog, common	*Rana temporaria*

Invertebrates

Beetle, heather	*Lochmaea suturalis*
Blackhead	*Histomonas meleagris*
Cleg or horse-fly	*Tabanus sulcifrons*
Coccidiosis	*Eimeria* spp
Crane-fly	*Molophilus ater, Nephrotoma* spp
Earthworm	*Allolobophora* spp
	Lumbricus spp
Greenbottle fly	*Lucilia sericata*
Liver fluke	*Fasciola hepatica*
Moth, vapourer	*Orgyia antiqua*
Moth, winter	*Operophtera brumata*
Mud snail (liver fluke vector)	*Lymnaea truncatula*
Round-worm	*Ascaridia compare*
Scab mite – sheep, foxes and dogs	*Psoroptes ovis* Sarcoptes scabiei*
Sciarid midge	*Lycoriella ingenua*
Sheep-louse or ked	*Melophagus ovis*
Spittle-bug	*Philaenus spumarius*
Tapeworm	*Hymenolepsis microps*
Tapeworm	*Paroniella urogalli*
Threadworm, caecal	*Trichostrongylus tenuis*
Tick, sheep	*Ixodes ricinus*
Warble fly – cattle red deer	*Hypoderma bovis* Hypoderma diana*

289

References

Amar, A., Anderson, G. D. A., Bolton, M., Bradbury, R. B., Eaton, M. A., Evans, A. D., Gibbons, D. W., Grant, M. C., Gregory, R. S., Hilton, G. M., Hirons, G. J. M., Hughes, J., Johnstone, I., Newbery, P., Peach, W. J., Ratcliffe, N., Smith, K. W., Summers, R. W., Walton, P. & Wilson, J. D. (2007). The predation of wild birds in the UK: a review of its conservation impact and management. *Research Report* **23**, RSPB, Sandy.

Andersen, R. & Sæther, B-E. (1996). *Elg i Norge: biologi, adferd og forvaltning.* Technologist forlag, Oslo.

Anderson, P. (2001). Heather restoration – the key factors. *The Heather Trust Annual Report* **17**, 28–9.

Anderson, P. W. & Yalden, D. W. (1981). Increased sheep numbers and loss of heather moorland in the Peak District, England. *Biological Conservation* **20**, 195–213.

Ash, J. S., Blank, T. H., Coleman-Cooke, J., Cranbrook, the Earl of, Fitter, R. S., Kimball, M. R., Lockie, J. D., Neal, E. G., Scott, W. N., Stuttard, R. M. & Wallace, R. E. (1967). *Predatory Mammals in Britain.* Council for Nature, London.

Baines, D., Aebischer, N., Brown, M. & Macleod, A. (2011). Analysis of capercaillie brood count data: long term analysis. *SNH Commissioned Report* **435**, Edinburgh.

Baines, D., Sage, R. B. and Baines, M. M. (1994). The implications of red deer grazing to ground vegetation and invertebrate communities of Scottish native pinewoods. *Journal of Applied Ecology* **31**, 776–83.

Baines, D., Moss, R. and Dugan, D. (2004). Capercaillie breeding success in relation to forest habitat and predator abundance. *Journal of Applied Ecology* **41**, 59–71.

Barov, B., Bennett, L., Booty, C., Bullock, D., Carus, H., Court, I., Crick, H., Cross, T., Everitt, J., Fern, R., Flynn, M., Hearn, R., Henry, M., Hughes, J., Jenkinson, S., Keating, G., Knott, J., Knox, C., Luxmoore, R., Murray, D., Sharp, C., Stirling-Aird, P., Ruddock, M., Thorpe, R., Waltho, C. & Whitehead, S. (2010). *Birds of Prey in the U. K.: On a Wing and a Prayer.* RSPB. Sandy.

Baynes, T. (2011). Wildlife Estates Scotland (WES). *Summer Review.* Bell Ingram, Perth.

Bell Ingram (2011). *Summer Review.* Bell Ingram, Perth.

Berdowski, J. J. M. (1987). Transition from heathland to grassland initiated by the heather beetle. *Vegetatio* **72**, 167–73.

Bevanger, K. & Brøseth, H. (2004). Impact of power lines on bird mortality in a sub-alpine area. *Animal Biodiversity and Conservation* **27** (2), 67–77.

BirdLife, 2011. BTO, Thetford.

Birks, H. J. B. (1988). Long-term ecological change in the British uplands. In: *Ecological Change in the Uplands* (eds M. B. Usher & D. B. A. Thompson), 17–56. Blackwell, Oxford.

Birse, E. L. & Robertson, J. S. (1976). *Plant Communities and Soils of the Lowland and Southern Upland Regions of Scotland.* The Macaulay Institute for Soil Research, Aberdeen.

Blackmore, A. (2011). *Why Should the RSPB Feign Ignorance?* Winter News Letter, Countryside Alliance, London.

Bobbink, R. & Heil, G. W. (1993). Atmospheric deposition of sulphur and nitrogen in heathland ecosystems. In: *Heathlands. Patterns and Processes in a Changing Environment* (eds R. Aerts, & G. W. Heil). Kluwer Academic Publishers, Dordrect.

British Association for Shooting and Conservation (2004). *Raptor Policy Paper.* BASC Scotland, Dunkeld.

Brown, R. (1997). The biology and ecology of bracken. *The Heather Trust Annual Report* **13**, 24– 6.

Brunsting, A. M. H. (1982). The influence of the dynamics of a population of herbivorous beetles on the development of vegetational patterns in a heathland system. In: *Proceedings of the 5th International Symposium on Insect-Plant Relationships*, 213–23. Pudoc, Wageningen.

Brunsting, A. M. H. & Heil, G. W. (1985). The role of nutrients in the interactions between an herbivorous beetle and some competing plant species in heathlands. *Oikos* **44**, 23–6.

Brys, R., Jacquemyn, H. & de Blust, G. (2005). Fire increases aboveground biomass, seed production and recruitment success of *Molinia caerulea* in dry heathland. *Acta Oecologica* **28**, 299–305.

Budianski, S. (1992). *The Covenant of the Wild.* Weidenfeld & Nicolson, London.

Cazenove, B. (1936). Grouse Shooting and Moor Management. *Country Life*, London.

Chaimberlain, D., Fuller, R., Bunce, R., Duckworth, J. & Shrubb, M. (2001). Changes in the abundance of farmland birds in relation to the timing of agricultural intensification in England and Wales. *Journal of Applied Ecology* **37** (5), 771–88.

Chapman, A. (1924). *The Borders and Beyond.* Gurney & Jackson, London.

Clarke, R. (1995). The Land of Northern Scotland: 200 Years of Sheep – 1795–1995. *The Heather Trust Annual Report* **11**, 47–9.

Clarke, R. (1996). The land of Northern Scotland: the next 200 years. *The Heather Trust Annual Report* **12**, 21–3.

Cobham Resource Consultants (1997). *Countryside Sports and their Economic, Social and Conservation Significance*. Standing Conference on Countryside Sports, Reading.

Colquhoun, J. (1840). *The Moor and the Loch*. W. Blackwood, Edinburgh & London.

Condry, W. M. (1966). *The Snowdonia National Park*. Collins, London.

Cox, R., Newborn, D., Baines, D., Thomas, C. J. & Sherratt, T. N. (2010). No evidence for resistance to fenbendazole in *Trichostrongylus tenuis*, a nematode parasite of the red grouse. *Journal of Wildlife Management* **74** (8), 1799–805.

Cunningham, J. M. M., Eadie, J., Maxwell, T. J. & Sibbald, A. R. (1978). Inter-relations between Agriculture and Forestry: An Agricultural View. *Scottish Forestry* **32** (3), 182–93.

Cunningham, R. 2010. Written answer – Scottish Parliament, 16th June, 2010.

DeGabriel, J. L., Albon, S. D., Fielding, D. A., Riach, D. J., Westaway, S. & Irvine, R. J. (2011). The presence of sheep leads to increases in plant diversity and reductions in the impact of deer on heather. *Journal of Applied Ecology* **48** (5), 1269–77.

Delahaye, R. (1992). Grouse disease: counts of hypobiotic larvae. *The Joseph Nickerson Reconciliation Project Annual Report* **8**, 51–2.

Delahaye, R. & Moss, R. (1994). Grouse disease. *The Heather Trust Annual Report* **10**, 52–3.

Denman, D. R., Roberts, R. A. & Smith, H. J. F. (1967). *Commons and Village Greens*. Leonard Hill, London.

de Smidt, J. T. (1995). The imminent destruction of north-west European heaths due to atmospheric nitrogen deposition. In: *Heaths and Moorland: Cultural Landscapes* (eds D. B. A. Thompson, A. J. Hester & M. B. Usher), 206–17. HMSO, Edinburgh.

Dickie, I., Hughes, J. & Esteban, A. (2006). *Watched Like Never Before. The Local Economic Benefits of Spectacular Bird Species*. RSPB, Sandy.

Dougall. J. D. (1875). *Shooting. Its Appliance, Practice and Purpose*. Samson Low, Marston, Low and Searle, London.

Duckworth-Chad, W. & Thomas, A. (2011). *Grouse Moor News*. Savills, London & Edinburgh.

Duncan, J. S., Reid, H. W., Moss, R., Phillips, J. D. P. & Watson, A. (1978). Ticks, louping ill and red grouse on moors in Speyside, Scotland. *Journal of Wildlife Management* **42**, 500–5.

Eadie, J. (1967). The nutrition of grazing hill sheep: utilisation of hill pastures. *H. F. R.O. 4th Report*, 38–45.

Eadie, J. (1969). Systems of hill sheep production. In: *Proceedings of 3rd Annual Conference*. Reading University Agricultural Club, Reading.

Eadie, J. (1978). Increased output in hill sheep farming. *Journal of the Royal Agricultural Society of England* **139**, 103–14.

Eaton, M. A., Brown, A. F., Noble, D. G., Musgrove, A. J., Hearn, R., Aebischer, N. J., Gibbons, D. W., Evans, A. & Gregory, R. D. (2009). Birds of conservation concern 3: the population status of birds in the United Kingdom, Channel Islands and the Isle of Man. *British Birds* **102**, 296–341.

Ellice, E. C. (1931). *Place-names in Glengarry and Glenquoich*. Routledge and Sons, London.

Erikstad, K. E. (1979). Effect of radio packages on reproductive success of willow grouse. *Journal of Wildlife Management* **43**, 170– 5.

Erikstad, K. E. & Andersen, R. (1983). The effect of weather on survival, growth rate and feeding time in different sized willow grouse broods. *Ornis Scandinavica* **14**, 249–52.

Erikstad, K. E. & Spidsø, T. K. (1982). The influence of weather on food intake, insect prey selection and feeding behaviour in willow grouse chicks in northern Norway. *Ornis Scandinavica* **13**, 176–82.

Etheridge, B., Summers, R. W. & Green, R. E. (1997). Effects of human persecution on the population dynamics of hen harriers *Circus cyaneus* in Scotland. *Journal of Applied Ecology* **34**, 1081–106.

Fantham, H. B. (1911). 'Grouse disease' – coccidiosis (ch 11). In: *The Grouse in Health and in Disease* (ed. A. S. Leslie), 235–72. Smith Elder & Co. London.

Fletcher, K. (2006). Predator control and ground-nesting waders. *Game Conservancy Trust – Review of 2005*, 80–3.

Fletcher, K., Aebischer, N. J., Baines, D., Foster, R. & Hoodless, A. N. (2010). Changes in breeding success and abundance of ground-nesting moorland birds in relation to the experimental deployment of legal predator control. *Journal of Applied Ecology* **47** (2), 263–72.

Fraser of Allander Institute for Research on the Scottish Economy (2010). *An Economic Study of Grouse Moors: an Update*. Game & Wildlife Conservation Trust, Scotland.

Fuller, R. J. & Gough, S. J. (1999). Changes in sheep numbers in Britain: implications for bird populations. *Biological Conservation* **91**, 73–89.

Galbraith, C. A., Stroud, D. A. & Thompson, D. B. A. (2003). Towards resolving raptor-human conflicts. In: *Birds of Prey in a Changing Environment* (eds D. B. A. Thompson, S. M. Redpath, A. H. Fielding, M. Marquiss & C. A. Galbraith), 527–35. HMSO, Edinburgh.

Gathorne-Hardy, A. E. (1900). *Autumns in Argyleshire*. Longmans Green, London.

Gilbert, L., Jones, L. D., Hudson, P. J., Gould, E. A. & Reid, H. W. (2000). Role of small mammals in the persistence of louping-ill virus: field survey and tick co-feeding studies. *Medical and Veterinary Entomology* **14**, 277–82.

Gimingham, C. (1972). *The Ecology of Heathlands*. Chapman & Hall, London.

Gladstone, H. C. (1922). *Record Bags and Shooting Records*. H. F. & G. Witherby, London.

Graham, K. (1908). *The Wind in the Willows*. Methuen & Co., London.

Grant, S. A. & Armstrong, S. M. (1993). Grazing ecology and the conservation of heather moorland: the development of models as aids to management. *Biodiversity & Conservation* **2**, 79–94.

Grant, S. A., Barthram, G. T., Lamb, W. I. C. & Milne, J. A. (1978). Effect of season and level of grazing on the utilisation of heather by sheep. 1. Responses of the sward. *Journal of British Grassland Society* **33**, 289–300.

Grant, S. A., Hunter, R. F. & Cross, C. (1963). The effect of muirburning *Molinia*-dominated communities. *Journal of the British Grassland Society* **18**, 249–57.

Gray, L. (2011). Grouse shooting helps rural economies, show new figures. 5th August. *Daily Telegraph*, London.

Grey Graham, H. (1937). *The Social Life of Scotland in the Eighteenth Century*. A. & C. Black, London.

Grimble, A. (1894). *Highland Sport*. Chapman & Hall, London.

Haldane, A. R. B. (1952). *The Drove Roads of Scotland*. T. Nelson & Sons Ltd, Edinburgh.

Haldane, A. R. B. (1962). *New Ways Through the Glens*. T. Nelson & Sons Ltd, Edinburgh.

Hall, R. (1885). *The Highland Sportsman and Tourist*. Simkin, Marshall & Co., London and John Menzies & Co., Edinburgh.

Halnan, E. T. & Garner, F. H. (1940). *The Principles and Practice of Feeding Farm Animals*. Longmans Green and Co., London.

Hamilton-Maxwell, W. (1832). *Wild Sports of the West*. EP Publishing Ltd, Wakefield.

Hardin, G. (1968). The Tragedy of the Commons. *Science* **162**, 1243–8.

Hare, C. E. (1939). The language of sport. *Country Life*, London.

Harrison, A., Newey, S., Gilbert, L., Haydon, D. T. & Thirgood, S. (2010). Culling wildlife hosts to control disease: mountain hares, red grouse and louping ill virus. *Journal of Applied Ecology* **47** (4), 926–30.

Hart, E. W. (2004). *Hefting in Practice. The Ancient Craft of Grazing the Open Hills*. The Farmers' Fund.

Hartley, S. (1993). Insect pests on heather moorland. *The Heather Trust Annual Report* **9**, 46–7.

Hartley, S. & Kerslake, J. (1995). The winter moth research programme: an update on current research. *The Heather Trust Annual Report* **11**, 63– 4.

Hartley, S., Kerslake, J. & Woodin, S. (1994). The winter moth research programme. *The Heather Trust Annual Report* **10**, 51.

Hartley, S. E. & Mitchell, R. J. (2005). Manipulation of nutrients and grazing levels on heather moorland: changes in *Calluna* dominance and consequences for community composition. *Journal of Ecology* **93**, 990–1004.

Haysom, S. (2011). Guardians of the Pinewood. *The Nature of Scotland* **12**, 34–5. SNH, Inverness.

Heather and Grass burning Code (2007). Natural England, Peterborough and DEFRA, London.

Heather Trust (1995). Raptors on moorland – the Heather Trust's view. *The Heather Trust Annual Report* **11**, 45.

Heinzel, H., Fitter, R. & Parslow, J. (1972). *The Birds of Britain and Europe with North Africa and the Middle East*. Collins, London.

Hely-Hutchinson, H, G. (1871). *Twenty Years Reminiscences of the Lews*. Horace Cox, London.

Hesketh Prichard, H. (1921). *Sport in Wildest Britain*. William Heinemann, London.

Hester, A. J. & Sydes, C. (1992). Changes in burning of Scottish heather moorland since the 1940s from aerial photographs. *Biological Conservation* **60**, 25–30.

Hewson, R. (1984). Mountain hare *Lepus timidus* bags and moor management. *Journal of Zoology* **204**, 563–5.

Hobbs, R. J. & Gimingham, C. H. (1987). Vegetation, fire and herbivore interactions in heathland. *Advances in Ecological Research* **16**, 87–173.

Hodgson, J. & Eadie, J. (1986). Animal production from the hills and uplands: vegetation resources and animal nutrition in hill areas. *Galway Symposium*, Republic of Ireland.

Hodgson, J. & Grant, S. A. (1981). Grazing animals and forage resources in the hills and uplands. *Proceedings of the British Grassland Society, Occasional Symposium No 12* (ed. J. Frame), 41–57.

Holland, J. P., Morgan-Davies, C., Waterhouse, T., Thomson, S., Midgley, A. & Barnes, A. (2011). An analysis of the impact on the natural heritage of the decline in hill farming in Scotland. *SNH Commissioned Report* **454**. Edinburgh.

Holloway, S. (1996). *The Historical Atlas of Breeding Birds in Britain and Ireland: 1875–1900.* Poyser, London.

Holstad, Ø., Karbøl, G. & Skorping, A. (1994). *Trichostrongylus tenuis* from willow grouse (*Lagopus lagopus*) and ptarmigan (*Lagopus mutus*) in northern Norway. *Bulletin of the Scandinavian Society for Parasitology* **4**, 9–13.

Hudson, P. J. (1986). *The Red Grouse: the Biology and Management of a Wild Gamebird.* Game Conservancy, Fordingbridge.

Hudson, P., Dobson, A. P. & Newborn, D. (1998). Prevention of population cycles by parasite removal. *Science* **282**, 2256–8.

Hudson, P. J. & Newborn, D. (1990). Brood defence in a precocial species: variations in the distraction displays of red grouse *Lagopus lagopus scoticus*. *Animal Behaviour* **40**, 254–61.

Hudson, P. J., Norman, R., Laurenson, M. K., Newborn, D., Gaunt, M., Jones, L., Reid, H., Gould, E., Bowers, R. & Dobson, A. (1995). Persistence and transmission of tick-borne viruses: *Ixodes ricinus* and louping ill virus in red grouse populations. *Parasitology* **111**, S49–S58.

Hunt, J. F. (2003). *Impacts of Wild Deer in Scotland – How Fares the Public Interest?* Report for WWF Scotland and RSPB Scotland.

Hutchinson, H. G. (1903). *Management of Grouse Moors.* In: *Shooting.* Country Life, London.

I C I Game Services, (1952). *Your Year's Work* **4**, 16–17. Fordingbridge.

I C I Game Services, (1953). *Your Year's Work* **5**, 6–20. Fordingbridge.

Jacquemyn, H., Brys, R. & Neubert, M. G. (2005). Fire increases invasive spread of *Molinia caerulea* mainly through changes in demographic parameters. *Ecological Applications* **15** (6), 2097–108.

Jenkins, D., Watson, A. & Miller, G. R. (1963). Population studies on red grouse *Lagopus lagopus scoticus* (Lath) in north-east Scotland. *Journal of Animal Ecology* **32**, 317–76.

Jenkins, D., Watson, A. & Miller, G. R. (1964a). Current research on red grouse in Scotland. *Bird Study* **3** (1), 3–13.

Jenkins, D., Watson, A. & Miller, G. R. (1964b). Predation and red grouse populations. *Journal of Applied Ecology* **1**, 183–93.

Jenkins, D., Watson, A. & Miller, G. R. (1967). Population fluctuations in the red grouse *Lagopus lagopus scoticus*. *Journal of Animal Ecology* **36**, 97–122.

Jenkins, D., Watson, A. & Miller, G. R. (1970). Practical results for management of red grouse. *Biological Conservation* **2** (4), 266–72.

Jones, L. D., Gaunt, M., Hails, R. S., Laurenson, K., Hudson, P. J., Reid, H., Henbest, P. & Gould, E. A. (1997). Transmission of louping ill virus between infected and uninfected ticks co-feeding on mountain hares. *Medical and Veterinary Pathology* **11**, 172–6.

Kaland, P. E. (1986). The origin and management of Norwegian coastal heaths as reflected by pollen analysis. In: *Anthropogenic indicators in Pollen Diagrams* (ed. K-E. Behre), 19–36. Blakema, Rotterdam.

Keatinge, T. H., Coupar, A. M. & Reid, E. (1995). Wet heaths in Caithness and Sutherland, Scotland. In: *Heaths and Moorlands: Cultural Landscapes* (eds D. B. A. Thompson, A. J. Hester & M. B. Usher), 20–7. HMSO, Edinburgh.

Keble Martin, W. (1965). *The Concise British Flora in Colour.* Ebury Press & Michael Joseph, London.

Kolb, H. H. & Hewson, R. (1980). A study of fox populations in Scotland from 1971–1976. *Journal of Applied Ecology* **17**, 7–19.

Lance, A. N. (1975). Egg production and chick development in a marginal red grouse population. *Congress of the International Union of Game Biologists* **12**, Tema 3, 1–10. Lisbon.

Lance, A. N. (1978). Territories and the food plant of individual red grouse. II. Territory size compared with an index of nutrient supply in heather. *Journal of Animal Ecology* **47**, 307–13.

Lance, A. N. (1983a). Selection of feeding sites by hen red grouse *Lagopus lagopus scoticus* during breeding. *Ornis Scandinavica* **14**, 78–80.

Lance, A. N. (1983b). Performance of sheep on unburned and serially burned blanket bog in western Ireland. *Journal of Applied Ecology* **20**, 767–75.

Lance, A. N. (1987). Estimating acceptable stocking levels for heather moorland. In: *Agriculture and Conservation in the Hills and Uplands.* (eds M. Bell & R. G. H. Bunce), 109–15. Institute of Terrestrial Ecology, Grange over Sands.

Lance, A. N. & Phillips, J. (1975). Responses of red grouse to serial heather burning in eastern Ireland. *Congress of the International Union of Game Biologists* **12**, Tema 3, 1–9. Lisbon.

Laurenson, M. K., Norman, R. A., Gilbert, L., Reid, H. W. & Hudson, P. J. (2003). Identifying disease reservoirs in complex systems: mountain hares as reservoirs of ticks and louping-ill virus, pathogens of red grouse. *Journal of Animal Ecology* **72**, 177–85.

Laurenson, M. K., Norman, R., Reid, H. W., Newborn, D. & Hudson, P. J. (2000). The role of lambs in louping-ill virus amplification. *Parasitology* **120**, 97–104.

Le Duc, M. (2000). Restoration of heath after bracken invasion. *The Heather Trust Annual Report* **16**, 29–31.

Lee, J. A., Caporn, S. J. M. & Reid, D. J. (1992). Effects of increasing nitrogen deposition and acidification on heathlands. In: *Acidification Research. Evaluation and Policy Applications* (ed. T. Schneider), 97–106. Elsevier, Colchester.

Lindström, E. R., Andrén, H., Angelstam, P., Cederlund, G., Hörnfeldt, B., Jäderberg, L., Lemnell, P.-A., Martinsson, B., Sköld, K. & Swenson, J. E. (1994). Disease reveals the predator: sarcoptic mange, red fox predation and prey populations. *Ecology* **75**, 1042–9.

Lindström, E. R., Brainerd, S. M., Helldin, J. O. & Overskaug, K. (1995). Pine marten/red fox interactions – a case of intraguild predation. *Annales Zoologici Fennici* **32**, 123–30.

Lloyd, H. G. (1980). *The Red Fox*. Batsford, London.

Loe, L. E., Mysterud, A., Stien, A., Steen, H., Evans, D. M. & Austrheim, G. (2007). Positive short-term effects of sheep grazing on the alpine avifauna. *Biology Letters* **3**, 109–11.

Lovat, Lord (1911). Moor Management (ch 17), Heather burning (ch 18) and Stock (ch 21). In: *The Grouse in Health and in Disease* (ed. A. S. Leslie), 372–413, 454–82. Smith, Elder & Co., London.

Mabbett, T. (2011). The Poisoned Heath. *Forestry Journal* **4** (11), 22–3.

MacColl, A. D. C., Piertney, S. B., Moss, R. & Lambin, X. (2000). Spatial arrangement of kin affects recruitment success in young male red grouse. *Oikos* **90**, 261–70.

McCreath, J. B. & Murray, R. D. (1954). *A survey of an Argyll hill farming district*. Report **26**. Economics Department, West of Scotland College of Agriculture, Ayr.

Macdonald, D. G. F. (1883). *Grouse Disease*. W. H. Allen & Co., London.

Macdonald, D. W. (1987). *Running with the Fox*. Hyman Unwin, London.

McGilvray, J. (1996). *An Economic Study of Scottish Grouse Moors*. University of Strathclyde, Glasgow.

McGilvray, J. & Perman, R. (1991). *Grouse Shooting in Scotland: Analysis of its Importance to the Economy and Environment*. University of Strathclyde. Report to The Game Conservancy, Fordingbridge.

McIntyre, D. (1926). *Days on the Hill*. Nisbet & Co., London.

Mackenzie, O. H. (1921). *A Hundred Years in the Highlands*. E. Arnold & Co., London.

Mackie, P. J. (1924). *The Keeper's Book*. Foulis & Co., London.

MacPherson, H. A. (1895). *The Grouse*. Fur & Feather Series. Longman, Brown & Green, London.

Malcolm, G. & Maxwell, A. (1910). *Grouse and Grouse Moors*. A. & C. Black, London.

Marcström, V., Kenward, R. E. & Engren, E. (1988). The impact of predation on boreal tetraonids during vole cycles: an experimental study. *Journal of Animal Ecology* **57**, 859–72.

Marrs, R. H., Le Duc, M. G., Mitchell, R. J., Goddard, D., Paterson, S. & Pakeman, R. J. (2000). The ecology of bracken: its role in succession and implications for control. *Annals of Botany* **85** (3), 3–15.

Marrs, R. H., Phillips, J. D. P., Todd, P. A., Ghorbani, J. & Le Duc, M. G. (2004). Control of *Molinia caerulea* on upland moors. *Journal of Applied Ecology* **41**, 398–411.

Martin, B. L. (1990). *The Glorious Grouse*. David & Charles, Newton Abbot.

Mather, A. S. (1978). The alleged deterioration of hill grazings in the Scottish Highlands. *Biological Conservation* **14**, 181–95.

Maxwell, H. (1922). *Memories of the Months*. Arnold & Co., London.

Maxwell, T. J., Grant, S. A., Milne, J. A. & Sibbald, A. R. (1986). Systems of sheep production on heather moorland, In: *Hill Land Symposium, Galway, 1985*. (ed. M. O'Toole), 188–211. Dublin: An Foras Taluntais.

Meikle, A., Paterson, S., Howse, P., Waterhouse, T. & Marshall, G. (1999). Management and geography: impact on the genetic structure of heather. *The Heather Trust Annual Report* **15**, 41.

Merricks, P. (2010). Lapwings, farming and environmental stewardship. *British Wildlife* **22** (1), 10–13. British Wildlife Publishing, Milton on Stour.

Miles, J. (1988). Vegetation and soil change in the uplands. In: *Ecological Change in the Uplands* (eds M. B. Usher & D. B. A. Thompson), 57–70. Blackwell Scientific Publications, Oxford.

Miller, G. R. (1968). Evidence for selective feeding on fertilised plots by red grouse, hares and rabbits. *Journal of Wildlife Management* **32**, 849–53.

Miller, G. R. (1979). Quantity and quality of the annual production of shoots and flowers by *Calluna vulgaris* in north-east Scotland. *Journal of Ecology* **67**, 109–29.

Miller, G. R., Jenkins, D. & Watson, A. (1966). Heather performance and red grouse populations. 1. Visual estimates of heather performance. *Journal of Applied Ecology* **3**, 313–26.

Miller, G.R., Jenkins, D. & Watson, A. (1970). Responses of red grouse populations to experimental improvement of their food. In: *Animal Populations in Relation to their Food Resources* (ed. A. Watson), 323–5. Blackwell Scientific Publications, Oxford.

Miller, G. R. & Miles, J. (1970). Regeneration of heather (*Calluna vulgaris* (L.) Hull) at different ages and seasons in north-east Scotland. *Journal of Applied Ecology* **7** (1), 51–60.

Miller, G. R. & Watson, A. (1978). Territories and the food plant of individual red grouse. 1. Territory size, number of mates and brood size compared with the abundance, production and diversity of heather. *Journal of Applied Ecology* **47**, 293–305.

Milligan, A. L., Marrs, R. H. & Putwain, P. D. (1997). Control of *Molinia caerulea* (L.) Moench in upland Britain. *British Crop Protection Conference* (Weeds), 679–80.

Milligan, A. L., Putwain, P. D., Cox, E. S., Ghorbani, J., Le Duc, M. G. & Marrs, R. H. (2003). Developing an integrated land management strategy for the restoration of moorland vegetation on *Molinia caerulea*-dominated vegetation for conservation purposes in upland Britain. *Biological Conservation* **119**, 371–85.

Milne, J. A. & Grant, S. A. (1978). Better use of heather hills for sheep production. *H.F.R.O. 7th Report, 1974–77*, 41–8.

Miquet, A. (1990). Mortality in black grouse *Tetrao tetrix* due to elevated cables. *Biological Conservation* **54**, 349–55.

Mitchell, R. J., Marrs, R. H., Le Duc, M. G. & Auld, M. H. D. (1999). A study of the restoration of heathland on successional sites: changes in vegetation and soil chemical properties. *Journal of Applied Ecology* **26**, 770–83.

Moorland Association (2010). *A Guide to Upland Policy Formation: Heather Moorland*. Moorland Association, Lancaster.

Moss, R. (1969). A comparison of red grouse (*Lagopus lagopus scoticus*) stocks with the production and nutritive value of heather (*Calluna vulgaris*). *Journal of Animal Ecology* **38**, 103–12.

Moss, R. (1972a). Food selection by red grouse (*Lagopus lagopus scoticus* Lath.) in relation to chemical composition. *Journal of Animal Ecology* **41**, 411–18.

Moss, R. (1972b). Effects of captivity on gut length in captive red grouse. *Journal of Wildlife Management* **36**, 99–104.

Moss, R. (1989). Grouse disease. *The Joseph Nickerson Reconciliation Project Annual Report* **5**, 39–40.

Moss, R. (2001). Second extinction of capercaillie (*Tetrao urogallus*) in Scotland? *Biological Conservation* **101**, 255–7.

Moss, R., Parr, R. & Lambin, X. (1995). Effects of testosterone on breeding density, breeding success and survival of red grouse (Vol. 258, p.175, 1994). *Proceedings of the Royal Society of London Series B, Biological Sciences* **260**, 373.

Moss, R., Shaw, J. L., Watson, A. & Trenholm, I. (1990a). Role of the caecal threadworm *Trichostrongylus tenuis* in the population dynamics of red grouse. In: *Red Grouse Population Processes* (eds A. N. Lance and J. H. Lawton), 62–71. RSPB, Sandy.

Moss, R., Trenholm, I., Watson, A. & Parr, R. (1990b). Parasitism, predation and survival of hen red grouse *Lagopus lagopus scoticus* in spring. *Journal of Applied Ecology* **59**, 631–42.

Moss, R. & Watson, A. (1984). Maternal nutrition, egg quality and breeding success of Scottish ptarmigan *Lagopus mutus*. *Ibis* **126**, 212–20.

Moss, R., Watson, A. & Parr, R. (1975). Maternal nutrition and breeding success in red grouse (*Lagopus lagopus scoticus*). *Journal of Animal Ecology* **44**, 233–44.

Moss, R., Watson, A. & Parr, R. (1988). Mate choice by hen red grouse *Lagopus lagopus scoticus* with an excess of cocks – role of territory size and food quality. *Ibis* **130**, 545–52.

Moss, R., Watson, A. & Parr, R. (1996). Experimental prevention of a population cycle in red grouse. *Ecology* **77**, 1512–30.

Moss, R., Watson, A., Rothery, P. & Glennie, W. W. (1981). Clutch size, egg size, hatch weight and laying date in relation to early mortality in red grouse *Lagopus lagopus scoticus* chicks. *Ibis* **123**, 450–62.

Moss, R., Watson, A., Trenholm, I. & Parr, R. (1993). Caecal threadworms *Trichostrongylus tenuis* in red grouse *Lagopus lagopus scoticus*: effects of weather and host density on estimated worm burdens. *Parasitology* **107**, 199–209.

Mougeot, F., Dawson, A., Redpath, S. M. & Leckie, F. (2005). Testosterone and autumn territorial behaviour in male red grouse *Lagopus lagopus scoticus*. *Hormones and Behavior* **47**, 576–84.

Muirburn Code (2011). Scottish Executive, Edinburgh.

Myrberget, S., Erikstad, K. E. & Spidsø, T. K. (1997). Variations from year to year in growth rates of willow grouse chicks. *Astarte* **10**, 9–14.

Nelson, K. (2010). Great expectations in the glen. *The Nature of Scotland* **10**, 26–31. SNH, Inverness.

Nethersole-Thompson, D. & Watson, A. (1974). *The Cairngorms*. Collins, London.

Newborn, D. (2004). *Strongylosis Control in Red Grouse*. Game Conservancy Trust, Fordingbridge.

Newborn, D. & Foster, R. (2002). Control of the parasite burdens in wild red grouse *Lagopus lagopus scoticus* through the indirect application of anthelmintics. *Journal of Applied Ecology* **39**, 908–14.

Newey, S., Iason, G. & Raynor, R. (2008). The conservation status and management of mountain hares. *SNH Commissioned Report* **287**, Edinburgh.

Nickerson, J. (1998). *A Shooting Man's Creed*. D. A. H. Grayling, Penrith.

Novoa, C., Hansen, E. & Ménoni, E. (1990). La mortalité de trois espèces de galliformes par collision dans les cables: resultants d'une enquête pyrénéene. *Bulletin Mensuel* **151**, 17–22. Office National de la Chasse, Paris.

Olstead, J. (ed.) (2011). Cloud cuckoo land. *Shooting and Conservation, May/June*, **106**, British Association for Shooting and Conservation, Wrexham.

Orchel, J. (1992). *Forest Merlins in Scotland*. Hawk & Owl Trust, Taunton.

PACEC (2006). *The Economics and Environmental Impacts of Sport Shooting*. 162-8, Regent Street, London.

Pakeman, R. J., Le Duc, M. G. & Marrs, R. H. (1998). An assessment of aerially applied asulam as a method of long-term bracken control. *Journal of Environmental Management* **53**, 255–62.

Palmer, S. C. F. (1997). Prediction of the shoot production of heather under grazing in the Uplands of Great Britain. *Grass and Forage Science* **52** (4), 408–24.

Park, K. J., Robertson, P. A., Campbell, S. T., Foster, R., Russell, Z. M., Newborn, D. & Hudson, P. J. (2001). The role of invertebrates in the diet, growth and survival of red grouse (*Lagopus lagopus scoticus*) chicks. *Journal of Zoology* **254**, 137–45.

Parker, E. (1929). *Shooting by Moor, Field and Shore*. The Lonsdale Library. Seeley Service & Co. Ltd, London.

Parker, H. (1985). Compensatory reproduction through renesting in willow ptarmigan. *Journal of Wildlife Management* **49**, 599–604.

Parr, R. (1975). Ageing red grouse chicks by primary moult and development. *Journal of Wildlife Management* **39**, 188–90.

Parr, R. A. (1990). *Moorland birds and their predators in relation to afforestation*. Institute of Terrestrial Ecology, Banchory.

Pearce-Higgins, J. W. & Grant, M. C. (2006). Relationships between bird abundance and the composition and structure of moorland vegetation. *Bird Study* **53**, 112–25.

Pearce-Higgins, J. W. & Yalden, D. W. (2003). Golden Plover *Pluvialis apricaria* breeding success on a moor managed for shooting red grouse *Lagopus lagopus*. *Bird Study* **50**, 170–7.

Pennant, T. (1772). *A Tour in Scotland and Voyage to the Hebrides*. Benjamin White, London.

Petty, S. J., Anderson, D. I. K., Davison, M., Little, B., Sherrat, T. N., Thomas, C. J. & Lambin, X. (2003). The decline of common kestrels *Falco tinnunculus* in a forested area of northern England: the role of predation by northern goshawks *Accipiter gentilis*. *Ibis* **145**, 472–83.

Phillips, D. (2011). An investigation into the success of the grey partridge (*Perdix perdix*) Restoration Project on The Northumberland Estates. M. Phil Thesis, Department of Land Economy, Cambridge.

Phillips, J. (1990a). The problems of 'white' moorland. *The Joseph Nickerson Reconciliation Project Annual Report* **6**, 40–1.

Phillips, J. (1990b). 'Buffer' feeding of raptors. *The Joseph Nickerson Reconciliation Project Annual Report* **6**, 46.

Phillips. J. (1990c). Cutting vs Burning. *The Joseph Nickerson Reconciliation Project Annual Report* **6**, 52.

Phillips, J. (1990d). The case for heather. *Journal of the Royal Agricultural Society of England* **151**, 96–102.

Phillips, J. (1991a). Heather burning and management. *The Joseph Nickerson Reconciliation Project Annual Report* **7**, 41–2.

Phillips, J. (1991b). Heather burning in September. *The Joseph Nickerson Reconciliation Project Annual Report* **7**, 51.

Phillips, J. (1991c). The Nickerson system of sheep grazing. *The Joseph Nickerson Reconciliation Project Annual Report* **7**, 53.

Phillips, J. & Moss, R. (1977). Effects of subsoil draining on heather moors in Scotland. *Journal of Range Management* **30**, 27–9.

Phillips, J., Steen, J. B., Raen, S. G. & Aalerud, F. (1992). Effects of burning and cutting on vegetation and on a population of willow grouse *Lagopus lagopus* in Norway. *Fauna Norvegica Series* C **15**, 37–42.

Phillips, J. & Watson, A. (1995). Key requirements for the management of heather moorland: now and for the future. In: *Heaths and Moorland: Cultural Landscapes* (eds D. B. A. Thompson, A. J. Hester & M. B. Usher), 344–61. HMSO, Edinburgh.

Phillips, J. & Wilson, K. (1987). The heather beetle (*Lochmaea suturalis*) and its effect on moorland. *The Reconciliation Project Annual Report* **3**, 31–3.

Phillips, J., Yalden, D. & Tallis, J. (1981). *Peak District Moorland Erosion Study, Phase 1 Report*. Peak Park Joint Planning Board, Bakewell.

Picozzi, N. (1968). Grouse bags in relation to the management and geology of heather moors. *Journal of Applied Ecology* **5**, 483–8.

Picozzi, N. (1971). Breeding performance and shooting bags of red grouse in relation to public access in the Peak District National Park. *Biological Conservation* **3**, 211–5.

Picozzi, N. (1975). Crow predation on marked nests. *Journal of Wildlife Management* **39** (1), 151–5.

Picozzi, N. (1980). Food, growth, survival and sex ratio of nestling hen harriers *Circus cyaneus* in Orkney. *Ornis Scandinavica* **11**, 1–11.

Picozzi, N. (1981). Common gull predation of winter moth larvae. *Bird Study* **28**, 68–9.

Pinder, P. S. & Hayes, A. J. (1986). An outbreak of vapourer moth (*Orgyia antiqua* L.: Lepidoptera Limnantridae) on Sitka spruce (*Picea stichensis* (Bong.) Carr) in Central Scotland. *Forestry* **59** (1), 97–106.

Potts, G. R. (1986). *The Partridge: Pesticides, Predation and Conservation*. Collins, London.

Ratcliffe, D. A. & Thompson, D. B. A. (1998). The British uplands: their ecological character and international significance. In: *Ecological Change in the Uplands* (eds M. B. Usher & D. B. A. Thompson), 9–36. Blackwell Scientific Publications. Oxford.

Red Deer Species Profile (2011). Trees for Life, Forres.

Redpath, S. M. & Thirgood, S. J. (1997). *Birds of Prey and Red Grouse*. HMSO, London.

Redpath, S. M. & Thirgood, S. J. (1999). Numerical and functional responses of general predators: hen harriers and peregrines on Scottish grouse moors. *Journal of Applied Ecology* **68**, 879–92.

Redpath, S. M., Thirgood, S. J. & Leckie, F. M. (2001). Does supplementary feeding reduce predation of red grouse by hen harriers? *Journal of Applied Ecology* **38**, 1157–60.

Reid, H. W. (1975). Experimental infection of red grouse with louping-ill virus (Flavivirus group). 1. The viraemia and antibody response. *Journal of Comparative Pathology* **85**, 223–9.

Reid, H. W., Duncan, J. S., Moss, R., Phillips, J. D. P. & Watson, A. (1978). Studies on louping-ill virus (Flavivirus group) on wild red grouse (*Lagopus lagopus scoticus*). *Journal of Hygiene* **81**, 321–9.

Reid, H. W. & Moss, R. (1980). The response of four species of birds to louping-ill virus. In: *Arboviruses in the Mediterranean Countries, Zbl. Bakt. Supplement* **9** (eds J. Vesenjak-Hirjan et al), 219–23. Gustav Fischer Verlag, Stuttgart & New York.

Reimers, E. (1989). *Villreinens verden*. H. Aschehoug & Co. (W. Nygaard), Oslo.

Richards, E. (1973). *The Leviathan of Wealth*. Routledge & Kegan Paul, Toronto.

Robertson, P. A., Park, K. J. & Barton, A. F. (2001). Loss of heather *Calluna vulgaris* in the Scottish uplands: the role of red grouse *Lagopus lagopus scoticus* management. *Wildlife Biology* **7**, 11–16.

Rochlitz, I., Pearce, G., Broom, D., Turner, J., Ross, S. & Harris, S. (2010). *The OneKind Report on Snaring*. OneKind & League Against Cruel Sports, Edinburgh.

Rodwell, J. S. (ed.) (1991). *British Plant Communities Vol. 2. Mires & Heaths*. Cambridge University Press.

Rosenberg, A. & Marrs, R. (2010). The heather beetle: a review. *Report to The Heather Trust*, Dumfries.

Sansom, A. (1998). Floods and sheep – is there a link? *The Heather Trust Annual Report* **14**, 33–5.

Savory, C. J. (1978). Food consumption in red grouse in relation to the age and productivity of heather. *Journal of Animal Ecology* **47**, 269–82.

Scandrett, E. & Gimingham, C. H. (1991). The effect of heather beetle *Lochmaea suturalis* on vegetation in a wet heath in N. E. Scotland. *Holarctic Ecology* **14**, 24–30.

Schoenian, S. (2008). Understanding anthelmintics. *Small Ruminant Information Sheet*. University of Maryland.

Scotland's Moorland Forum (2003). *Principles of Moorland Management*. SNH, Perth.

Scott, Lord George (1937). *Grouse Land and the Fringe of the Moor*. H. F. & G. Witherby, London.

Scott, Lord George (1940). *The Fleeting Opportunity*. H. F. & G. Witherby, London.

Scottish Natural Heritage (2003). Muirkirk and North Lowther Uplands Management Scheme. ISSN 1479-7798-2000203.

Scottish Natural Heritage (2010)a. Galloway Kite Trail. *Nature of Scotland* **10**, 18. SNH, Inverness.

Scottish Natural Heritage (2010)b. *Diversionary Feeding of Hen Harriers on Grouse Moors: A Practical Guide*. SNH, Inverness.

Scrope, W. (1838). *The Art of Deer Stalking*. J. Murray, London.

Small, R. J. & Rusch, D. H. (1985). Backpacks vs ponchos: survival and movements of radio-marked ruffed grouse. *Wildlife Society Bulletin* **13**, 163–5.

Smedshaug, C. A., Selås, V., Lund, S. E. & Geir, A. S. (1999). The effect of a natural reduction of red fox *Vulpes vulpes* on small game hunting bags in Norway. *Wildlife Biology* **5**, 157–66.

Smith, A. (2005). Are tick mops effective in Scotland? *Game Conservancy Trust Annual Review of 2005*, 86–7, Fordingbridge.

Soulbury, C. D., Iossa, G., Baker, P. J., Cole, N. C., Funk, S. M. & Harris, S. (2007). The impact of sarcoptic mange *Sarcoptes scabiei* on the British red fox *Vulpes vulpes* population. *Mammal Review* **37**, 278–96.

Speedy, T. (1886). *Sport in the Highlands and Lowlands of Scotland*. W. Blackwood, Edinburgh & London.

Steen, J. B. (1989). *Ryper, Ryperliv og Ryperjakt*. Gyldendal Norsk Forlag, Oslo.

Steen, J. B., Steen, H., Stenseth, N. C., Myrberget, S. & Marcström, V. (1988). Microtin density and weather as predictors of chick production of willow ptarmigan *Lagopus lagopus*. *Oikos* **51**, 367–73.

Stevenson, A. C. & Thompson, D. B. A. (1993). Long-term changes in the extent of heather moorland in upland Britain and Ireland: paleoecological evidence for the importance of grazing. *The Holocene* **3**, 70–6.

St John, C. (1845). *Sketches of the Wild Sports and Natural History of the Highlands*. J. Murray, London.

St John, C. (1882). *Natural History and Sport in Moray*. D. Douglas, Edinburgh.

Storch, I. & Leidenberger, C. (2003). Tourism, mountain huts and distribution of corvids in the Bavarian Alps. *Wildlife Biology* **9**, 301–8.

Straker-Smith, P. & Phillips, J. (1994). The logistics and arithmetic of heather burning. *The Heather Trust Annual Report* **10**, 7–19.

Stuart-Wortley, A. J. (1895). *The Grouse. Fur & Feather Series*. Longman, Brown & Green, London.

Sutherland, J. (1997). The rabbit's role in heather degeneration. *The Heather Trust Annual Report* **13**, 39–40.

Tharme, A. P., Green, R. E., Baines, D., Bainbridge, I. P. & O'Brien, M. (2001). The effect of management for red grouse shooting on the population density of breeding birds on heather-dominated moorland. *Journal of Applied Ecology* **38**, 439–57.

Thirgood, S., Redpath, S., Newton, I. & Hudson, P. (2000). Raptors and red grouse: conservation conflicts and management solutions. *Conservation Biology* **14**, 95–104.

Thomas, L. H. & McDiarmid, A. (2011). Badger Status. *The Field* **317**, (7281), 18.

Thompson, D. B. A. (1992). Scottish Natural Heritage: what the future holds. *The Heather Trust Annual Report* **8**, 46–7.

Thorley, J. (1998). Traditional sheep farming and the heather moor. *The Heather Trust Annual Report* **14**, 37–8.

Thomson, J. A. (1921). *Mountains and Moorland*. MacMillan, London.

Thornton, T. (1804). *A Sporting Tour through the Northern Parts of England and Greater Part of the Highlands of Scotland*. Vernor & Hood, London.

Todd, P. A., Phillips, J. D. P., Putwain, P. D. & Marrs, R. H. (2000). Control of *Molinia caerulea* on moorland. *Grass and Forage Science* **55**, 181–91.

Tucker, G. (2003). Review of the impacts of heather and grassland burning in the uplands on soils, hydrology and biodiversity. *Research Report* **550**. English Nature, Peterborough.

Turner, J. (2010). The impact of a snare-free Scotland on biodiversity. *Onekind Report*. Submission to Scottish Parliament.

Usher, M. B. & Thompson, D. B. A. (1993). Variations in the upland heathlands of Great Britain: conservation importance. *Biological Conservation* **66**, 69–81.

Waddington, R. (1958). *Grouse Shooting and Moor Management*. Faber & Faber, London.

Walker, J. (1808). *Economical History of the Hebrides and Highlands*. **1**, 158–60. Edinburgh University Press.

Wallace, R. (1917). *Heather and Moor Burning for Grouse and Sheep*. Oliver & Boyd, Edinburgh.

Walters, J. (ed.) (2011a). Project tackles harrier decline. *Nature of Scotland* **11**, 18. SNH, Inverness.

Walters, J. (ed.) (2011b). Extra rangers help capercaillie. *Nature of Scotland* **13**, 22. SNH, Inverness.

Warren, P. & Baines, D. (2007). Dispersal distances of juvenile radio-tagged red grouse *Lagopus lagopus scoticus* on moors in northern England. *Ibis* **149**, 758–62.

Watson, A. (1967a). Population control by territorial behaviour in red grouse. *Nature* **213**, 1274–5.

Watson, A. (1967b). Social status and population regulation in the red grouse (*Lagopus lagopus scoticus*). *Proceedings of the Royal Society Population Study Group* **2**, 22–30. The Royal Society, London.

Watson, A. (1979). Bird and mammal numbers in relation to human impact at ski lifts on Scottish hills. *Journal of Applied Ecology* **16**, 753–64.

Watson, A. (1989). Recent influences on grouse habitat from increases in red deer. *The Joseph Nickerson Reconciliation Project Annual Report* **5**, 45–6.

Watson, A. (1996), Human-induced increases in carrion crows and gulls on Cairngorms plateaux. *Scottish Birds* **18**, 17–19.

Watson, A., Hewson, R., Jenkins, D. & Parr, R. (1973). Population densities of mountain hares compared with red grouse on Scottish moors. *Oikos*, **24**, 225–30.

Watson, A. & Jenkins, D. (1968). Experiments on population control by territorial behaviour in red grouse. *Journal of Animal Ecology* **37**, 595–614.

Watson, A. & Miller, G. R. (1976). *Grouse Management*. The Game Conservancy, Fordingbridge.

Watson, A., Miller, G. R. & Green, F. H. W. (1966). Winter browning of heather (*Calluna vulgaris*) and other moorland plants. *Transactions of the Botanical Society of Edinburgh* **40** (2), 195–203.

Watson, A. & Moss, R. (1990). Spacing behaviour and winter loss in red grouse. In: *Red Grouse Population Processes* (eds A. N. Lance & J. H. Lawton), 35–52. RSPB, Sandy.

Watson, A. & Moss, R. (1993). Red Grouse in Marginal Areas. *The Heather Trust Annual Report* **9**, 39–42.

Watson, A. & Moss, R. (2008). *Grouse*. T. & A. D. Poyser. Berkhampstead.

Watson, A., Moss, R. & Parr, R. (1984a). Effects of food enrichment on numbers and spacing behaviour of red grouse. *Journal of Animal Ecology* **53**, 663–78.

Watson, A., Moss, R. & Parr, R. (1984b). Grouse increase on Mull. *Landowning in Scotland* **207**, 6.

Watson, A., Moss, R., Phillips, J. & Parr, R. (1977). The effect of fertilizers on red grouse stocks on Scottish moors grazed by sheep, cattle and deer. In: *Écologie du petit Gibier et Aménagement des Chasses* (ed. P. Pesson), 193–212. Gaulthier-Villars, Paris.

Watson, A. & O'Hare, P. J. (1979). Red grouse populations on experimentally treated and untreated Irish bog. *Journal of Applied Ecology* **16**, 433–52.

Watson, A. & O'Hare, P. J. (1980). Dead sheep and scavenging birds and mammals on Mayo bog. *Irish Birds* **1**, 487–91.

Watson, A., Phillips, J. & Moss, R. (1995). Moorland history and grouse declines in Ireland, with an outline policy for conservation and recovery. In: *Canada Life Irish Red Grouse Conference* (ed. R. O'Dwyer), 26–100. Irish Field Trials Grouse Ground Committee, Cappagh, Waterford.

Watson, D. (1977). *The Hen Harrier*. T. & A. D. Poyser, Berkhampstead.

Watson Lyall, J. (1912). *The Sportsman's and Tourist's Guide*. J. Watson Lyall & Co., London.

Webb, N. R. (1989). The invertebrates of heather and moorland. *Botanical Journal of the Linnean Society* **101**, 307–12.

Welch, D. (1984). Studies in the grazing of heather moorland in north-east Scotland. ll. Responses of heather. *Journal of Applied Ecology* **1**, 197–207.

Welch, D. (1998). Response of bilberry *Vaccinium myrtillus* stands in the Derbyshire Peak District to sheep grazing and implications for conservation. *Biological Conservation* **83** (2), 155–64.

Welch, D., Scott, D., Moss, R. & Bayfield, N. G. (1994). Ecology of blaeberry and its management in British moorlands. Institute of Terrestrial Ecology, Cumbria.

Wildlife Scotland (2011). Visitscotland.com.

Yalden, D. W. (1972). The red grouse (*Lagopus lagopus scoticus*, L.) in the Peak District. *Naturalist*, Hull. **922**, 89–102.

Index

Page numbers in *italic* refer to illustrations; page numbers in **bold** refer to tables